PRINCIPLES OF THE SURFACE TREATMENT OF STEELS

Charlie R. Brooks
Professor of Metallurgical Engineering
Materials Science and Engineering Department
The University of Tennessee
Knoxville, Tennessee

LANCASTER · BASEL

Principles of the Surface Treatment of Steels

a **TECHNOMIC**® publication

Published in the Western Hemisphere by
Technomic Publishing Company, Inc.
851 New Holland Avenue
Box 3535
Lancaster, Pennsylvania 17604 U.S.A.

Distributed in the Rest of the World by
Technomic Publishing AG

Printed in the United States of America
10 9 8 7 6 5 4 3 2 1

Main entry under title:
Principles of the Surface Treatment of Steels

A Technomic Publishing Company book
Bibliography: p.
Includes index p. 283

Library of Congress Card No. 91-67902
ISBN No. 87762-796-7

CONTENTS

PREFACE

This book is designed to present the physical metallurgy principles of the surface treatment of steels, as the title states. It should serve as a useful review for metallurgists and materials engineers involved in commercial treatments of steel surfaces, as well as a source of background information for engineers involved in using surface treated steel components. The extensive appendices provide a convenient source of related information.

The only prerequisite to the use of the book is an introductory course in metallurgy or materials science, or an equivalent industrial experience.

Since the properties are controlled by the microstructure, the book contains many micrographs to illustrate the effect of processing on microstructure. In carburizing, the chemical thermodynamics of the gas reactions is of prime importance, and thus I have given a treatment of the principles of the chemical thermodynamics of the gas equilibrium involved in the carburizing reactions. However, I have written this section so that a background in chemical thermodynamics is not required to appreciate the application of the principles to carburizing.

Because of the extensive literature data cited, I have not attempted to convert any of the data into SI units. Conversion tables are given in Appendices 2–5.

ACKNOWLEDGEMENTS

This book makes extensive use of information from the literature, and I want to express my appreciation to the many authors and publishers who have allowed me to use their data. They are acknowledged in the figure captions. Original prints were kindly supplied by Mr. J. M. Williams and Drs. W. C. Oliver, J. R. Conrad, G. K. Hubler, R. H. Shay and T. Ericsson.

I used considerable information from sources from ASM International (formerly American Society for Metals), and I thank Timothy Gall for his assistance in this.

I am especially indebted to Dr. Len Samuels for allowing me to make extensive use of his excellent micrographs from his book *Optical Microscopy of Carbon Steels* [American Society for Metals, Metals Park, Ohio (1980)]. He not only allowed me the use of this information, but kindly supplied original glossy prints of the micrographs.

I thank my wife Sue for editorial assistance and Mr. Yee Chung Lin for making the drawings.

CHAPTER 1 **Introduction**

The surfaces of steels are treated by a variety of methods for a variety of reasons. Some examples of surface treatments and the reasons for them are as follows: painting to prevent or minimize corrosion; plastic deformation by shot-peening to develop a harder surface and favorable residual stresses; heat treatment to austenitize just the surface layers, followed by quenching to form a layer of hard martensite and to produce favorable residual stresses; and processing to alter the chemistry of the surface in order to improve the mechanical properties and to develop favorable residual stresses. In this book, it is important to note that the term *surface* does not mean simply the most superficial layer of the steel, but the region below it as well.

The variety of surface treatments is too great for this book to cover all of them in detail. Instead, emphasis has been placed on four methods—thermal processing, carburizing, nitriding, and ion implantation—that have been chosen both because they are widely used and because they involve basic principles that can be applied to the study of other surface treatment processes. The first of these, thermal processing, involves the manipulation of temperature alone—for example, heating just the surface layer to form austenite, and then cooling the steel rapidly. Later chapters bring various chemical processes into play as well. Carburizing involves alteration in the chemistry of steel surfaces through the addition of carbon. Nitriding also involves chemical alterations, although the effects of nitrogen addition differ significantly from those of carbon addition. Ion implantation is covered at the end of the book as an example of emerging technology for altering the surface chemistry of steel.

One of the most important developments in this field in recent years is the great advancement in the computer modeling of steel surface treatment processes. For example, computer programs are available that will calculate the carbon gradient from the surface to the core of a steel part during carburizing. The gradient can be calculated even if the surface carbon content changes (controlled, for example, by the gas composition) or if the temperature changes with time. Such approaches are very important in industrial processes since, if the temperature or gas composition inadvertently changes greatly, the computer program allows calculation of a new set of operating parameters in order to recover the required carbon gradient. Also, computer programs are available to calculate the microstructure and the accompanying residual stress distribution for different heat treatments. In this book, reference is made to some of the results of computer programs, although the rapid rate of advancement in programming technology makes it difficult to include the latest developments.

CHAPTER 2 **Surface Treatments Involving Only Thermal Processing**

In this section we examine surface treatments that involve only the conversion of the surface layer of the steel part to austenite, followed by cooling to form martensite. There is no change in the chemistry of the surface layer. One important advantage of this type of heat treatment is that the core properties can be set by prior heat treatment, allowing only the surface layer to be affected during the heating. For example, a tough core can be obtained, and the surface can be austenitized and quenched to form hard martensite.

Surface-heating can be accomplished by a number of methods, such as by the use of a flame (flame-hardening), high frequency current (induction-hardening), lasers, solar radiation, and electron beams. The method used, and its application, control the heating and cooling cycle and the depth of heating. However, the principles of the response of the steel can be examined without invoking the details of the method used to cause the heating, and that is what is done in the first sections. In the last three sections, characteristics of three surface-heating methods used for surface hardening—induction-heating, flame-heating, and laser-heating—are examined.

2a. MICROSTRUCTURAL EFFECTS

Consider a plain carbon steel containing 0.4% C that has its surface heated for a certain time so that the temperature–time profiles obtained at different depths are those shown schematically in Figure 1. Also, assume that the affected depth is small compared to the total depth, so that once the energy source is removed (time t_0), the heated regions cool relatively rapidly by conduction of heat into the cooler central region of the part. The layers below the surface also lose heat to the surface, at a rate depending upon the way the surface itself is cooled once the energy source is removed. For example, the surface may cool in air, or it may be sprayed with water.

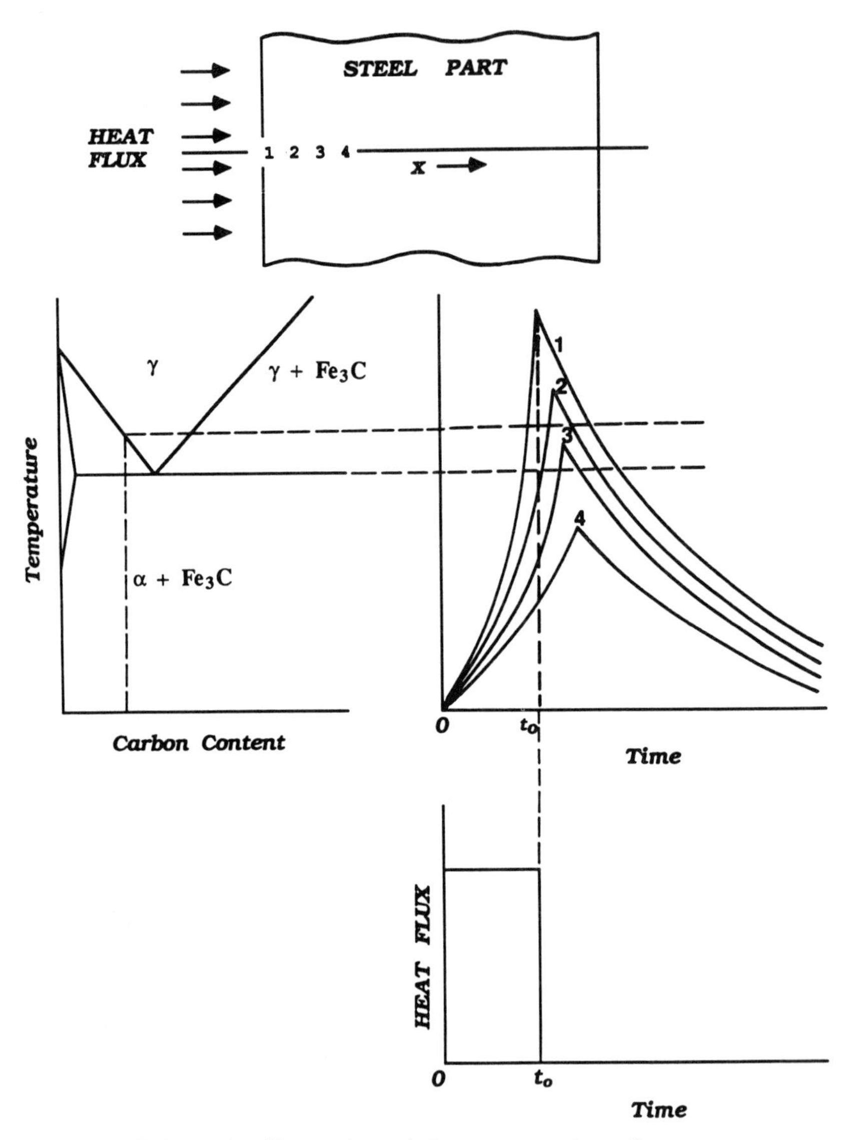

FIGURE 1. Schematic illustration of the temperature–time curves as a function of depth for surface-heating.

In Figure 2 are shown schematically the microstructures developed upon heating. Note that at the surface the steel will attain the most austenitization, in terms of austenitizing time and temperature. This is the region in which the austenite grain size will be largest, in principle, and care must be taken in the heating not to allow development of excessively large grains. As the distance from the heated surface increases, a location is reached (x_2) where all austenite is just formed, since this location was above the A_{c3} temperature (see Figure 3) sufficiently long for the transformation to occur. Further from the heated surface, there is a region (x_3) in which not all austenite formed. For a hypoeutectoid steel, this region will

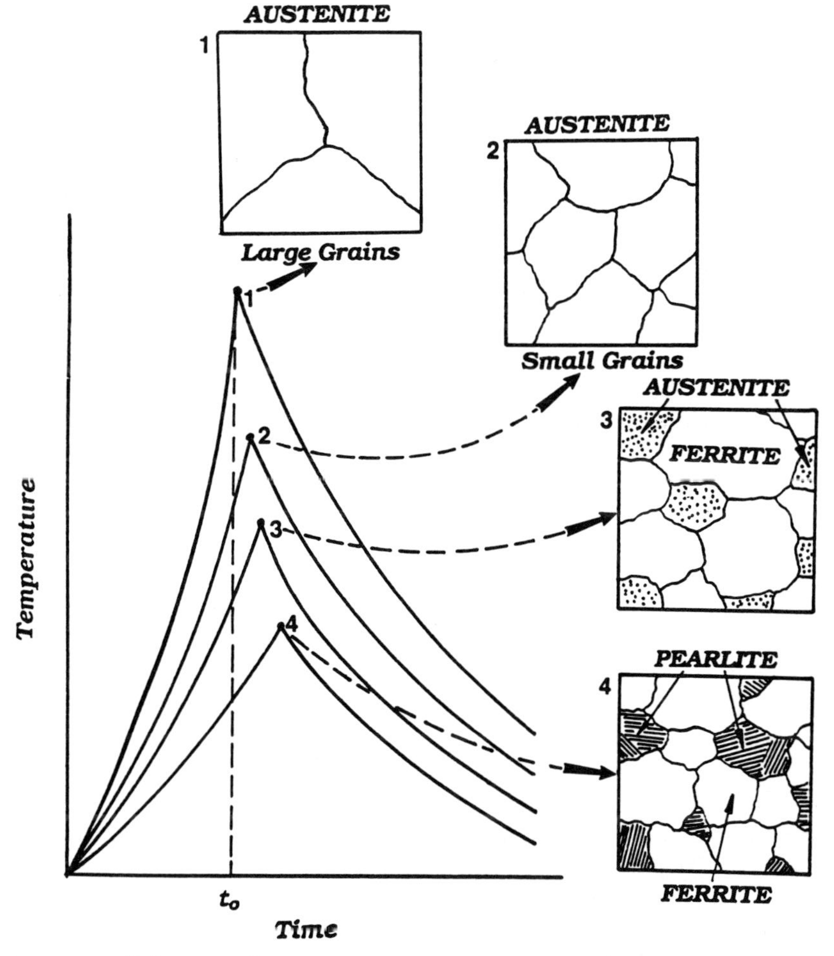

FIGURE 2. Schematic illustration of the formation of austenite in a surface-hardened 0.4% C steel.

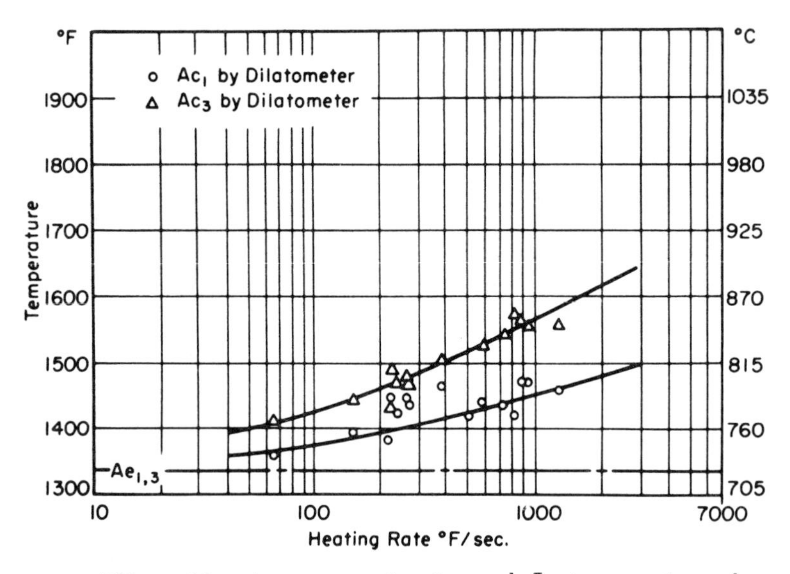

FIGURE 3. Effect of heating rate on the A_{c1} and A_{c3} temperatures for an annealed 1080 steel (from Feuerstein, W. J. and W. K. Smith. 1954. *Trans. ASM*, 46:1270).

contain ferrite and austenite. Even further removed from the surface (x_4), the steel will not have formed any austenite (although the temperature may have reached above A_{e1}; see Figure 3). The structure in this region will depend upon the starting microstructure. If this is primary ferrite and pearlite (as depicted in Figure 2), no change will be apparent because the time will be too short to have allowed any spheroidization. If the starting microstructure is martensite, then in this region the martensite will be tempered.

The microstructure developed upon cooling at a given depth depends upon the hardenability at this location and on the cooling rate. Figure 4 shows schematically the probable structure, because in most surface treatments of the type being considered here the cooling rate is high. The hardenability generally will be lower than it would be for the same steel conventionally austenitized, since in surface-hardening the heating time is, perforce, necessarily short. Thus, the austenite may not be homogeneous. The effect on hardenability of the shorter austenitizing time obtained upon induction heating is reflected in the Jominy curves in Figure 5. The furnace-heated samples were conventionally austenitized (40 min at 870°C), then end-quenched. The induction-heated samples were heated to 870°C in a few seconds, held there for about 30 sec, then end-quenched. The more heterogeneous structure causes a lower Jominy curve (lower

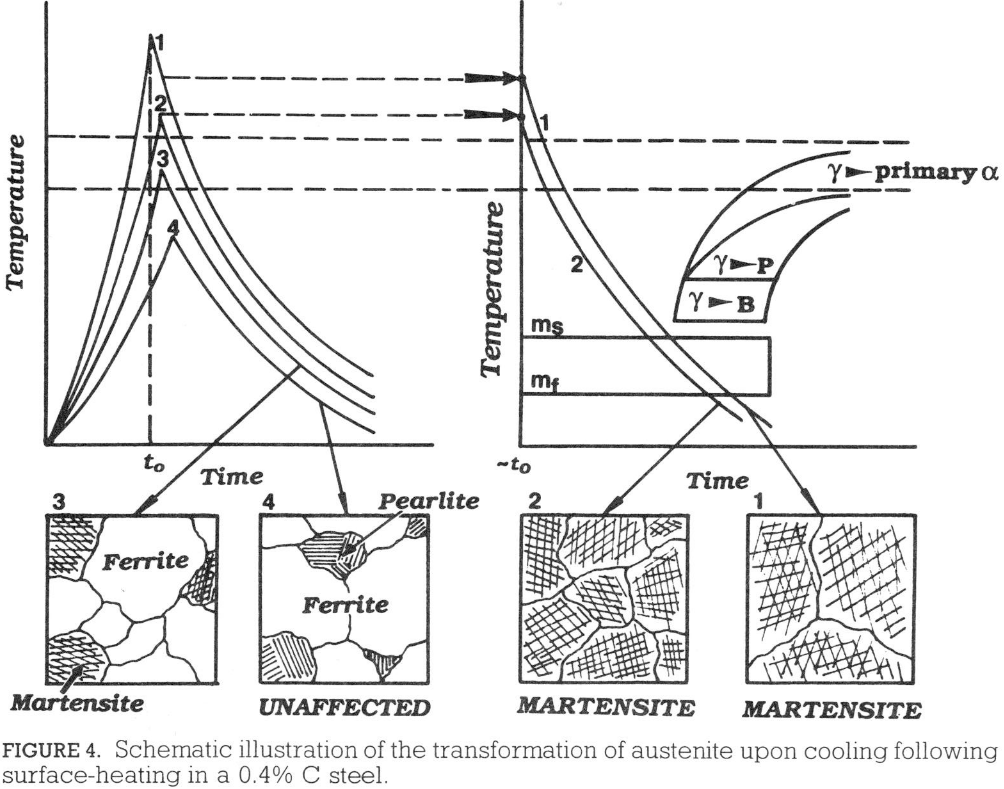

FIGURE 4. Schematic illustration of the transformation of austenite upon cooling following surface-heating in a 0.4% C steel.

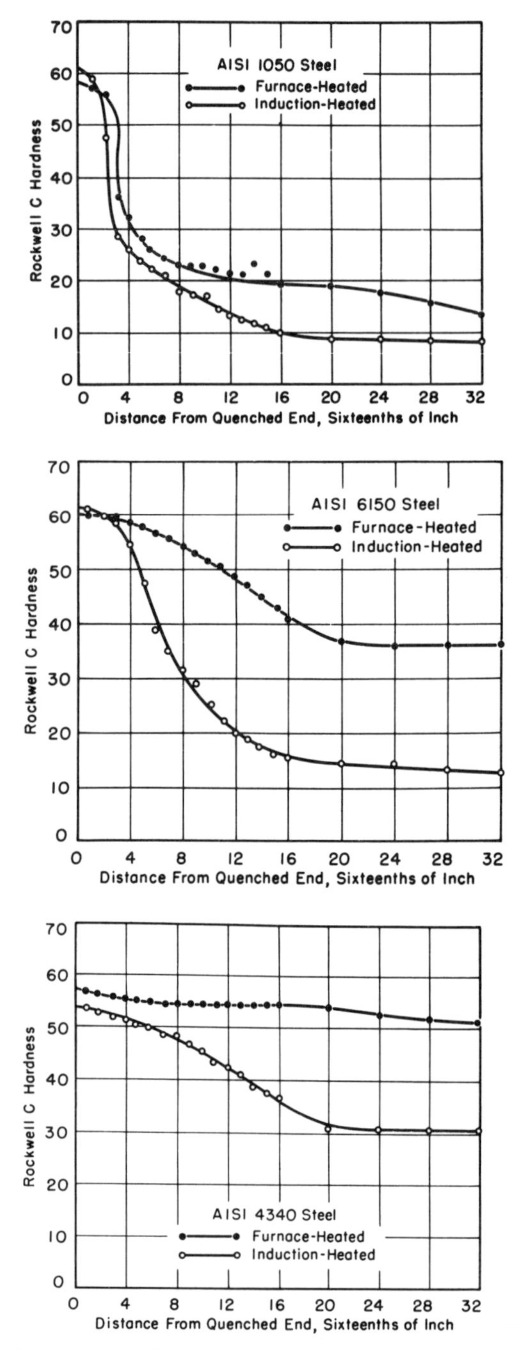

FIGURE 5. Jominy curves of steels that have been conventionally austenitized, then end-quenched, and that have been austenitized for short times by induction-heating, then end-quenched (from Libsch, J. F., W.-P. Chuang and W. J. Murphy. 1950. *Trans. ASM*, 42:121).

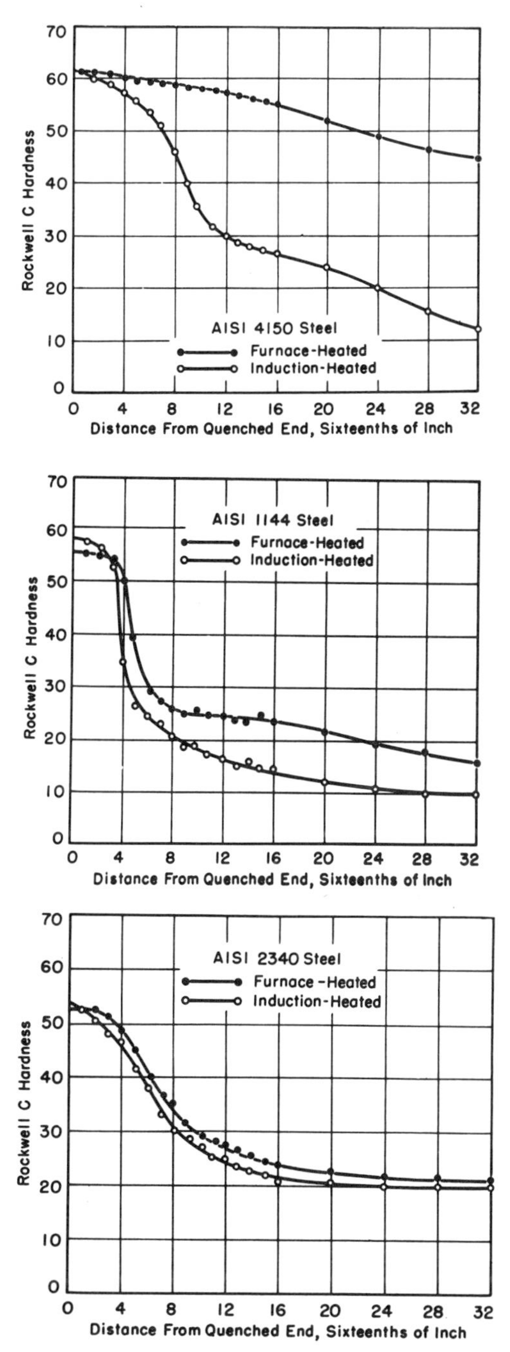

FIGURE 5 (continued). Jominy curves of steels that have been conventionally austenitized, then end-quenched, and that have been austenitized for short times by induction-heating, then end-quenched (from Libsch, J. F., W.-P. Chuang and W. J. Murphy. 1950. *Trans. ASM*, 42:121).

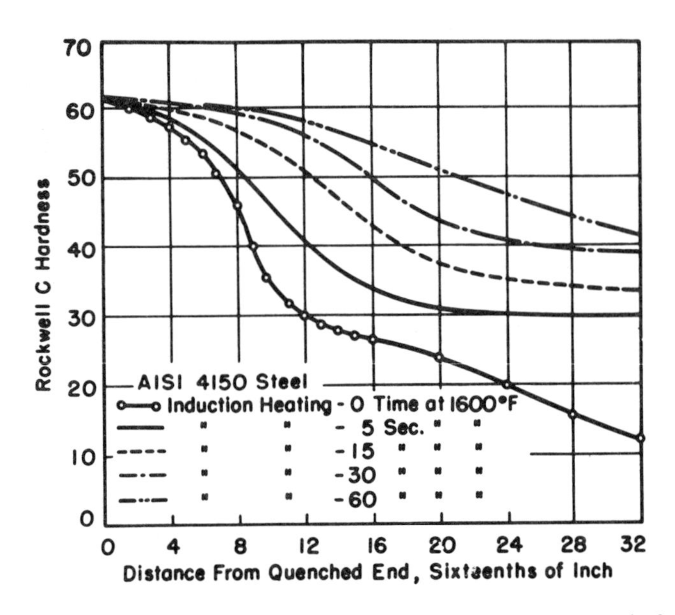

FIGURE 6. Effect of time at temperature during austenitizing by induction-heating on the Jominy curves of a 4150 steel (from Libsch, J. F., W.-P. Chuang and W. J. Murphy. 1950. *Trans. ASM*, 42:121).

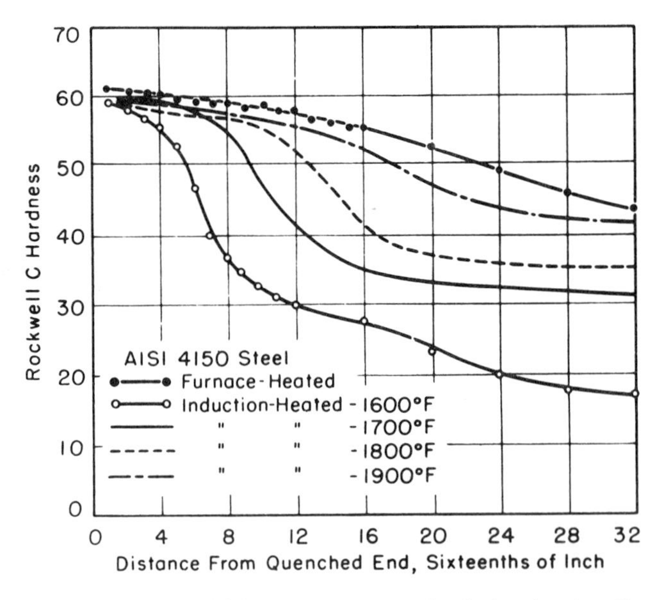

FIGURE 7. Effect of austenitizing temperature by induction-heating on the Jominy curves of a 4150 steel (from Libsch, J.F., W.-P. Chuang and W. J. Murphy. 1950. *Trans. ASM*, 42:121).

hardenability). The effect of the induction-heating time on the hardenability is shown in Figure 6, and the effect of austenitizing temperature in Figure 7. The longer time and higher temperature allow a more homogeneous structure to form and hence produce a higher hardenability.

As described above, at a given location in the surface layers of the steel part, the microstructure depends upon hardenability and the cooling rate. As the distance from the surface increases, the cooling rate decreases and the hardenability decreases because of the lower austenitizing temperature. Eventually a depth is reached where complete austenite will not form. Instead, a two-phase structure of austenite plus ferrite (in hypoeutectoid steels) or austenite plus carbide (in hypereutectoid steels) will be present. The austenite will contain a different carbon content from that of the steel, and it will transform to a structure that depends upon the cooling rate and the hardenability of this austenite region. (Note that these microstructural effects are the same as those encountered in the heat-affected zone of a weld; see Figure 4.)

In surface-heating, the energy input rate to the surface is necessarily high, since only the surface layers are to be austenitized. Thus the heating

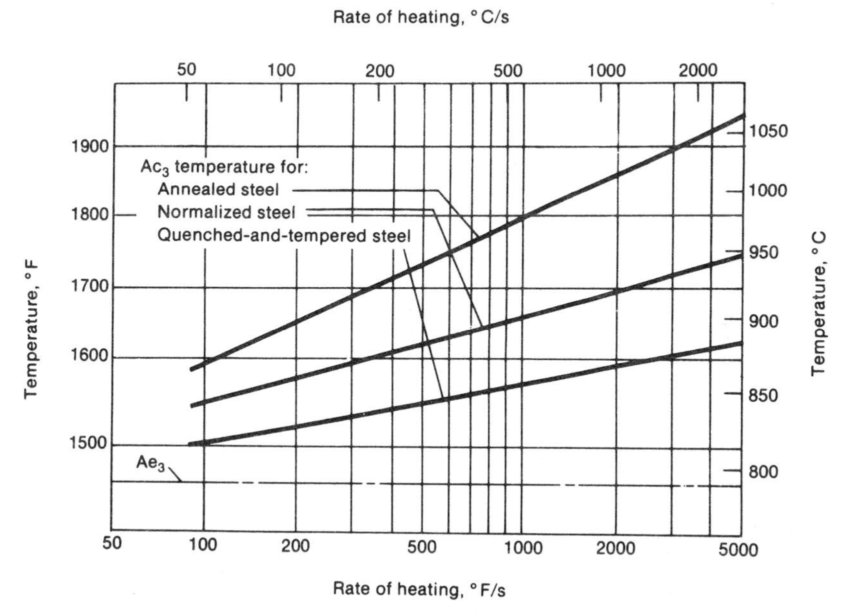

FIGURE 8. Effect of initial microstructure and heating rate on the A_{c3} temperature for 1042 steel (adapted from Feuerstein, W. J. and W. K. Smith. 1954. *Trans. ASM,* 46:1270, as given in Semiatin, S. L. and D. E. Stutz. 1986. *Induction Heat Treatment of Steel.* Metals Park, Ohio: American Society for Metals).

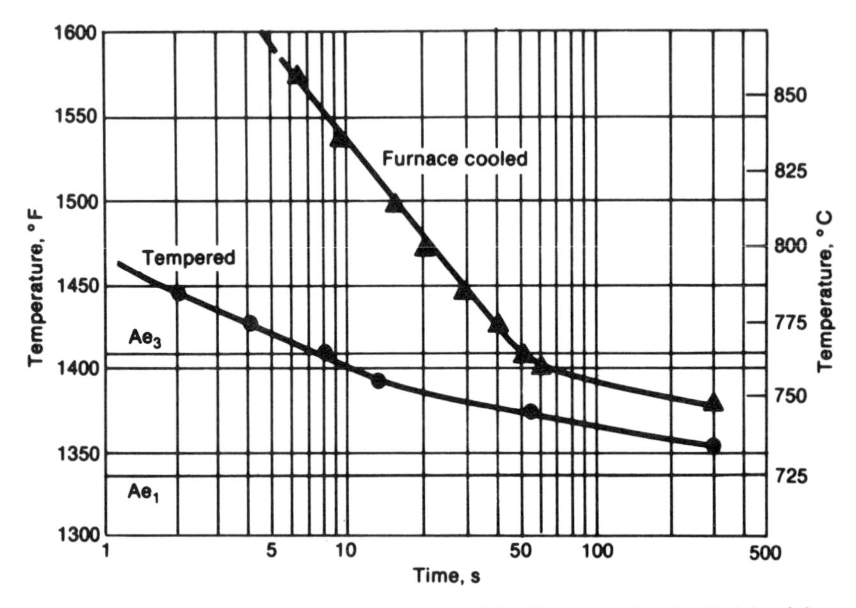

FIGURE 9. Comparison of the isothermal TTT diagrams for the finish of the formation of austenite for a 1050 steel (from Spenser, T. H. et al. 1964. *Induction Hardening and Tempering*. Metals Park, Ohio: American Society for Metals; and Martin, D. L. and W. G. Van Note. 1946. *Trans. ASM*, 36:210, as adopted from S. L. Semiatin and D. E. Stutz. 1986. *Induction Heat Treatment of Steel*. Metals Park, Ohio: American Society for Metals).

rate in this region is relatively high and, as mentioned above, superheating of the starting microstructure is more pronounced than in conventional austenitizing. The effect of heating rate on the critical temperatures of some plain carbon steels was shown in Figure 3. The effect of the starting microstructure on the nucleation of austenite is of more concern when short austenitizing times are used. The data in Figure 8 show that the coarser structure (e.g., coarse pearlite) requires more superheating than does a finer structure (e.g., tempered martensite) in which austenite can nucleate more easily. This effect is also illustrated by the data in Figure 9, which shows the time for complete austenitization for two different beginning microstructures.

Also, as mentioned above, the austenite grain size in the surface-heated layers is of importance. The higher the austenitizing temperature, the greater the possibility of obtaining undesirable, large grains. This temperature depends upon the time. In conventional austenitizing, the austenitizing time typically is one hour. In surface-heating, the austenitizing times are much shorter, so that higher austenitizing temperatures can be used without the danger of forming large austenite grains. Figure 10 shows the effect of austenitizing temperatures and short austenitizing times on the austenite grain size for two steels. Note that in the alloy steel, a grain size of about ASTM 6–8 can be attained upon using an austenitizing temperature of 1000°C.

We now examine some induction-heated steels to find examples of the microstructures that form. In Figure 11 we see the microstructure of a 0.80% C steel that had a pearlite structure prior to heating [Figure 11(a)]. When the surface was induction-heated at about 300°C/sec to 775°C, then water-quenched, the structure formed at the surface was as shown in Figure 11(b). This structure consists of martensite and undissolved ferrite and carbides, as seen by the remnants of pearlite present. The hardness at this location is quite high (Figure 12), indicating a large amount of martensite. In Figure 11(c) we see the structure in the region where not all

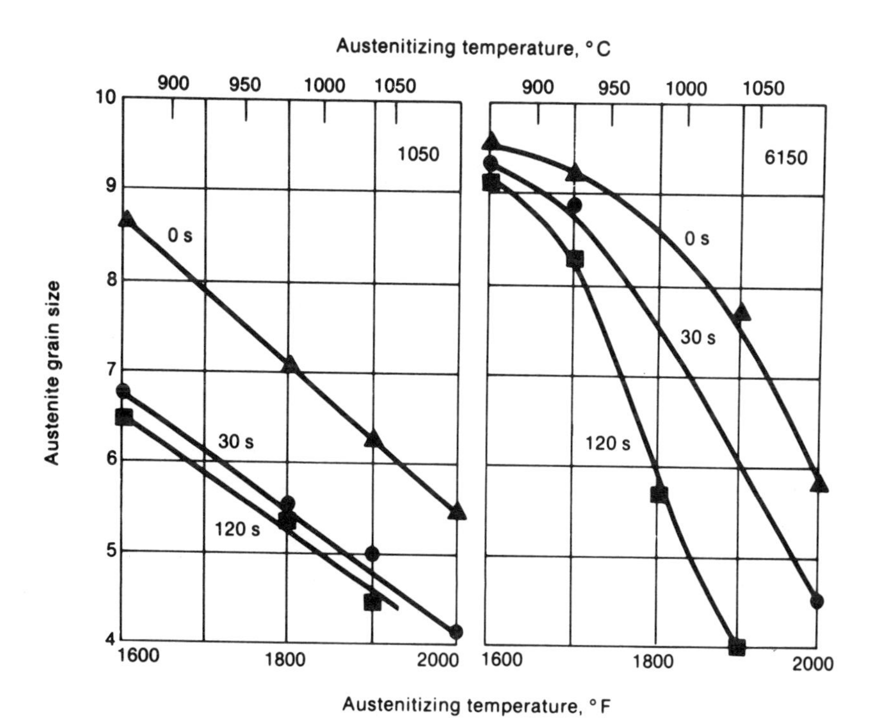

FIGURE 10. Effect of austenitizing temperature and time on austenite grain size in induction-hardened 1050 and 6150 steel (from Spenser, T.H. et al. 1964. *Induction Hardening and Tempering.* Metals Park, Ohio: American Society for Metals; and Martin, D. L. and W. C. Van Note. 1946. *Trans. ASM,* 36:210, as adopted from S. L. Semiatin and D. E. Stutz. 1986. *Induction Heat Treatment of Steel.* Metals Park, Ohio: American Society for Metals).

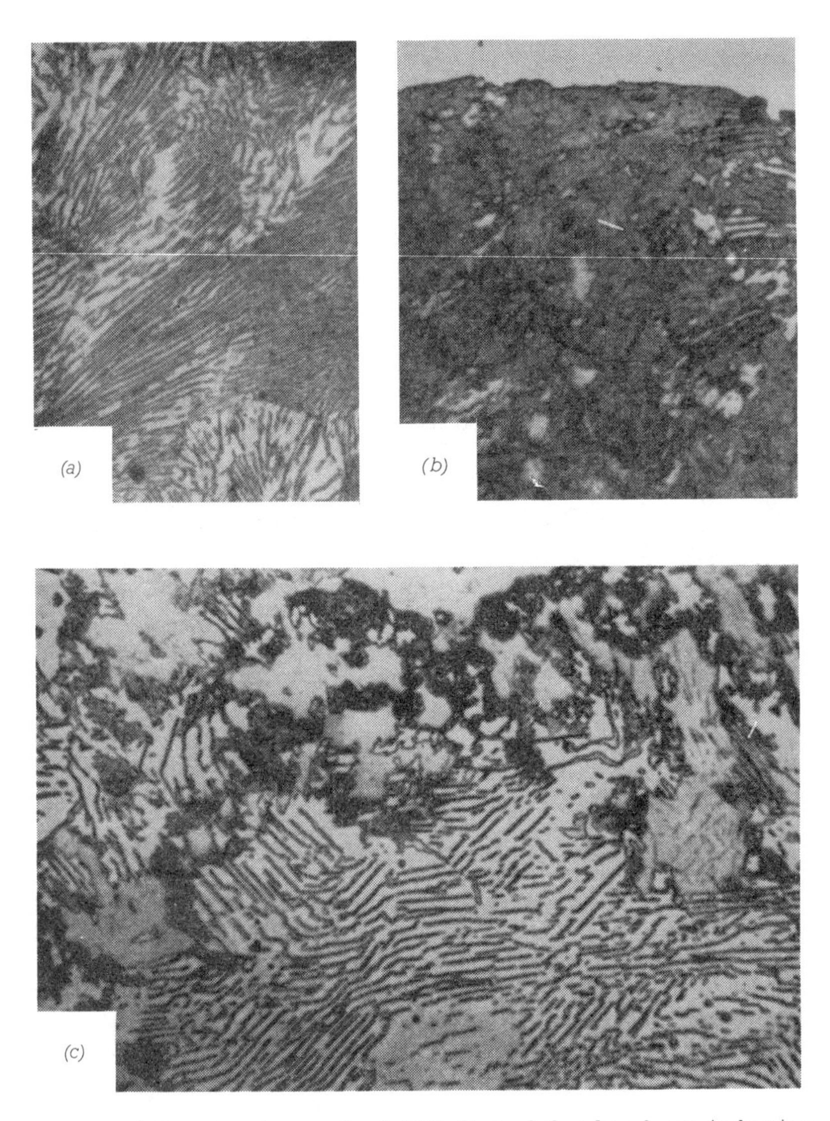

FIGURE 11. Microstructures of a 0.80% C steel that has been induction-hardened with a maximum surface temperature of 775°C. (a) The initial pearlitic microstructure. (b) Microstructure at the edge (shown at top): martensite, ferrite and pearlite. (c) Area near transition zone: martensite and pearlite (at bottom). All micrographs at 1000× (from Martin, D. L. and F. E. Wiley. 1945. *Trans. ASM*, 34:351).

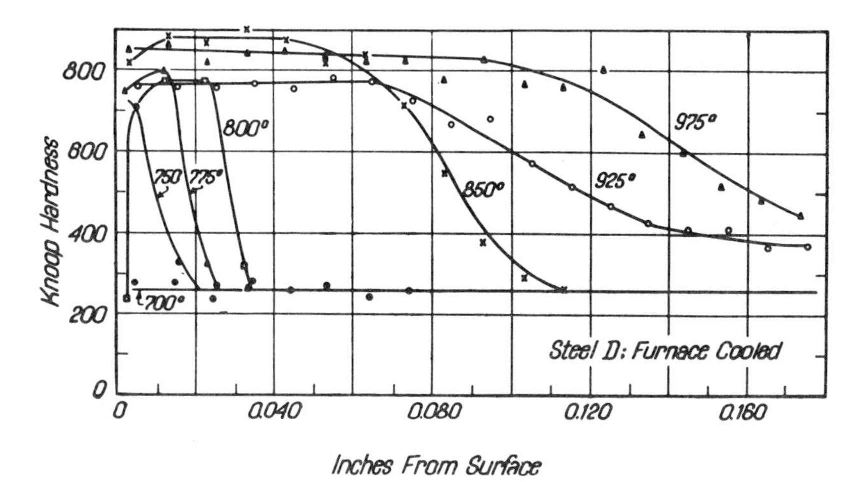

FIGURE 12. Hardness profile for an induction-hardened 0.80% C steel, for different maximum surface temperatures (from Martin, D. L. and F. E. Wiley. 1945. *Trans. ASM,* 34:351).

austenite formed. Note the coarse pearlite, and the regions that clearly have begun to form austenite, but that after quenching consist of a few undissolved carbides and martensite. Figure 13 shows the microstructure of this steel at the surface after induction-heating to 975°C, followed by quenching. Note that the undissolved pearlite seen in Figure 11(b) is not present. Figure 12 shows that the hardness is high at the surface.

The difficulty in dissolving carbides is shown by the microstructures in Figure 14 for a 1.10% C steel. The starting microstructure [Figure 14(a)] was a spheroidized structure. After induction-heating to 800°C, followed by quenching, the structure consisted of undissolved carbides and martensite [Figure 14(b)].

Figure 15 shows microstructures for a 0.30% C steel. The starting microstructure [Figure 15(a)] was primary ferrite and pearlite. When the maximum surface temperature was 850°C, the structure at the time of removal of the induction power, and prior to quenching, was ferrite and austenite. Then, after quenching, the structure was ferrite and martensite [Figure 15(b)]. Further from the surface, even more ferrite is present, as shown in Figure 15(c). If this steel is induction-heated to 975°C, the surface structure appears to be all martensite [Figure 15(d)]. The hardness profiles corresponding to these treatments are shown in Figure 16. Note that the hardness profile is about the same for heating to 850 and to 970°C, in spite of the presence of some primary ferrite for the lower temperature.

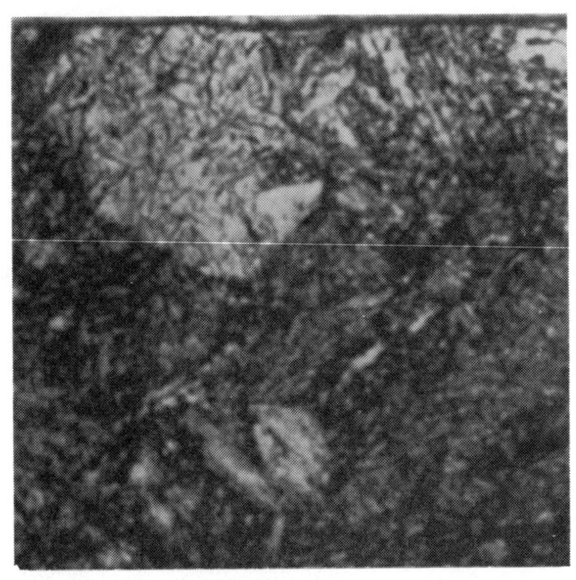

FIGURE 13. Microstructure (all martensite) of the edge (at top) of a 0.80% C steel that was induction-hardened with a surface temperature of 975°C (from Martin, D. L. and F. E. Wiley. 1945. *Trans. ASM*, 34:351).

2b. HARDNESS PROFILES

The hardness profile depends upon the heating cycle and the cooling cycle. Typical hardness profiles are shown in Figures 17 and 18 for laser-hardening and flame-hardening, respectively. Typical profiles for induction-hardening are shown in Figures 12 and 16. The factors that affect these profiles will be discussed in subsequent sections.

The depth of hardening is called the *case depth*. It can be defined in several ways, one of which is illustrated in Figure 19. It is an important parameter in setting up heat treatments, as it can be correlated to mechanical behavior of surface-hardened steel components.

2c. RESIDUAL STRESSES

The development of residual stresses during the formation of martensite in the surface layers is illustrated schematically in Figure 20. When the surface layers form the close-packed austenite, a contraction occurs that places these layers in tension and the center of the cylinder in compression [see Figure 20(b)]. However, the austenite is at a high temperature and has a relatively low strength compared to the center, which is at 25°C. Also

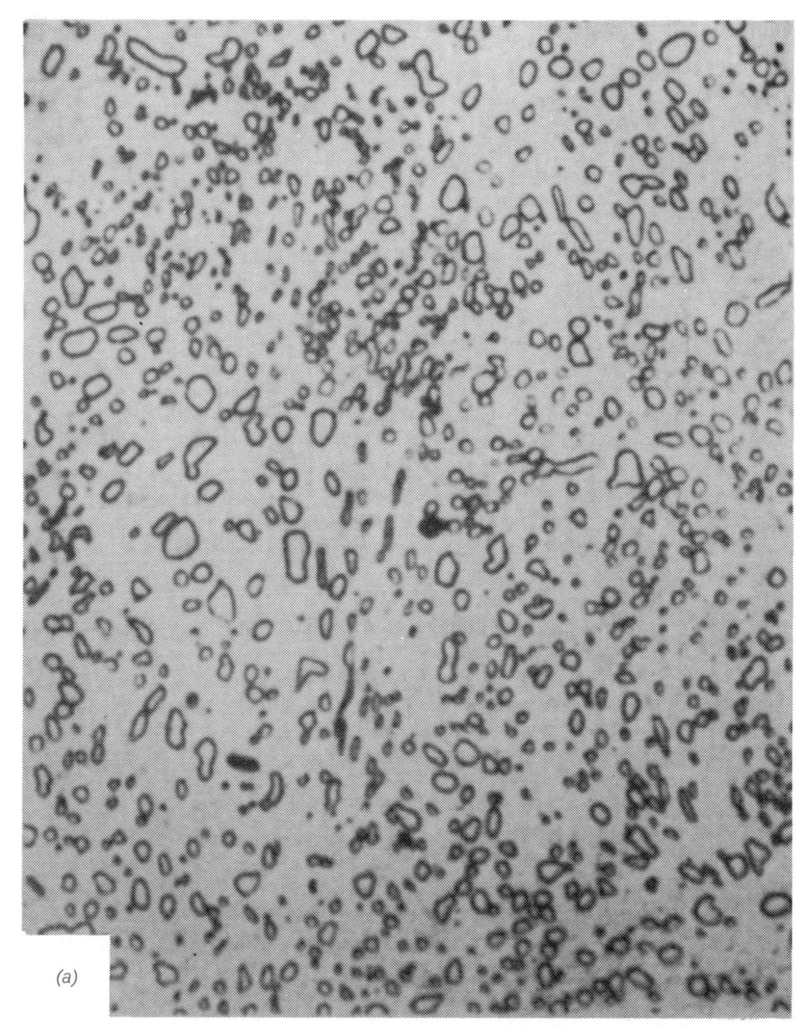

FIGURE 14. The microstructure at the surface of a 1.10% C steel heated to 800°C by induction-heating. The initial spheroidized microstructure is shown in (a).

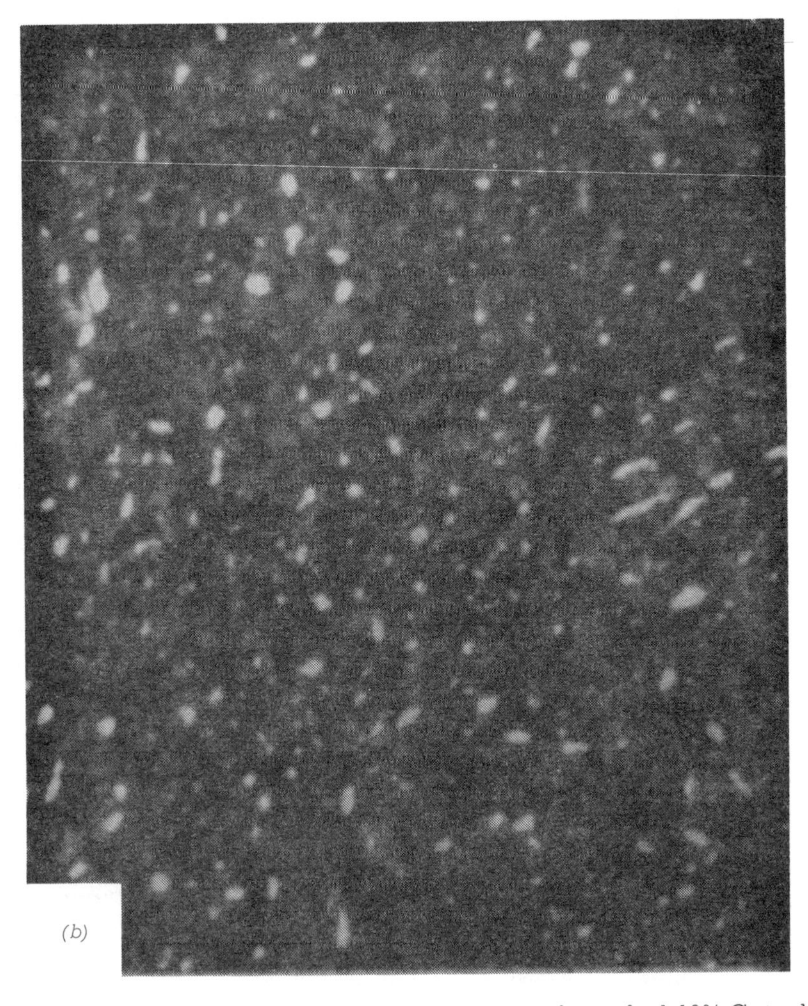

FIGURE 14 (continued). The microstructure at the surface of a 1.10% C steel heated to 800°C by induction-heating. In (b) is shown the microstructure after induction-heating, showing undissolved carbides (white spheroids) and martensite (1000×) (from Martin, D. L. and F. E. Wiley. 1945. *Trans. ASM*, 34:351).

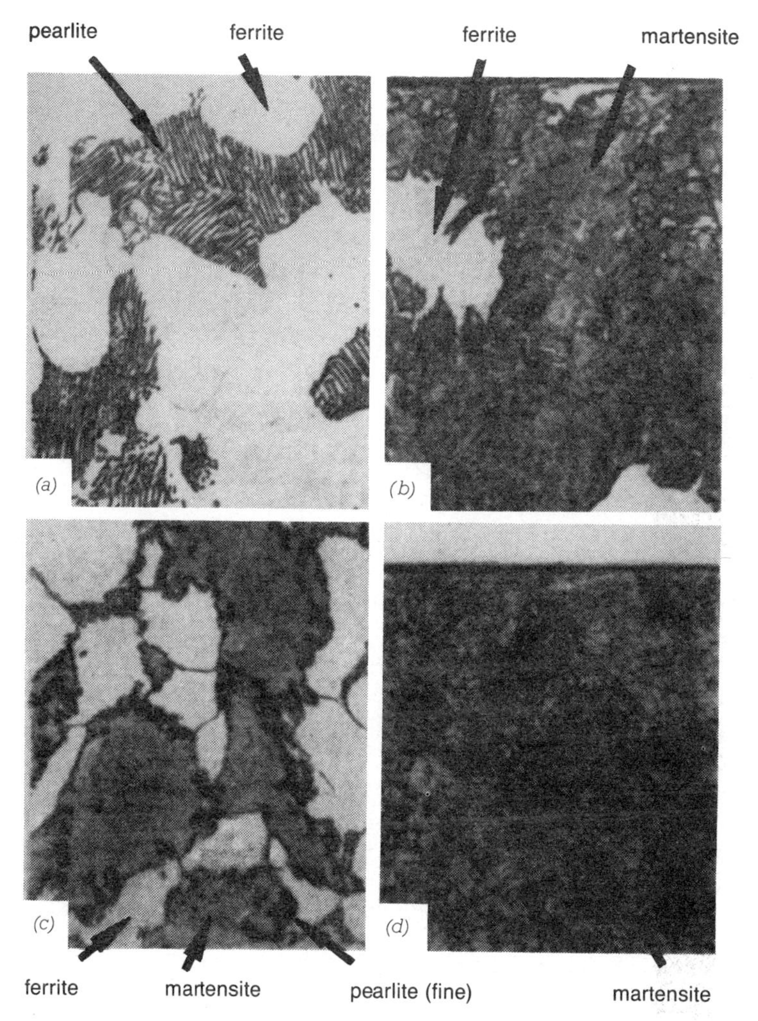

FIGURE 15. Microstructures of an induction-hardened 0.30% C steel. (a) Original furnace-cooled structure: primary ferrite and pearlite. (b) Heated to 850°C: martensite and ferrite. Taken at the surface. (c) Same sample as (b): martensite, ferrite and low carbon pearlite. Taken in the transition zone. (d) Heated to 975°C: martensite. Taken at the surface. All micrographs at 1000× (from Martin, D. L. and F. E. Wiley. 1945. *Trans. ASM,* 34:351).

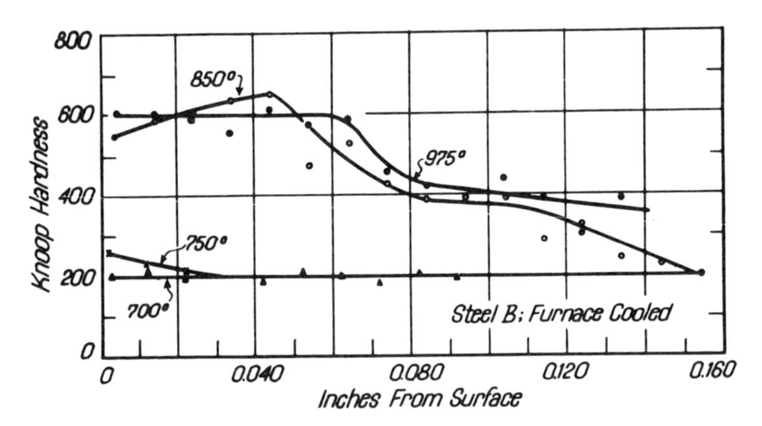

FIGURE 16. Hardness profiles for an induction-hardened 0.30% C steel (from Martin, D. L. and F. E. Wiley. 1945. *Trans. ASM*, 34:351).

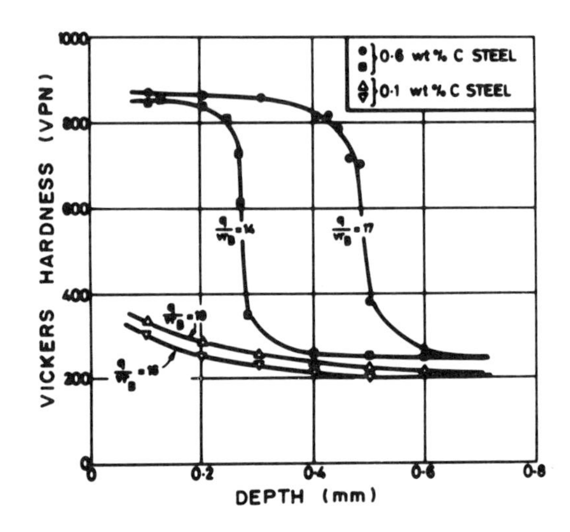

FIGURE 17. Hardness profiles for two steels that have been laser-hardened (reprinted with permission from Ashby, M. F. and K. E. Easterling. *Acta Metallurgica*, 32:1935, © 1984, Pergamon Press).

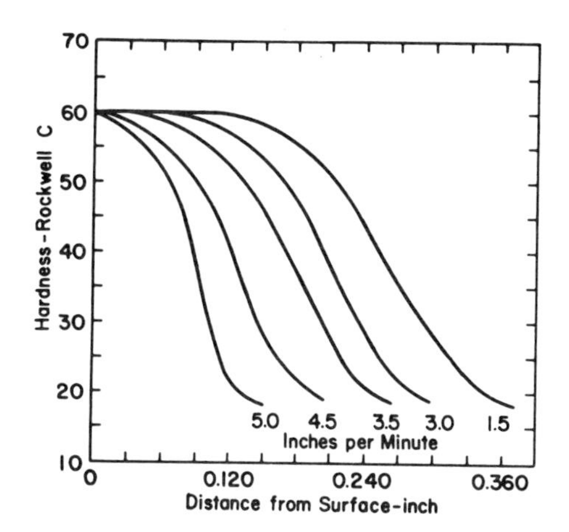

FIGURE 18. Hardness profiles of a flame-hardened 1050 grade steel forging (reprinted with permission from Harvey, P. D, editor. 1979. *Surface Hardening*. Metals Engineering Institute, Metals Park, Ohio: American Society for Metals).

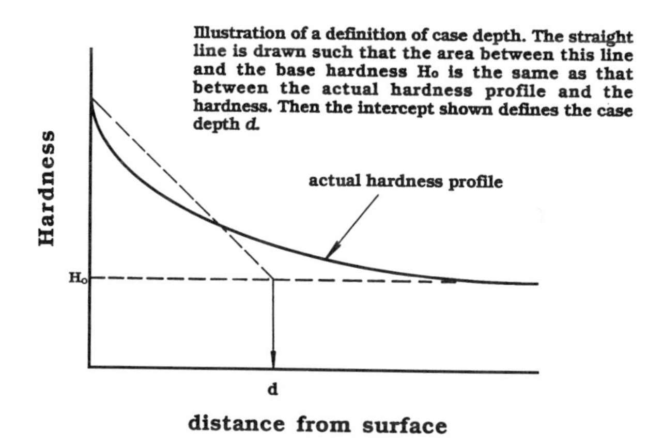

FIGURE 19. Schematic illustration of a definition of case depth.

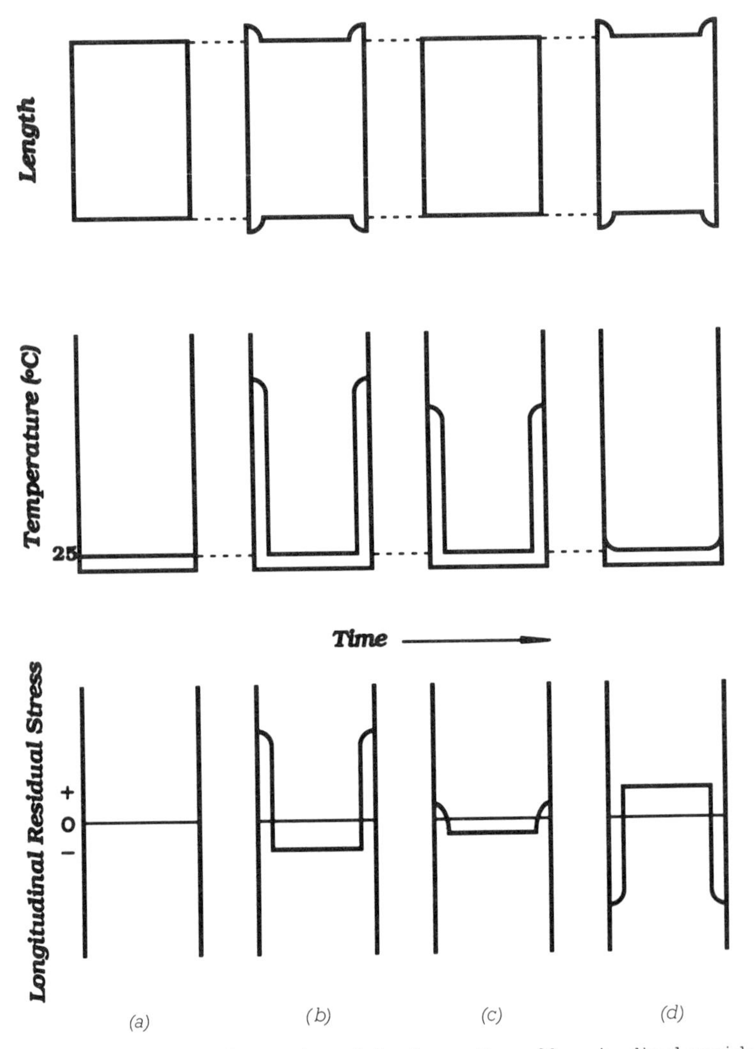

FIGURE 20. Schematic illustration of the formation of longitudinal residual stresses in a cylinder when surface-hardened.

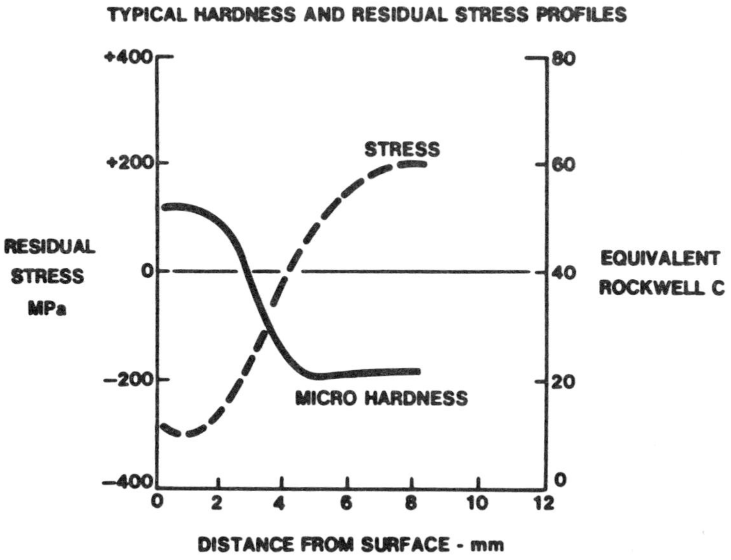

FIGURE 21. Typical longitudinal residual stress distribution in steel surface-hardened by induction-heating (reprinted with permission from Walker, E. D. 1981. In *Residual Stress for Designers and Metallurgists*, L. J. Vande Walle, editor, Metals Park, Ohio: American Society for Metals, as adapted from *SAE Handbook Series*, J784a. 1971. Warrendale, PA: Society for Automotive Engineers).

the center region is larger than the thin surface layer. Thus the surface undergoes plastic deformation and stress relaxation, and the residual stresses are reduced to low values [see Figure 20(c)]. Upon removal of the heat source, the surface layers cool. When martensite forms, the surface layers expand. This places the center in tension and the surface in compression [Figure 20(d)].

Figure 21 shows residual stresses for an induction-hardened surface. Also shown is the hardness profile. Note the high longitudinal compressive residual stress at the surface, as predicted in the description above.

2d. INDUCTION-HARDENING

Heating by induction produces temperature profiles that are different from those produced by techniques that supply the energy only to the surface. In flame- and laser-hardening, the surface is the only location of energy input to the steel. However, in induction-heating, high-frequency electromagnetic radiation is coupled to the steel part, producing induced

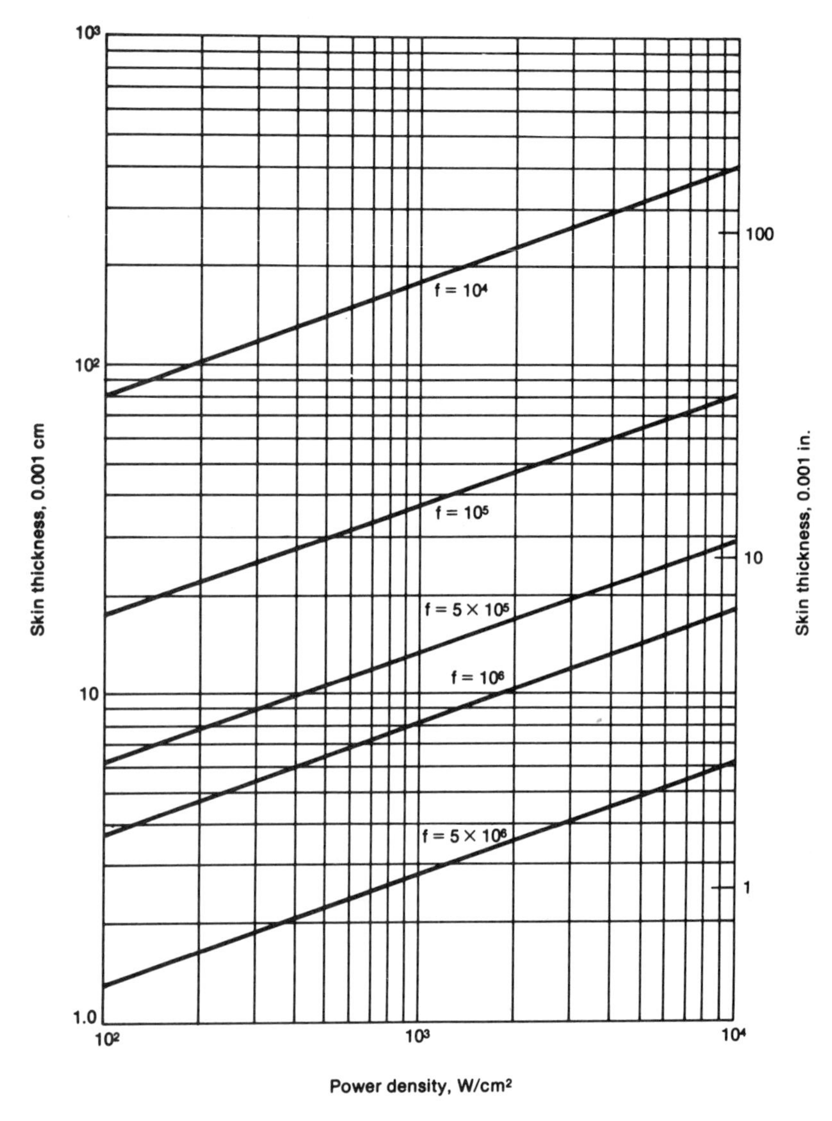

FIGURE 22. Skin depth of steel as a function of power density and frequency (reprinted with permission from Semiatin, S. L. and D. E. Stutz. 1986. *Induction Heat Treatment of Steel*. Metals Park, Ohio: American Society for Metals, as adapted from Brown, G. H., C. N. Hoyler and R. A. Bierwirth. 1947. *Theory and Application of Radio-Frequency Heating*. New York: Van Nostrand).

currents that cause the heating in the steel. Thus, heating of the steel does not occur just on the surface, but also in the surface layers. A key to the induction-hardening method is the correlation of the radiation's depth of penetration to its frequency. As the frequency increases, the current is concentrated more in the surface layers. The relation between the approximate depth of the affected layer, called the skin depth, is

$$d = 5000\ (\varrho/\mu f)$$

where d is in cm, ϱ is the electrical resistivity of the steel (in ohm-cm), μ is the relative magnetic permeability and f is the frequency of the alternating magnetic field (in hertz). Here d is the distance from the surface at which the induced field strength and current are $1/e$, or 37%, of that at the surface.

The depth of heating depends upon the frequency and the power density, as shown in Figure 22. In this plot, specific values of ϱ and μ typical of steels have been used. The depth is determined by the temperature dependence of these parameters (see Figure 23). The power required depends upon the temperature desired, and this depends upon the heat transfer of the process. The steel surface loses heat by radiation to the environment, and heat is conducted from the heated layers into the colder center of the part. Thus, the power required depends upon the size and thermal conductivity of the steel and upon the surface radiation properties.

The hardness profile, usually indicated by the case depth, is controlled by the parameters of the induction-heating process. (These are treated in detail in references that deal specifically with induction-heating. See, for example, Semiatin, S. L., D. E. Stutz and I. L. Harry. 1986. *Induction Heat Treatment of Steel.* Metals Park, Ohio: American Society for Metals.) Figure 24 shows the relationship between the heating time to attain a given case depth, the frequency used, and the surface power density. Note that for a given heating time, the case depth is greater the lower the frequency.

It was mentioned earlier that austenitizing for surface-hardening uses a relatively short austenitizing time, so that the nucleation and growth of austenite is of more concern than in conventional austenitizing where the time is long (e.g., 1 hour). This is illustrated by the data in Figure 25, which show hardness profiles for three steels as a function of the maximum surface temperature attained before quenching. (Note that for surface-hardening, only the lower temperature curves are applicable; the curve, for example, for 900°C corresponds to hardening almost to the center of the bar.) For a maximum temperature of 790°C, the case depth is about the same (0.012 inches) for the 1350 and 2350 steel, as is the surface hardness. But for the higher alloy 4160 steel, the case depth is considerably less

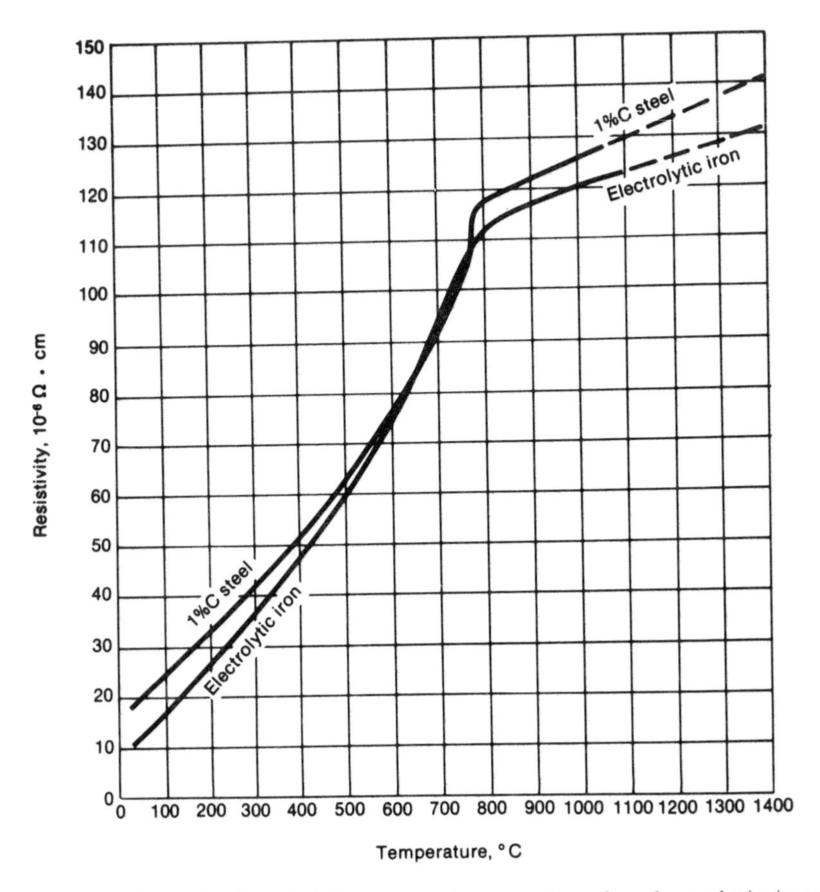

FIGURE 23. Electrical resistivity versus temperature for electrolytic iron and for a 1% C steel (adapted with permission from Stansel, N. R. 1949. *Induction Heating*. New York: McGraw-Hill).

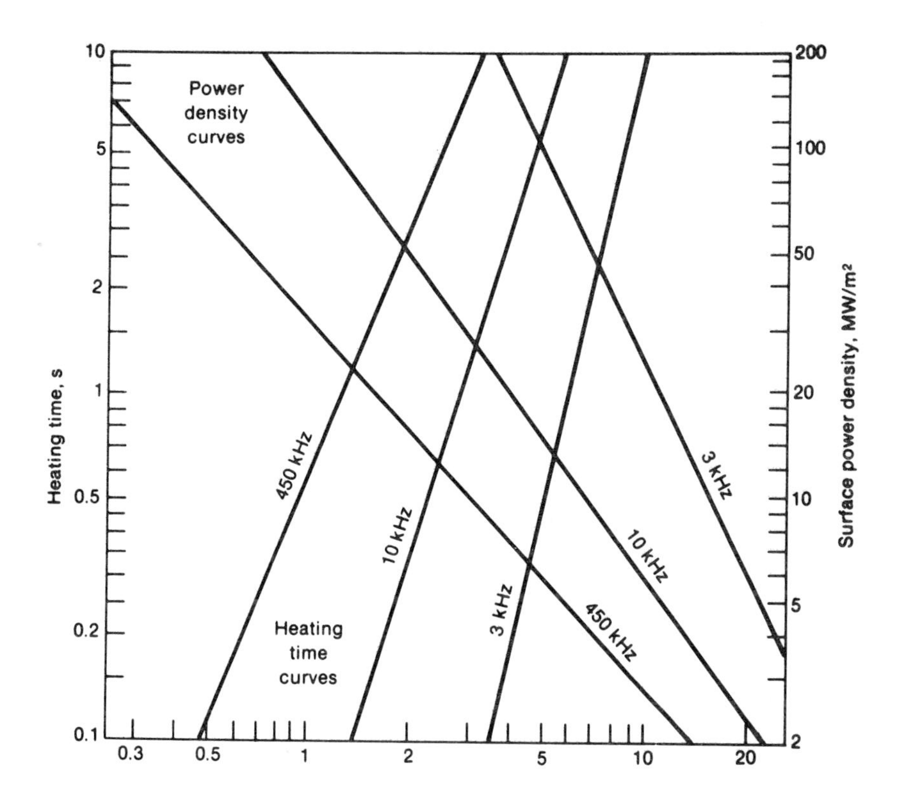

FIGURE 24. Interrelationship among heating time, surface power density and hardened depth for various induction generator frequencies (adapted with permission from Lazinskii, M. G. *Industrial Applications of Induction Heating*, © 1969, Pergamon Press PLC).

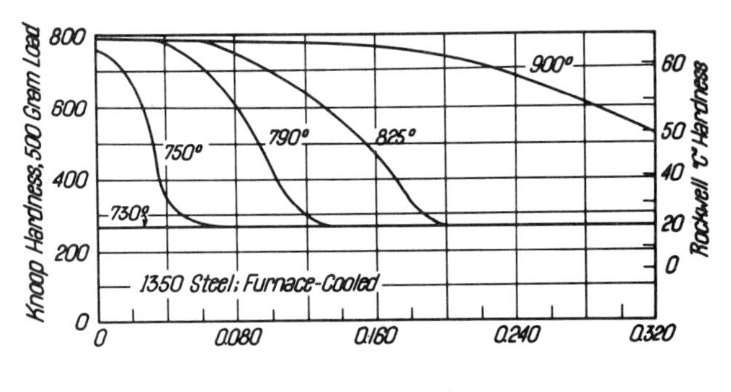

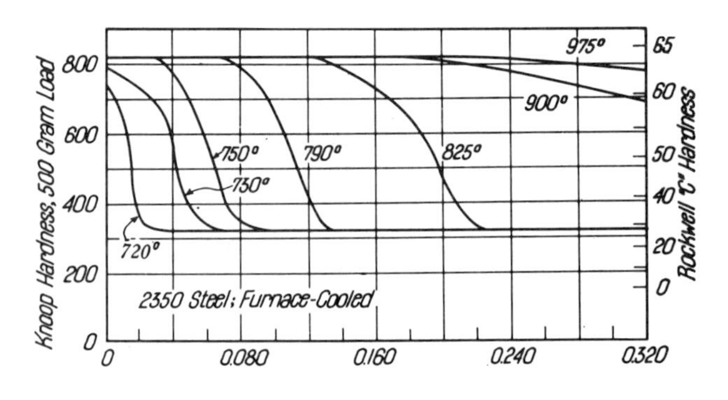

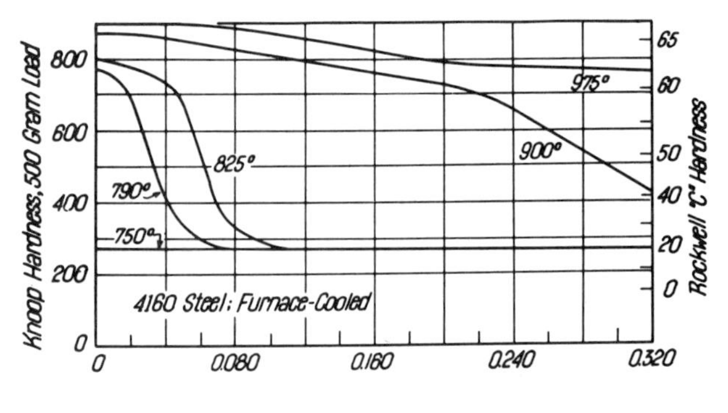

FIGURE 25. Hardness profiles for three steels which have been induction-hardened (from Martin, D. L. and W. G. Van Note. 1946. *Trans. ASM*, 36:210).

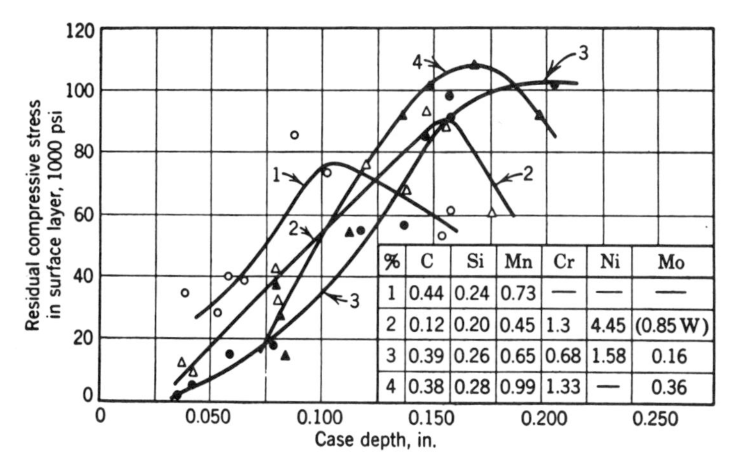

%	C	Si	Mn	Cr	Ni	Mo
1	0.44	0.24	0.73	—	—	—
2	0.12	0.20	0.45	1.3	4.45	(0.85 W)
3	0.39	0.26	0.65	0.68	1.58	0.16
4	0.38	0.28	0.99	1.33	—	0.36

FIGURE 26. Longitudinal residual stresses in the surface layers of induction-hardened cylinders as a function of case depth (reprinted with permission from Horger, O. J. In *Handbook of Experimental Stress Analysis*, M. Hetenyi, editor, © 1950, John Wiley & Sons, Inc., p. 459).

(0.007 inches), as is the surface hardness. This reflects the difficulty of forming austenite in the higher alloy steel.

The residual stresses depend upon the depth of hardening, as is illustrated by the data in Figure 26. If the depth is too great, then the surface heat treatment begins to approach conventional through-hardening. The advantage of induction-hardening in improvement of fatigue life is well established. Figure 27 shows fatigue curves for an unhardened steel, and for induction-hardened samples of the same steel for different case depths. The importance of case depth is clearly indicated.

The advantage of induction-hardening an actual steel component is shown by the data for tractor axles in Figure 28. Here the comparison is between conventionally quenched and tempered axles and those that were induction-hardened. Note that the fatigue life is approximately doubled by induction-hardening. This effect is also shown in Figure 29.

Another advantage accrued by induction-hardening is a somewhat higher "martensite" hardness. Usually, the hardness of martensite is considered to be dependent only on the carbon content of the martensite, and independent of the alloy content and prior austenite grain size. However, in steels austenitized for short times, then quenched, the hardness is sometimes harder than for martensite formed from homogeneous austenite. This effect is sometimes referred to as superhardness and is shown in Figure 30. It may be due to an unusually fine prior austenite grain size that transforms to a very fine martensite, and to undissolved carbides.

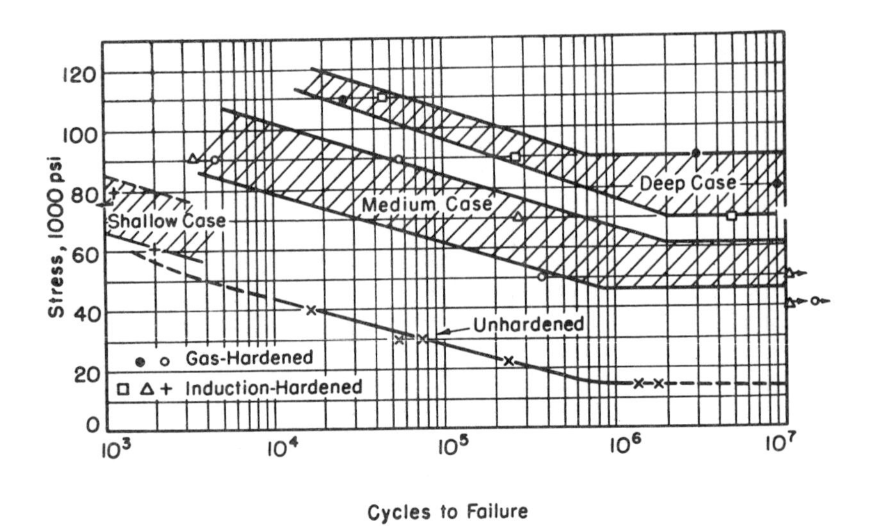

FIGURE 27. Fatigue curves from reversed bending tests of unhardened and of surface-hardened steels of different case depths (from Case, S. L., J. M. Berry and H. J. Grover. 1952. *Trans. ASM*, 44:667).

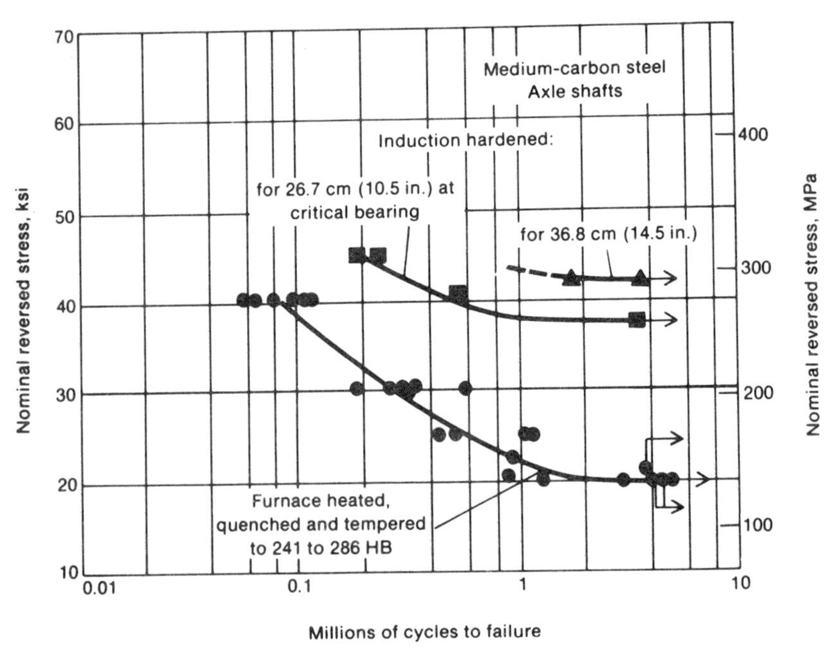

FIGURE 28. Bending fatigue response of furnace-hardened and of induction-hardened medium-carbon steel tractor axles (reprinted with permission from Spencer, T. H. et al. 1964. *Induction Hardening and Tempering*. Metals Park, Ohio: American Society for Metals).

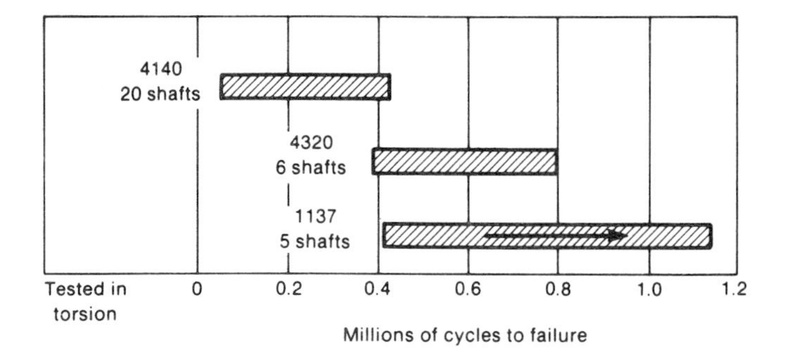

Steel	Surface hardness, HRC	Method of hardening
4140	36 to 42	Through-hardened
4320	40 to 46	Carburized to 1.0 to 1.3 mm (0.040 to 0.050 in.)
1137	42 to 48	Induction hardened to 3.0-mm (0.120-in.) min effective depth and 40 HRC

Arrow in lower bar (induction hardened shafts) indicates that one shaft had not failed after testing for the maximum number of cyles shown.

FIGURE 29. Comparison of fatigue life of induction surface-hardened transmission shafts with that of through-hardened and carburized shafts (reprinted with permission from Spencer, T.H. et al. 1964. *Induction Hardening and Tempering.* Metals Park, Ohio: American Society for Metals).

Figure 31 shows temperature–time curves for induction-heating, followed by cooling, of a roll of a cold rolling mill. Note that at 5 mm from the surface the maximum temperature was about 1000°C, and this location was in the austenite region (from about 800 to 1000°C) for about 400 sec. The microstructure in the surface layers and the hardness profile for this roll are shown in Figure 32.

Tempering following induction-hardening can be carried out using induction-heating. The effect on hardness profiles is illustrated in Figure 33.

The effect of prior structure on the hardness profile produced by induction-hardening is illustrated by the following example. Axles 4.1 cm in diameter of a 1041 steel were austenitized, then air-cooled prior to surface-hardening. One axle was cooled such that the structure was 10% primary ferrite–90% pearlite, and the hardness was Rockwell C 22. Another axle was cooled more slowly, producing 30% primary ferrite–70% pearlite, with a hardness of 15 Rockwell C. (Equilibrium cooling will produce 50% primary ferrite–50% pearlite.) Then the surface of the axles

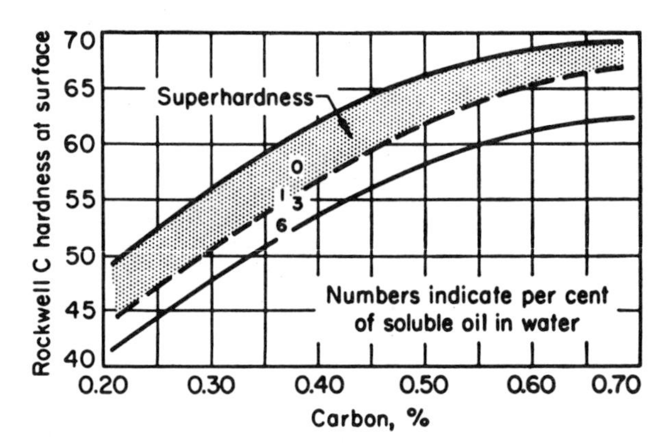

FIGURE 30. Superhardness produced by induction-hardening compared to the hardness produced by conventional furnace-hardening (reprinted with permission from 1964. *Metals Handbook, 8th Edition, Vol. 2.* Metals Park, Ohio: American Society for Metals).

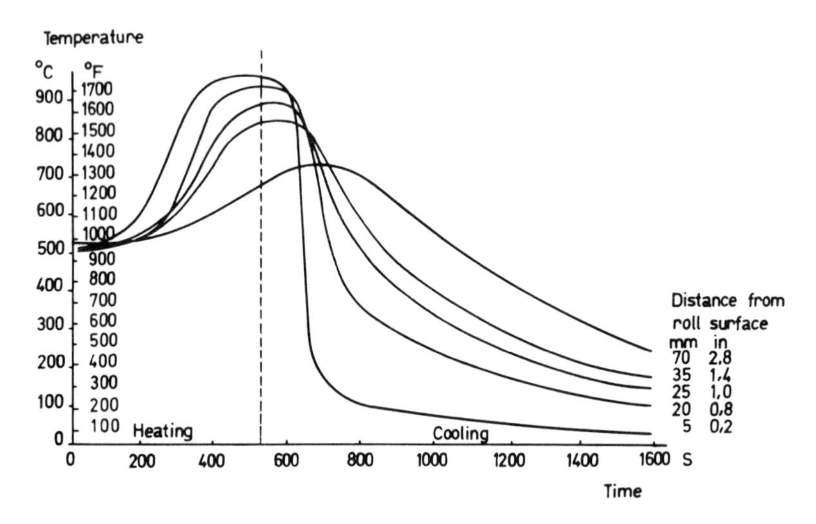

FIGURE 31. Temperature–time curves for induction-hardening of a 30 cm diameter cold rolling roll. The roll had been preheated to 500°C prior to induction-heating (reprinted with permission from Thelning, K.-E. 1975. *Steel and Its Heat Treatment.* London: Butterworths).

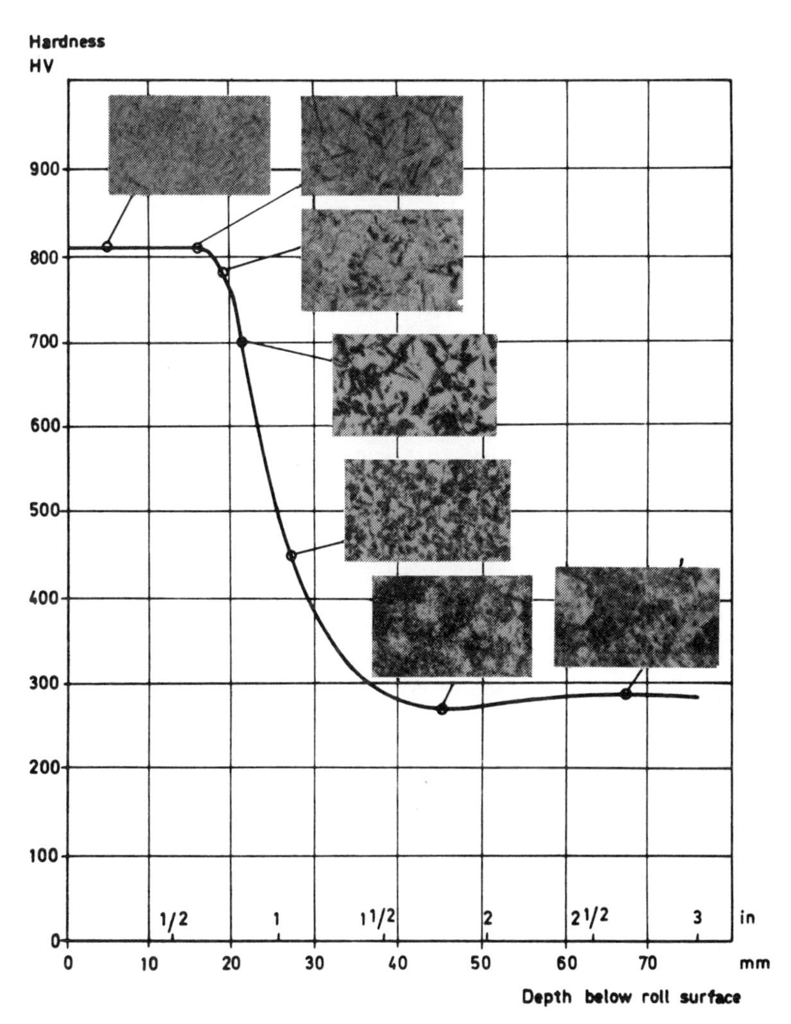

FIGURE 32. Hardness and microstructure in the surface zone of an induction-hardened roll for cold rolling (reprinted with permission from Thelning, K.-E. 1975. *Steel and Its Heat Treatment.* London: Butterworths).

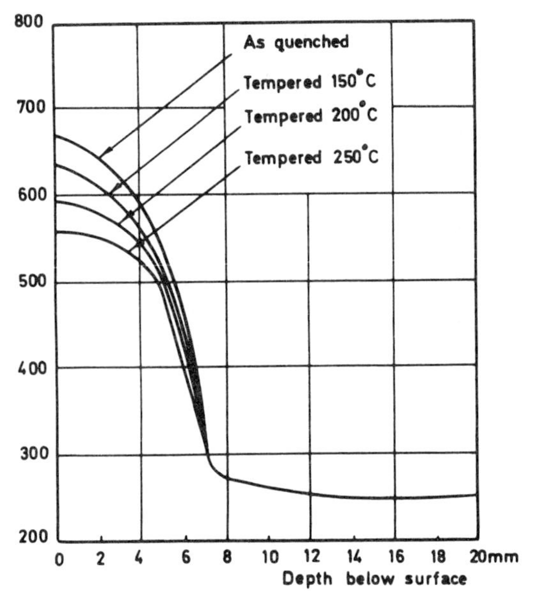

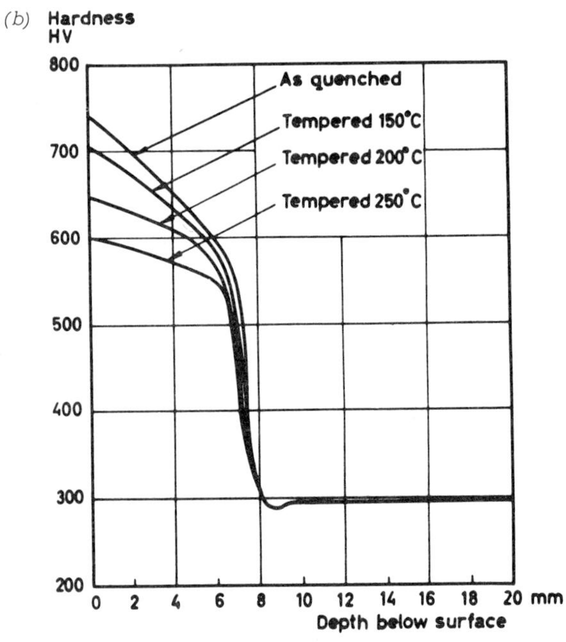

FIGURE 33. Hardness profiles of induction-hardened steels and after tempering. The steels were (a) AISI 4135 (Bofors R O 752) and (b) AISI 4142 (Bofors R O 952). The samples were induction-heated (frequency 10,000 Hz) to 900°C for 25 sec, and subsequently tempered at the temperatures shown (reprinted with permission from Thelning, K.-E. 1975. *Steel and Its Heat Treatment.* London: Butterworths).

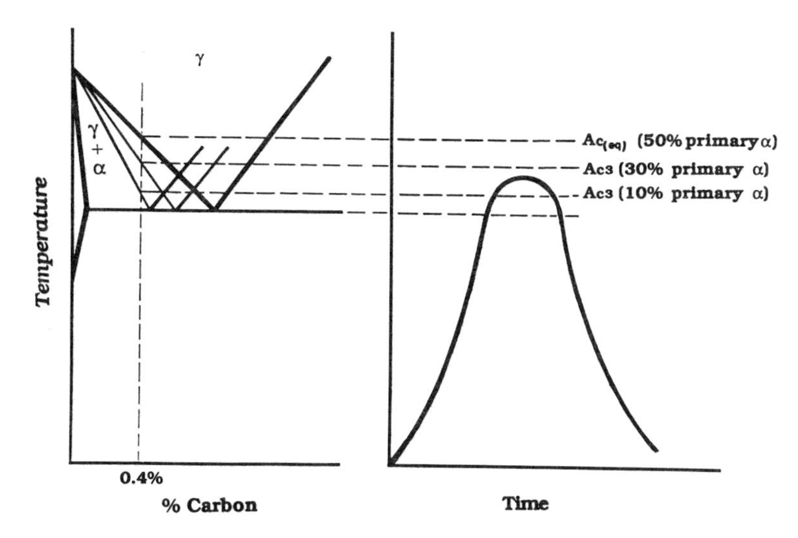

FIGURE 34. Schematic diagram illustrating the effect of beginning microstructure on the temperature above which austenite can form in a 0.4% C steel.

was heated at about 110°C/sec. As shown in Figure 34, the eutectoid composition is different for the two structures, and hence the A_{c3} temperatures are different. Thus at 6 mm below the surface, even though both axles had the same heating and cooling curves, the axle with a structure of 10% ferrite will be in the austenite region, and hence will form all martensite upon quenching, whereas that with 30% ferrite will be in the ferrite–austenite region, and will form a structure containing only 70% martensite. Thus, the axle having a beginning structure of 30% primary ferrite–70% pearlite prior to surface-heating will have a lower hardness at this 6 mm location. Figure 35(a) shows the hardness profiles for the two axles. At the 6 mm depth, the hardness values for the two beginning structures differ by about 9 Rockwell C. This can be a significant difference, depending upon the case depth criteria.

The above description neglects the effect of heating rate on the A_{c3} temperature. However, this can be taken into account by using Figure 8, which shows the effect of heating rate and prior structure on the A_{c3} temperature. At a heating rate of 110°C/sec, the A_{c3} temperature for the normalized 1042 steel is about 870°C, whereas it is about 900°C for the annealed steel. The annealed steel will contain 50% primary ferrite–50% pearlite. The amount of primary ferrite–pearlite in the normalized steel depends upon the cooling rate, but here it is not specified. However, we can use these data to analyze the response of the axles to surface-hardening. The maximum temperature as a function of depth is shown in Figure 35(b). The A_{c3}

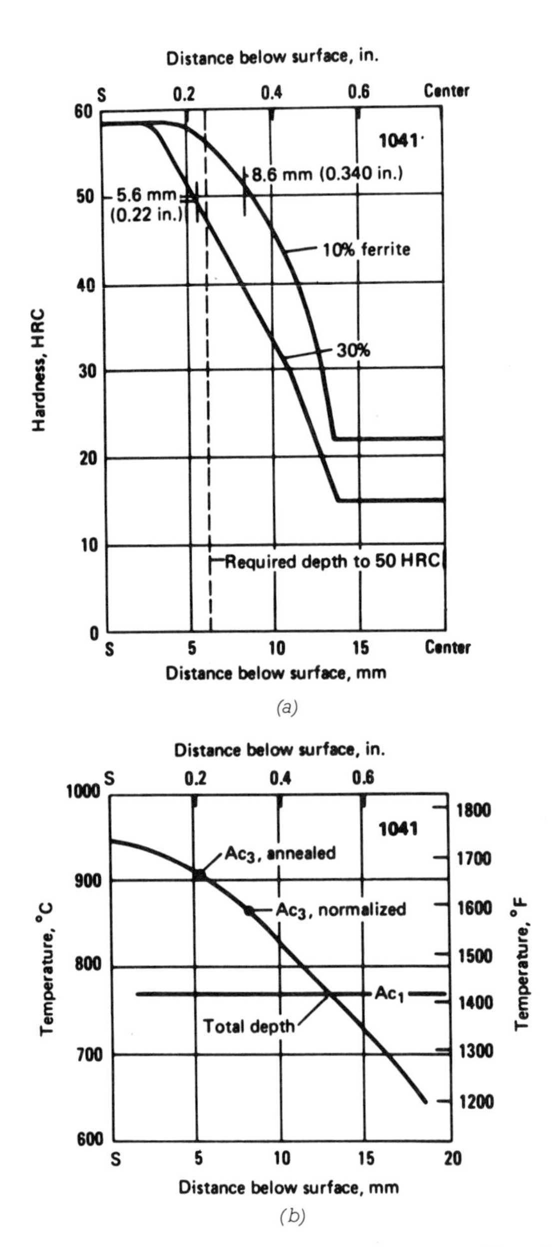

FIGURE 35. Hardness profiles and temperature profiles for induction-hardened axles of 1041 steel (reprinted with permission from *Metals Handbook, 9th Edition, Vol. 4, Heat Treating.* 1981. Metals Park, Ohio: American Society for Metals).

temperatures from Figure 8 are noted. For the annealed structure, all austenite (and hence all martensite) cannot be formed below 5.6 mm from the surface; for the normalized structure, this is 8.6 mm. (The location at which all martensite actually forms is less than these values, because of the finite time required to convert the beginning structure to all austenite.)

The effect of prior structure is also illustrated by Figure 36. Shafts made of 1070 steel were heat-treated (prior to surface-hardening) by annealing, by normalizing, and by quenching and tempering. Then the shafts were surface-hardened using a heating rate of about 950°C/sec, followed by water quenching. The hardness profiles are shown in Figure 36(a). The maximum temperature as a function of depth is shown in Figure 36(b). The A_{c3} temperatures are labeled on this curve, obtained from data such as in Figure 8. As above, the finer structure of tempered martensite formed more martensite, and thus this prior treatment produced a surface-hardened steel with a greater case depth.

The sequence of the phase transformations upon cooling during surface-hardening can be estimated by calculation. Figure 37 shows the temperature–time curves at different depths for a 30 cm diameter cylinder of a 0.74% C steel when induction-heated. The beginning structure was all pearlite. The cylinder was preheated to 550°C prior to induction-heating. After 400 sec, it was cooled in air for 80 sec, then quenched with water for 1 hr. The location 7 mm below the surface followed almost the same heating and cooling curve as that at the surface. The amount of microconstituents present as a function of depth after cooling to 15°C is shown in Figure 38. Note that there is some retained austenite present. Below about 50 mm, the steel did not attain a sufficiently high temperature to form austenite. The calculated longitudinal stresses as a function of depth and of cooling time are shown in Figure 39. (The times chosen for the calculation were not short enough to show the initial formation of tensile stresses at the surface.)

2e. LASER-HARDENING

In laser-hardening of steels, a laser beam is used to heat the surface for a short time, then this layer cools rapidly by conduction into the underlying cold metal, producing a hard surface layer. This method is distinct from induction-hardening and flame-hardening in that the energy is put into the steel in a very short time, with the result that the surface layer affected is quite thin.

There are a number of advantages in laser-hardening. Since only a thin layer is formed, the total energy input is relatively small. The heat-affected zone is small, and less distortion is accrued. The laser beam can be con-

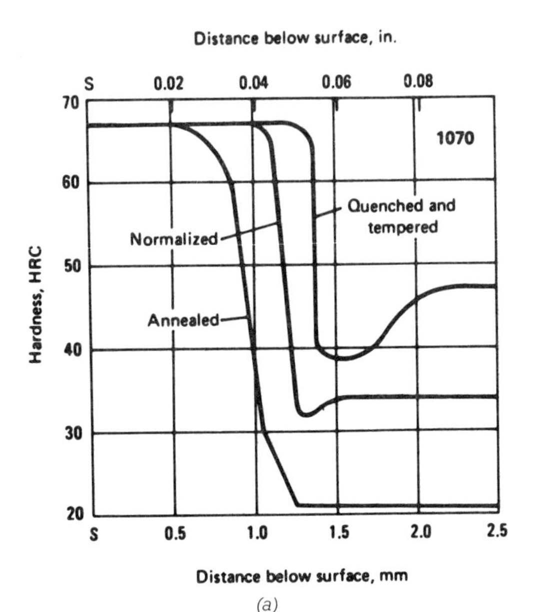

(a)

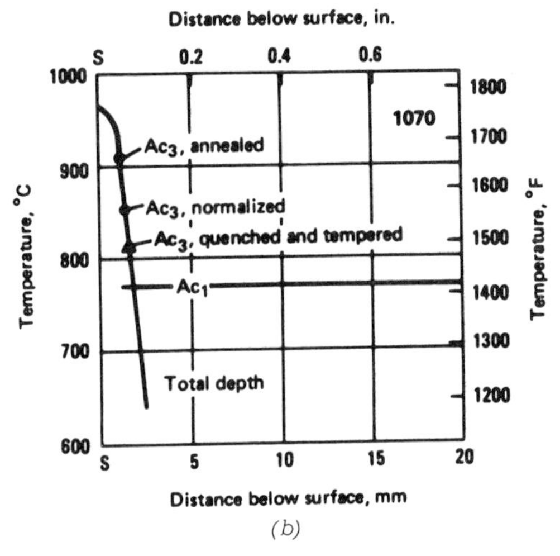

(b)

FIGURE 36. Hardness profiles and temperature profile for induction-hardened shafts of 1070 steel (reprinted with permission from *Metals Handbook, 9th Edition, Vol. 4, Heat Treating.* 1981. Metals Park, Ohio: American Society for Metals).

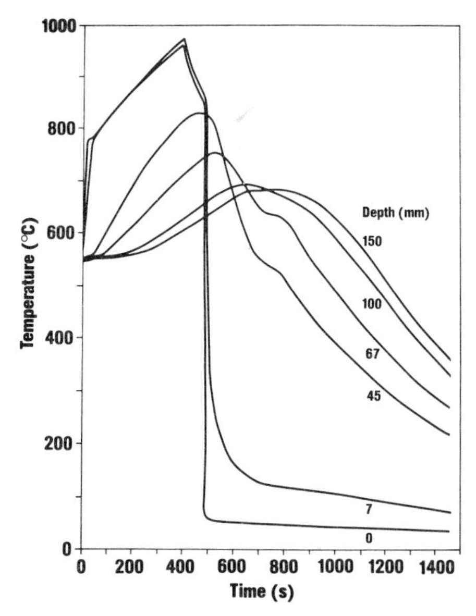

FIGURE 37. Calculated temperature–time curves in an induction-hardened cylinder (30 cm diameter) made of a 0.74% C steel. The cylinder was preheated to 550°C prior to induction-heating. The different curves describe the temperature at the distance from the surface (reprinted with permission from Ericsson, T. and B. Hildenwall. 1981. In *Residual Stress and Stress Relaxation*, E. Kula and V. Weiss, editors, New York: Plenum).

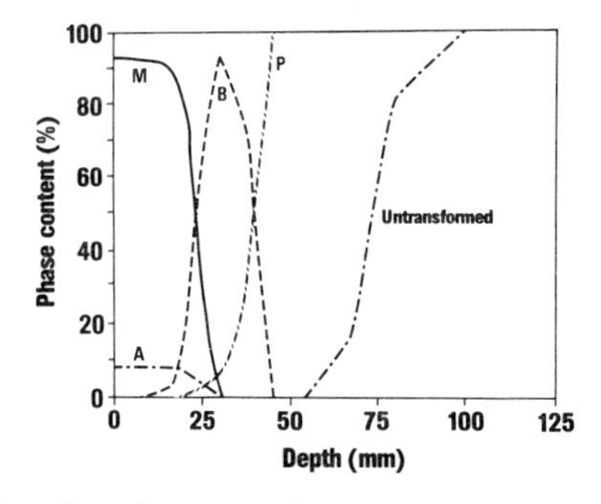

FIGURE 38. Calculated microconstituent content versus depth after induction-hardening of a 30 cm diameter, 0.74% C cylinder. The original pearlite (to the right) remains untransformed (reprinted with permission from Ericsson, T. and B. Hildenwall. 1981. In *Residual Stress and Stress Relaxation*, E. Kula and V. Weiss, editors, New York: Plenum).

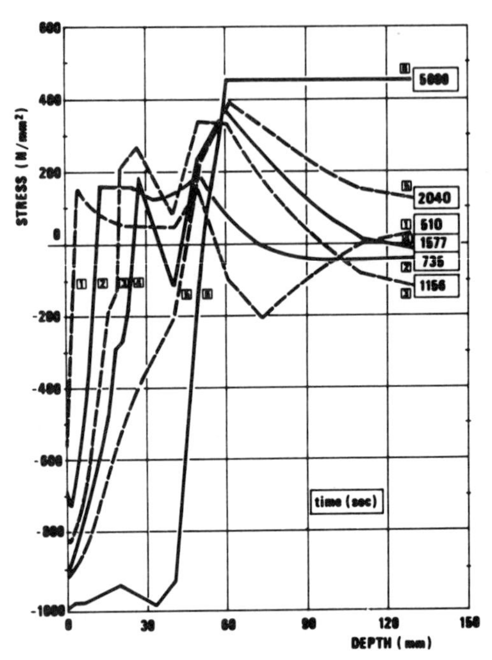

FIGURE 39. Calculated longitudinal residual stresses as a function of depth and time during quenching an induction-heated 30 cm diameter, 0.74% C cylinder (reprinted with permission from Ericsson, T. and B. Hildenwall. 1981. In *Residual Stress and Stress Relaxation*, E. Kula and V. Weiss, editors, New York: Plenum).

trolled well, so that good control of the hardening is achievable. Selective surface-hardening can be conducted by control of the beam movement. Since only a thin layer is affected, self-quenching by the cold center of the component is used for cooling, and thus no other quench is required. An important disadvantage to the use of the laser for surface treatment, however, is that the surface usually has to be covered with an energy-absorbing coating, such as graphite. Also there is the high capital cost of the laser and ancillary equipment.

In laser-hardening, the depth of the affected region is about 1 mm, so that experimentally it is difficult to measure the temperature profiles. Instead, these are obtained by solution of the appropriate heat transfer equations. (In the following section, the treatment described by Amende, cited in Figure 40, is given.) Consider a flat surface heated with a laser beam of average power density q_b for time t_c. Limiting attention to only heat flow normal to the surface in direction Z, the equation to be solved to obtain the temperature T as a function of depth Z and time t is

$$dT/dt = (\alpha)\,(d^2T/dZ^2)$$

where K is the thermal diffusivity of the steel. Using an average value for K, and assuming a fixed value for the absorptivity of the laser beam on the surface, the following solution is obtained:

$$T(Z,t) = \frac{2\epsilon q_b}{K}\sqrt{\alpha t_c} \cdot i \text{ erf } c\left(\frac{Z}{2\sqrt{\alpha t_c}}\right)$$

Here i erf c is the integrated complementary error function, given by

$$i \text{ erf } cx = \int_x^{\infty} \text{erf} \times dy = \int_x^{\infty} (1 - \text{erf } x)dy$$

where the error function is

$$\text{erf } x = \frac{2}{\sqrt{\pi}} \int_o^x e^{-Y^2} dy$$

K is the thermal conductivity, ϵ is the absorptivity, q_b the power density, and t_c the heating time. During cooling, after the laser beam is removed at time t_c, solving the appropriate equation yields the relation

$$T(Z,t) = \frac{2\epsilon q_b}{K}\sqrt{\alpha}\left[\sqrt{t} \cdot i \text{ erf } c\left(\frac{Z}{2\sqrt{\alpha t}}\right) - \sqrt{t - t_c} \cdot i \text{ erf } c\left(\frac{Z}{2\sqrt{\alpha(t - t_c)}}\right)\right]$$

By setting $Z = 0$, the surface temperature T_0 during heating is

$$T_0(t) = \frac{2\epsilon q_b}{K}\sqrt{\frac{\alpha t}{\pi}}$$

Figure 40 shows the calculated heating and cooling curves for gray cast iron using a power density of 1500 watts/cm^2. The laser power was on for 0.8 sec. These curves illustrate three important points about laser hardening. One is that the heating rate is quite high, about 1000°C/sec. Another is that the depth that is heated sufficiently to attain austenite is about 1 mm. The other is that the time to cool from the austenite region to about 500°C (to about the nose of the CCT diagram) is only about 1 sec.

The description above is for a stationary laser beam. However, in most applications, it is desirable to harden an area considerably larger than that

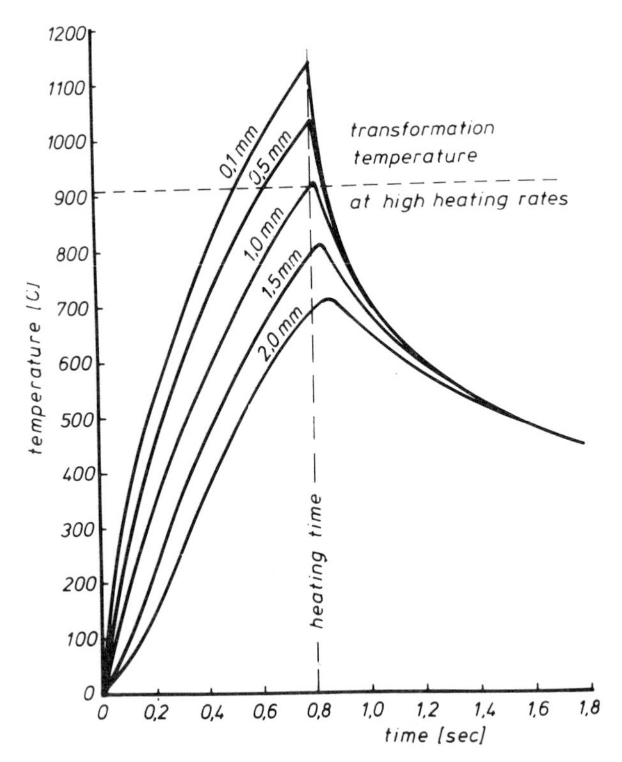

FIGURE 40. Calculated temperature–time curves for a laser-heated pearlitic gray cast iron (reprinted with permission from Amende, W. 1984. In *Industrial Applications of Lasers*, H. Koebner, editor, © John Wiley & Sons, Ltd., p. 79).

which can be covered by a stationary beam. Thus, the steel component must be moved relative to the beam. In this situation, the temperature as a function of location and time depends also on the velocity of movement. Thus, the temperature at a given point in the steel will have a different time response than when a stationary laser beam is used. This is illustrated in Figure 41. To calculate the temperature–time relations, the heat transfer equation used above for a stationary beam is modified to take into account the movement of the beam. The laser beam is assumed to have dimensions such as a and b, as shown in Figure 42.

The mathematical solution to this complicated problem has been obtained, in which the temperature dependence of the thermal conductivity, the heat capacity, and the surface absorptivity has been taken into account. The heat loss from the surface by both radiation and convection after the laser beam has passed a given position was found to be negligible (less than 1% of the total heat absorbed).

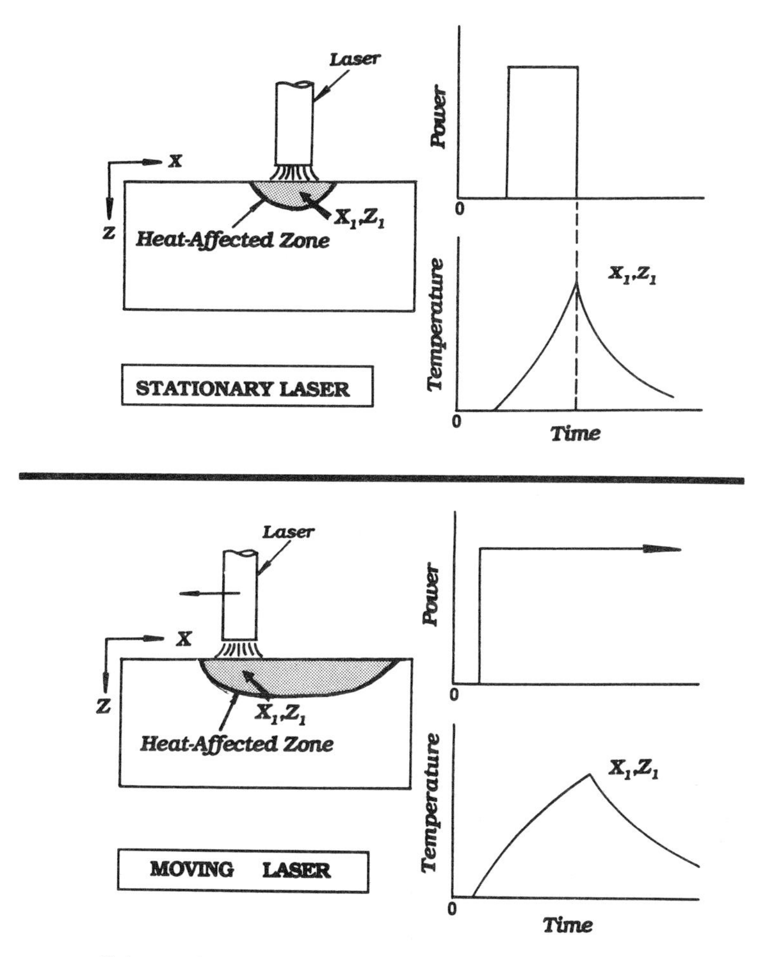

FIGURE 41. Schematic illustration of the difference in the temperature-time curve for surface-heating using a stationary laser and a moving laser.

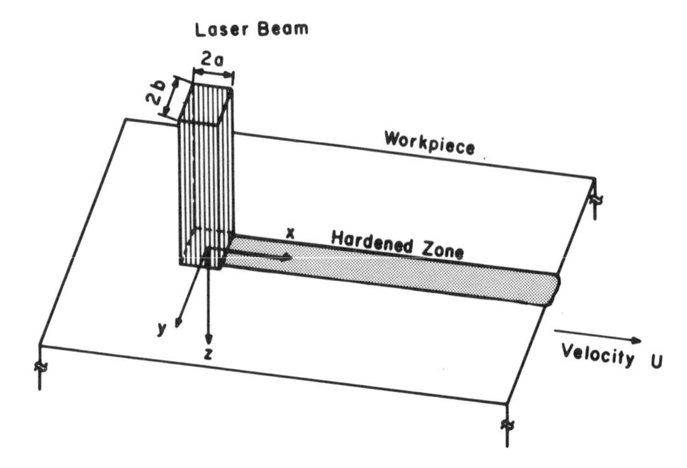

FIGURE 42. Schematic sketch showing the model used to calculate the temperature–time response in laser-heating (reprinted with permission from Kou, S., D. K. Sun and Y. P. Le. 1983. *Met. Trans.*, 14A:643, a publication of The Minerals, Metals and Materials Society, Warrendale, PA).

Figure 43 shows the temperature as a function of depth and distance. The plot is in terms of the dimensionless quantities

$$T^* = [K\alpha(T_0 - T)/(QU)]$$
$$x^* = xU/\alpha$$
$$y^* = yU/\alpha$$
$$z^* = zU/\alpha$$

Here α is the thermal diffusivity, K is the thermal conductivity, Q is the power absorbed by the sample, T_0 is the initial temperature and U is the sample velocity. At a given location, the peak temperature occurs soon after the beam passes.

A very important consideration in laser-hardening is that melting be avoided. Using a beam that is too powerful will cause this. Figure 44 shows the influence of beam power on the depth of the heat affected zone. The sharp change in slope corresponds to melting. Thus, in this case, to avoid laser melting the power would have to be below about 5 Kw.

Figure 45 shows the temperature as a function of distance x and time. Note that these two quantities are coupled, as shown on the abscissa, through the travel speed (U). The shape of the curves is similar to that shown in Figure 40 for a stationary beam. Also in this case the time to attain maximum temperature is less than one second, and the time to cool to about 500°C from the peak temperature at the surface is about 1 sec.

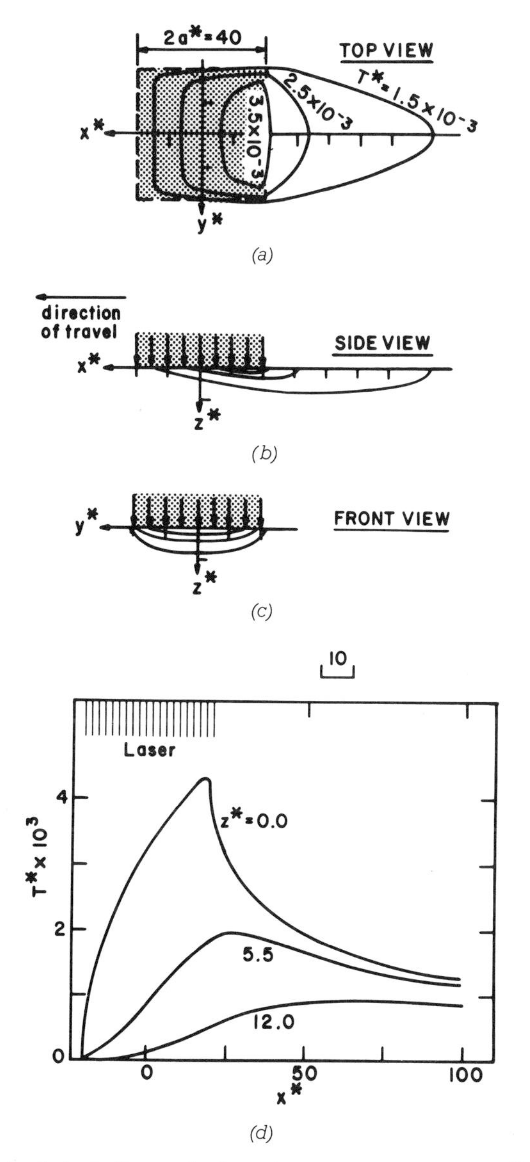

FIGURE 43. Calculated dimensionless temperature versus dimensionless distance curves (reprinted with permission from Kou, S., D. K. Sun and Y. P. Le. 1983. *Met. Trans.*, 14A:643, a publication of The Minerals, Metals and Materials Society, Warrendale, PA).

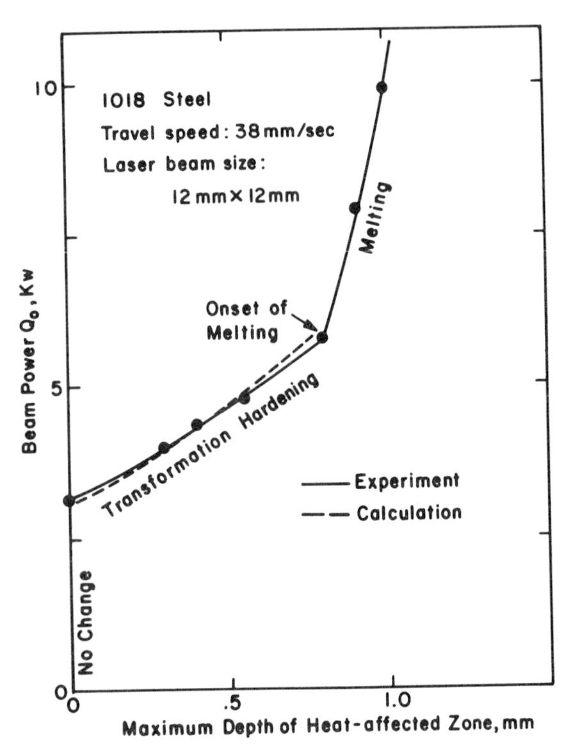

FIGURE 44. The effect of beam power on the depth of the heat-affected zone in laser-heating (reprinted with permission from Kou, S., D. K. Sun and Y. P. Le. 1983. *Met. Trans.*, 14A:643, a publication of The Minerals, Metals and Materials Society, Warrendale, PA).

Figure 46 shows calculated temperature curves for a laser beam moving along a cylinder. The projection of these is shown in Figure 47; also shown are isotherms.

The effects of some of the important parameters in laser-hardening are summarized in Figure 48. If the surface is traversed slowly (low value of U), the surface does not reach as high a temperature as is attained for higher velocity [Figure 48(a)]. This is because upon slower movement, more of the energy input at the surface has time to diffuse into the center. Thus the hardness profile will be affected because the cooling rate will be less for the slower hardening [Figure 48(b)]. If the velocity of the beam is constant, a lower energy input (Q) will not heat the surface to as high a temperature [Figure 48(c)], and the cooling rate will be lower. Thus the depth of hardening will be less [Figure 48(d)]. The effects of Q and U on the maximum surface temperature are shown in Figure 48(e) and Figure 48(f). Increasing Q increases the maximum temperature, but increasing U decreases it.

The laser beam size and shape must also be considered. For given Q and U values, the maximum temperature at the surface and the depth of hardening decrease with beam size. This is illustrated by the curves in Figure 49.

Some of these characteristics are reflected in the hardness data in Figure 50. In these experiments, decreasing beam velocity [Figure 50(b)] increased the depth of hardening, as was depicted in Figure 48(b). Increasing power density [Figure 50(a)] increased the depth of hardening initially, as depicted in Figure 48(d), but then decreased it. This decrease may have been due to melting.

If laser-hardening is accomplished by multiple passes, the effect on the hardness of the previous pass by the new pass must be considered. The subsequent pass will austenitize part of the first pass, forming martensite upon cooling. However, there is a region in the first pass that develops a

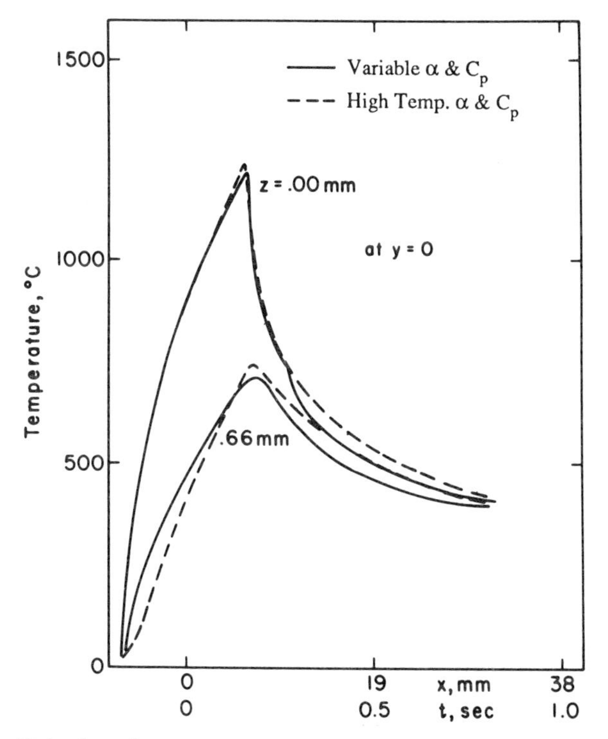

FIGURE 45. Calculated temperature–time (and distance) curves for laser-heating. α is the thermal diffusivity and C_p is the heat capacity. The curves were calculated using either variable α and C_p values or high temperature (>800°C) values (reprinted with permission from Kou, S., D. K. Sun and Y. P. Le. 1983. *Met. Trans.*, 14A:643, a publication of The Minerals, Metals and Materials Society, Warrendale, PA).

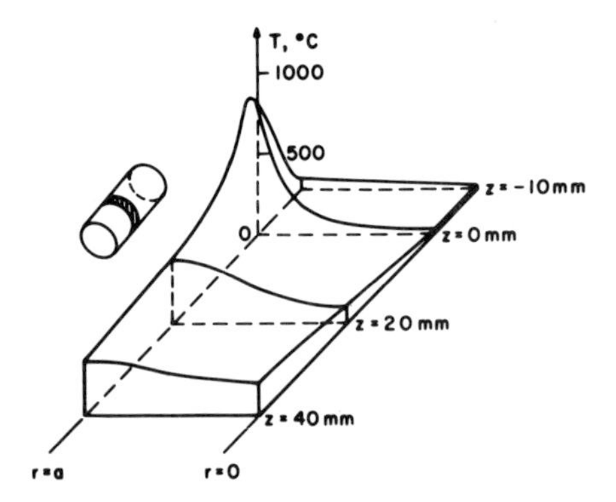

Three-dimensional view of temperature distribution in a solid cylinder of 4140 steel. Q = 9020 W, U = 2.54 mm/sec, a = 28.6 mm, and b = 3.8 mm.

FIGURE 46. Calculated temperature profiles for laser-heating a cylinder (reprinted with permission from Kou, S. and D. K. Sun. 1983. *Met. Trans.*, 14A:1859, a publication of The Minerals, Metals and Materials Society, Warrendale, PA).

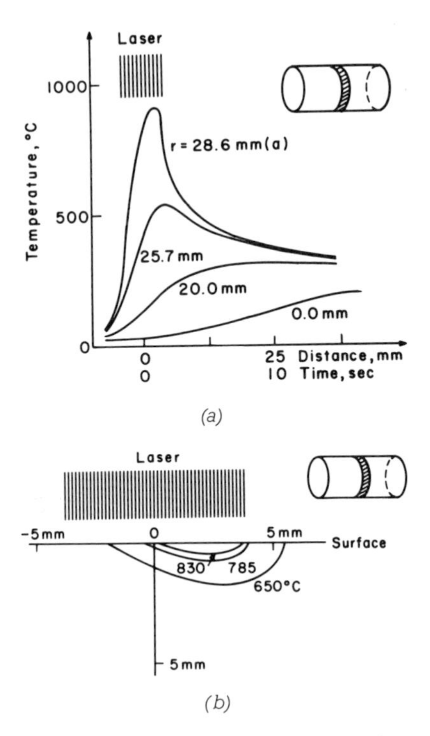

FIGURE 47. (a) Projections of the isotherms from the surface shown in Figure 46, and (b) isotherms on an axial plane of the cylinder (reprinted with permission from Kou, S. and D. K. Sun. 1983. *Met. Trans.*, 14A:1859, a publication of The Minerals, Metals and Materials Society, Warrendale, PA).

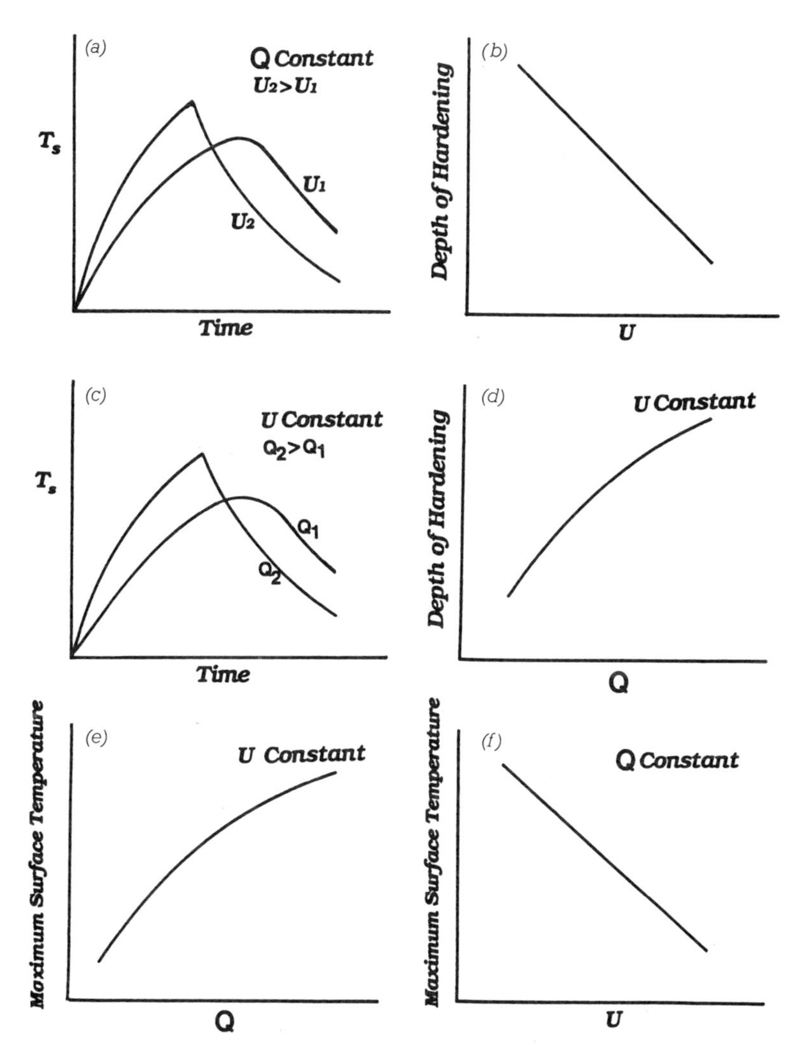

FIGURE 48. Schematic illustration of the effect of some of the parameters in laser-heating of the temperature–time curves and the depth of hardening.

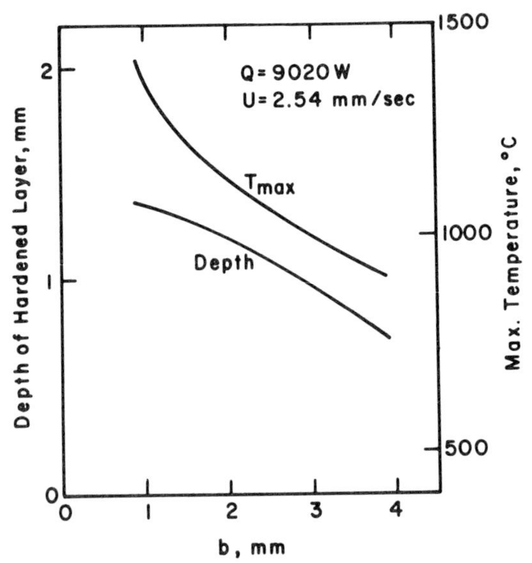

Effect of the width of laser beam on depth of hardened layer and maximum surface temperature of solid cylinder of 4140 steel. Q = 9020 W, U = 2.54 mm/sec, and a = 28.6 mm.

FIGURE 49. Effect of the laser beam size on the depth of hardening (reprinted with permission from Kou, S. and D. K. Sun. 1983. *Met. Trans.*, 14A:1859, a publication of The Minerals, Metals and Materials Society, Warrendale, PA).

heat-affected zone from the second pass, causing tempering and softening. This is illustrated by the data in Figure 51.

Figure 52 shows the microstructures in a cross section through a laser-hardened surface of a 1018 steel, obtained by using a moving laser beam and a single pass. The beginning microstructure was primary ferrite and pearlite. For this laser treatment, the calculated temperature–time curves at the surface and at the location of the beginning of the heat-affected zone are shown in Figure 53. Again, note the rapid heating and cooling. The heat-affected zone begins at location (D) (Figure 52), as revealed by the higher magnification micrograph. At the bottom of this micrograph is primary ferrite and pearlite (black regions), identical to the unaffected zone [see micrograph (E)], and at the top the regions that were pearlite are mixed dark and white, due to untransformed pearlite and to martensite formed on cooling. As the surface is approached, regions of increasing austenitization are reached which, after quenching, contain increasing amounts of martensite. Micrograph (B), from the surface, indicates incomplete austenitization because some untransformed ferrite appears to be present.

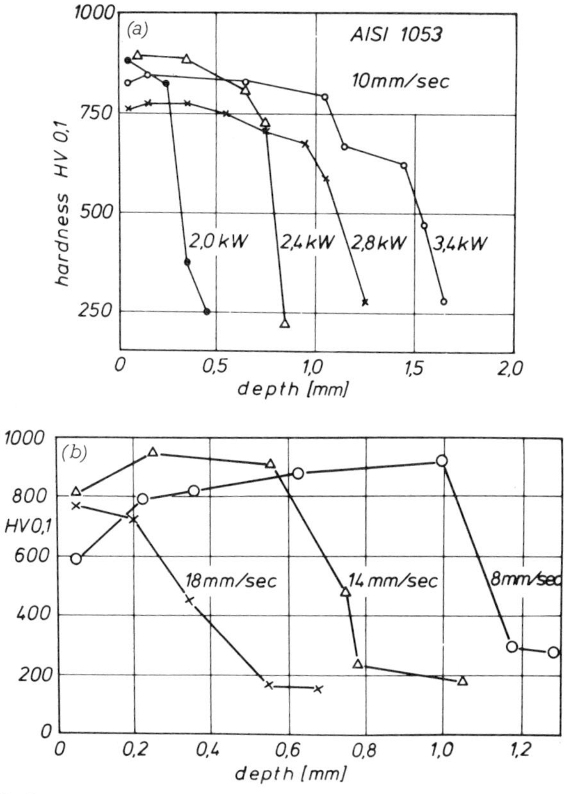

FIGURE 50. Influence of (a) power density and (b) beam velocity (3 kW power density) on the hardness profiles of a 1053 steel (reprinted with permission from Amende, W. 1984. In *Industrial Applications of Lasers*, H. Koebner, editor, © John Wiley & Sons, Ltd., p. 79).

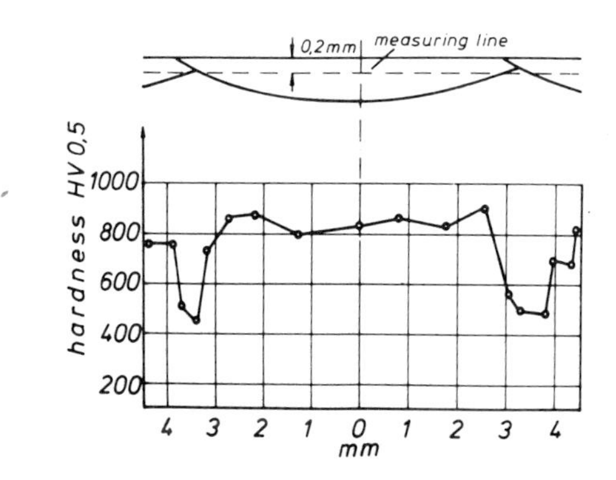

FIGURE 51. Hardness profile just below the surface of a multi-pass laser-hardened surface (reprinted with permission from Amende, W. 1984. In *Industrial Applications of Lasers*, H. Koebner, editor, © John Wiley & Sons, Ltd., p. 79).

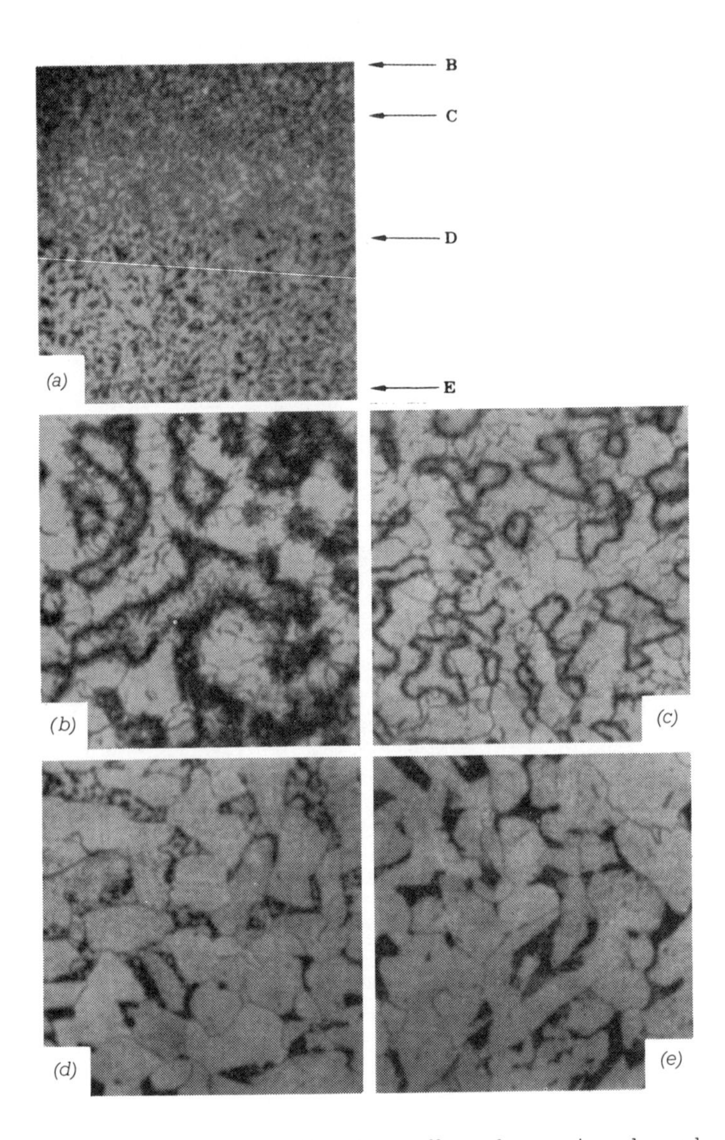

FIGURE 52. Microstructures of the heat-affected zone in a laser-heated 1018 steel. The temperature–time curves are shown in Figure 53 (reprinted with permission from Kou, S., D. K. Sun and Y. P. Le. 1983. *Met. Trans.*, 14A:643, a publication of The Minerals, Metals and Materials Society, Warrendale, PA).

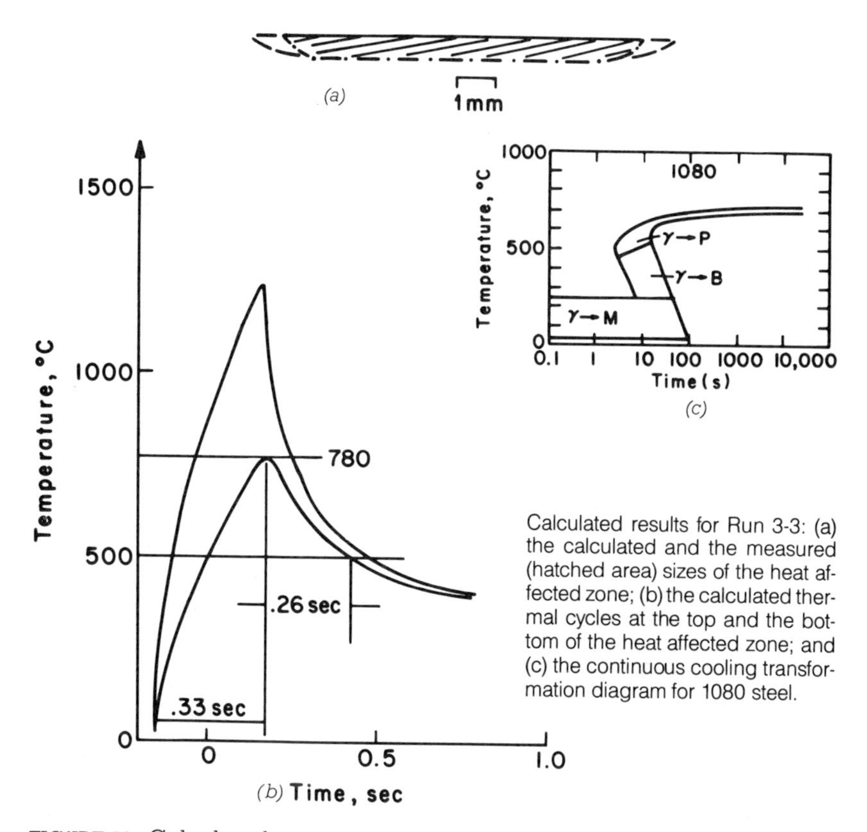

Calculated results for Run 3-3: (a) the calculated and the measured (hatched area) sizes of the heat affected zone; (b) the calculated thermal cycles at the top and the bottom of the heat affected zone; and (c) the continuous cooling transformation diagram for 1080 steel.

FIGURE 53. Calculated temperature–time curves for laser heating a 1018 steel (reprinted with permission from Kou, S., D. K. Sun and Y. P. Le. 1983. *Met. Trans.*, 14A:643, a publication of The Minerals, Metals and Materials Society, Warrendale, PA).

Figure 54 shows calculated temperature–time curves for laser-heating the same steel, using the same beam velocity but a higher beam power. Note the higher maximum temperature attained at the surface. The microhardness traverse is shown in Figure 55. The microstructure near the surface is shown in Figure 56, showing a martensitic structure. The hardness here was 435 Knoop, which agrees well with an all-martensite structure for this carbon content (0.18%).

Figure 57(a) shows another example of the microstructure of a surface-hardened steel. It contained about 0.45% C and the original structure consisted of tempered martensite. The fine martensitic structure at the surface is clearly revealed. The hardness distribution is shown in Figure 57(b).

Figure 58 shows some of the parameters measured in determining the response of steels to laser-hardening. These data are for a 1045 steel, and

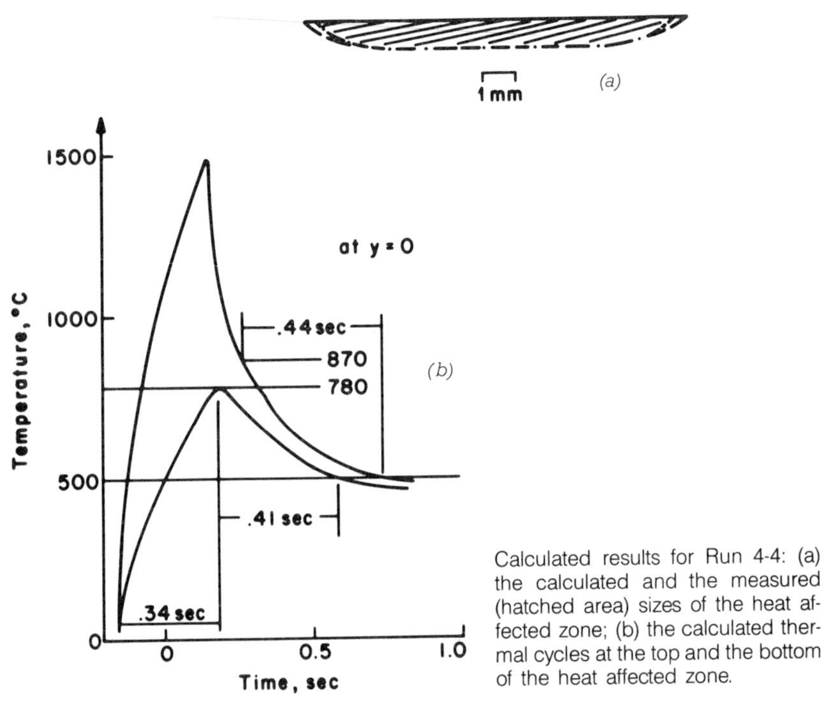

FIGURE 54. Calculated temperature–time curves for laser heating a 1018 steel, using a higher power density than for the case shown in Figure 53 (reprinted with permission from Kou, S., D. K. Sun and Y. P. Le. 1983. *Met. Trans.*, 14A:643, a publication of The Minerals, Metals and Materials Society, Warrendale, PA).

show the effects of laser power on penetration depth and surface hardness as a function of beam velocity. For a given velocity, if the power becomes too high, melting occurs at the point at which each curve intersects the curve denoted as melting:no melting.

The microstructure of the 1045 steel after laser-hardening is shown in Figure 59. The beginning microstructure was primary ferrite and pearlite. This micrograph shows just the hardened surface, the thin white layer at the top; it had a hardness of about 715 DPH, which is about the value expected for a homogeneous martensite of this carbon content. The coarser layer below this contained some undissolved ferrite, and had a hardness of about 610 DPH.

The dependence of surface hardness and penetration depth on laser power for 1045 and 1095 steels is compared in Figure 60. The higher carbon content allows a greater penetration depth and a higher surface hard-

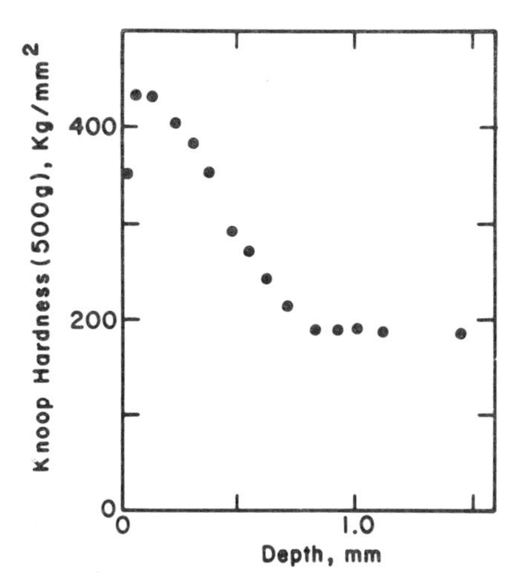

FIGURE 55. Hardness profile for the temperature-time curves shown in Figure 54 (reprinted with permission from Kou, S., D. K. Sun and Y. P. Le. 1983. *Met. Trans.*, 14A:643, a publication of The Minerals, Metals and Materials Society, Warrendale, PA).

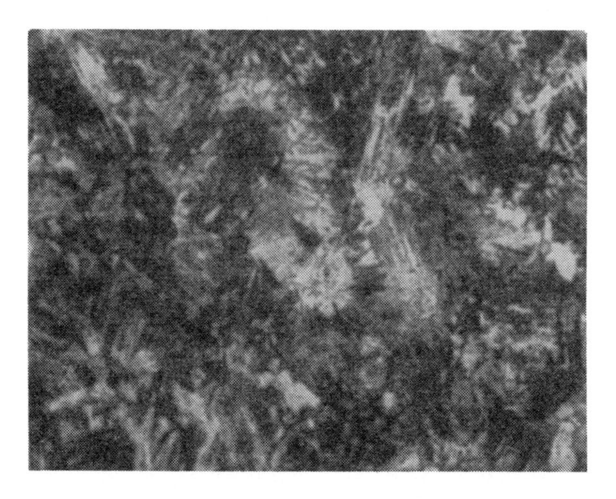

The microstructure near the top of the heat affected zone of Run 4-4. The Knoop hardness is 435 kg/mm² (500 grams). Magnification 370 times.

FIGURE 56. The microstructure near the surface of the laser-hardened 1018 steel corresponding to the hardness profile in Figure 55 (reprinted with permission from Kou, S., D. K. Sun and Y. P. Le. 1983. *Met. Trans.*, 14A:643, a publication of The Minerals, Metals and Materials Society, Warrendale, PA).

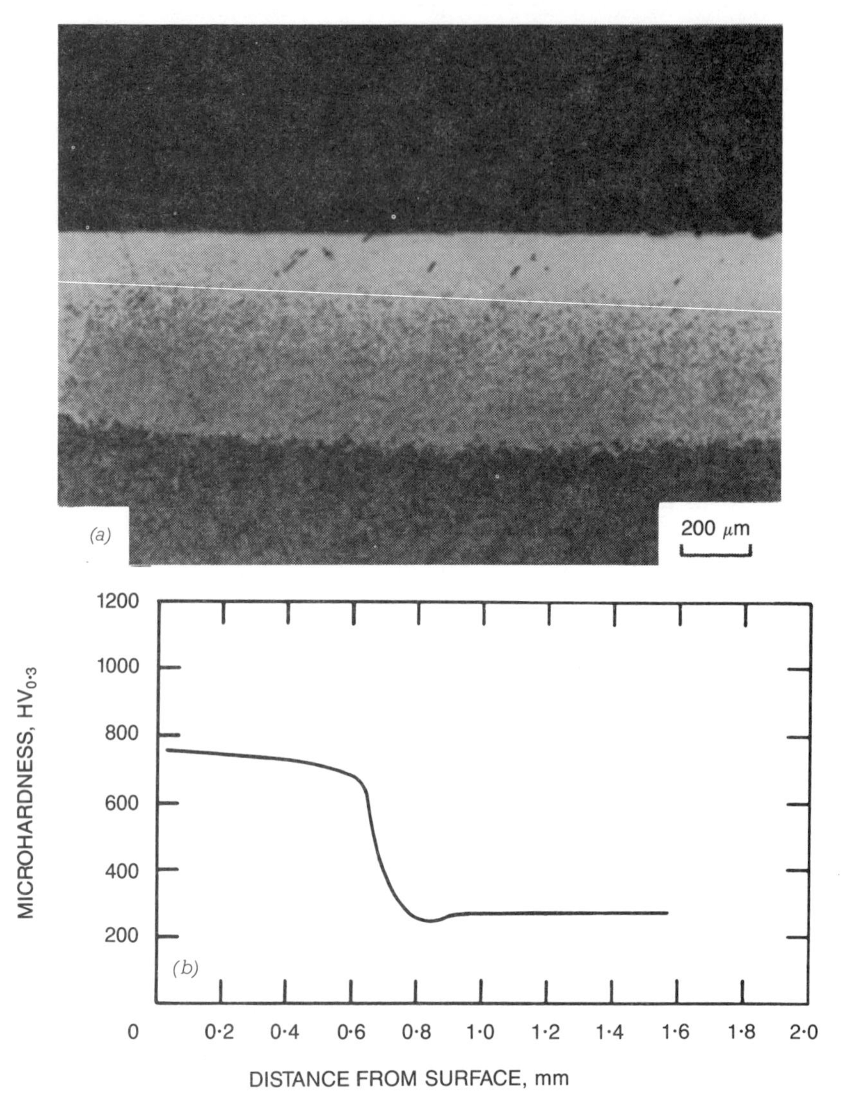

a **Micrograph showing the laser-hardened case formed on quenched and tempered En 8 steel; IPD 5×10^3 W/cm^2, coverage rate 30 cm^2/min, etched in picral;** *b* **microhardness profile after treatment**

FIGURE 57. Microstructure and hardness profile of a 0.45% C laser-hardened steel (from Trafford, D. N. H., T. Bell, J. H. P. C. Megaw and A. S. Bransden. 1980. In *Heat Treatment '79*. London: The Metals Society, p. 32; by permission of The Institute of Metals).

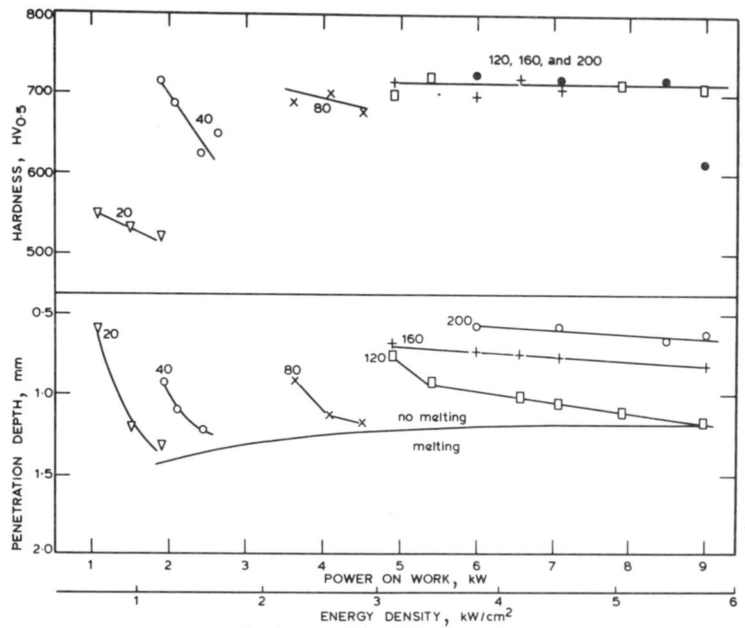

speed of work travel, cm/min, indicated for each curve; transverse oscillation 450 Hz; axial oscillation 120 Hz; pattern 1·27 × 1·27

Hardness and depth of penetration obtained with oscillating optics (F105) on AISI 1045 steel using a manganese phosphate coating

FIGURE 58. Surface hardness and depth of penetration of laser-hardened 1045 steel. A manganese phosphate coating was used (from Bonelo, L. and M. A. H. Howes. 1980. In *Heat Treatment '79*. London: The Metals Society, p. 39; by permission of The Institute of Metals).

hardness: area of fine needles, ~715 HV; area of coarse needles ~610 HV × 150

FIGURE 59. Micrograph of the surface layer of a 1045 steel hardened by a laser (from Bonelo, L. and M. A. H. Howes. 1980. In *Heat Treatment '79*. London: The Metals Society, p. 39; by permission of The Institute of Metals).

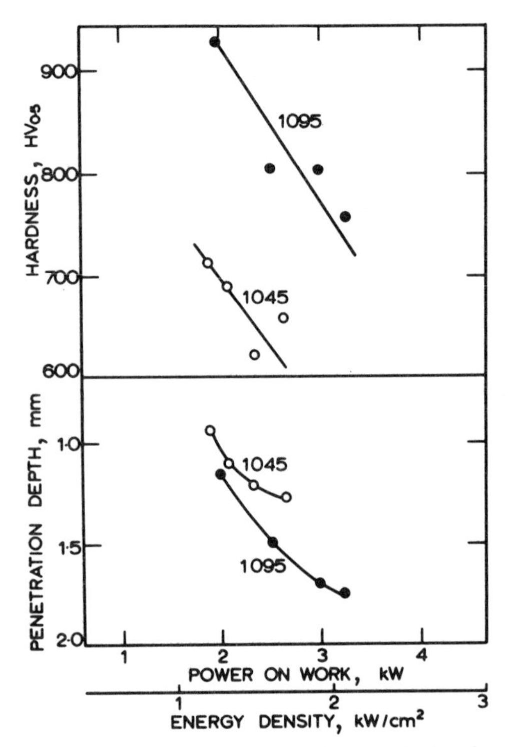

Comparative hardness and depth of hardening for two steels laser treated at 40 cm/min

FIGURE 60. Comparison of the dependence of the penetration depth and the surface hardness on the laser power of laser-hardened 1045 and 1095 steels (from Bonelo, L. and M. A. H. Howes. 1980. In *Heat Treatment '79*. London: The Metals Society, p. 39; by permission of The Institute of Metals).

ness. However, the hardness profiles in Figure 61 show that retained austenite is present, since sub-zero cooling after laser-hardening increased the surface hardness.

One study has examined the effect of laser surface-hardening on the fatigue strength of a 1045 steel orginally in the annealed condition. The samples were rotated (at about 400 rpm) while the laser beam moved along the sample axis. Figure 62 shows the cross-sectional microstructures for one sample. Note that the heat-affected zone increased in thickness in the beam movement direction. The microstructure shown in (b) was irradiated by five overlapping passes of the laser beam. The microstructure at the surface is shown at higher magnification in Figure 63. The core is shown at the bottom, consisting of primary ferrite and pearlite. The hardness in the core was from about 270 to 340 Knoop. As the surface is approached, a finer structure is found, with a higher hardness. The layer within about

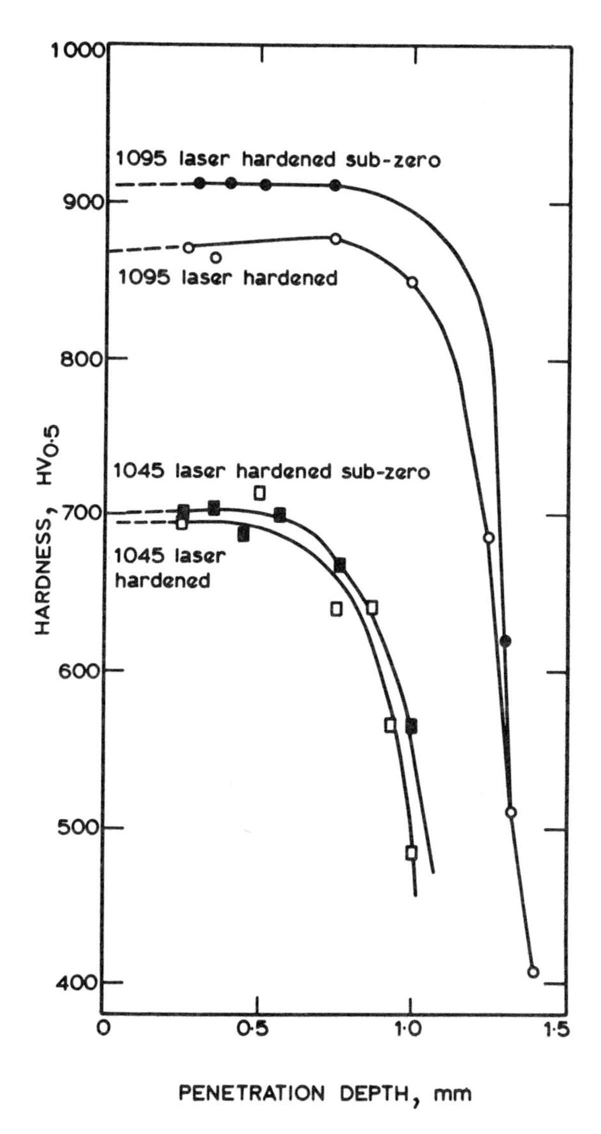

FIGURE 61. Hardness profiles for laser-hardened 1045 and 1095 steels, as-hardened and also after sub-zero cooling (from Bonelo, L. and M. A. H. Howes, 1980. In *Heat Treatment '79*. London: The Metals Society, p. 39; by permission of The Institute of Metals).

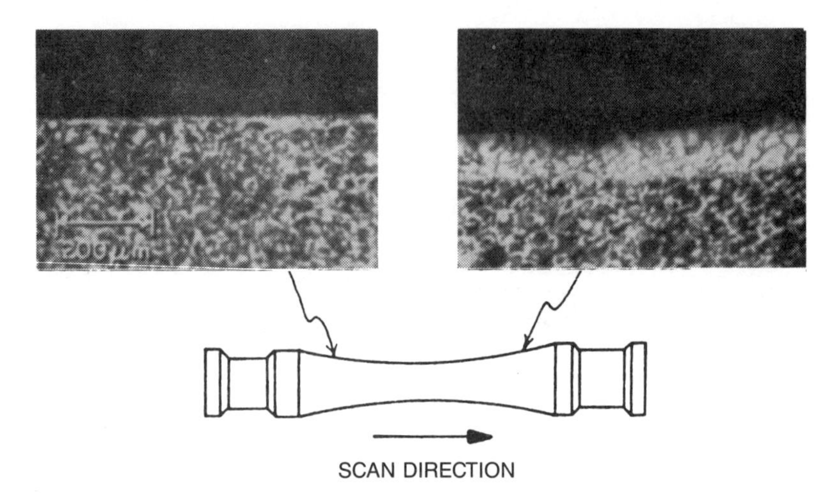

FIGURE 62. Low magnification micrographs of the microstructure of a laser-hardened 1045 steel, with a beginning microstructure of primary ferrite and pearlite (reprinted with permission from Singh, H. B., S. M. Copley and M. Bass. 1981. *Met. Trans.*, 12A:138, a publication of The Minerals, Metals and Materials Society, Warrendale, PA).

10 μm of the surface was martensitic, and the hardness was between about 650 to 800 Knoop.

Figure 64 shows the fatigue data from this study. Note that the fatigue strength (fatigue limit) is increased from about 50,000 psi to about 70,000 psi by laser-hardening. (A brief treatment of laser surface-hardening and of the equipment used is given by Sandven, O. 1981. In *Metals Handbook, 9th Edition, Heat Treating*. Metals Park, Ohio: American Society for Metals.)

2f. FLAME-HARDENING

The main distinction between flame-hardening and laser-hardening is that the rate of energy transfer to the surface is lower in flame-hardening. Thus, any attempt to attain similar energy input rates with flames results in melting. Therefore, the heating rates in flame-hardening are less than in laser-heating, as shown in Figure 65. Note that the heating time at the surface is about 60 sec. This results in a larger heat-affected zone, and slower cooling if only self-quenching is relied upon. Thus in flame-hardening, it is customary to water-quench the surface following heating to the desired temperature. Figure 66 shows the effect of not using any water-quench, but allowing the heated region to cool to the air and to the cooler inside of the

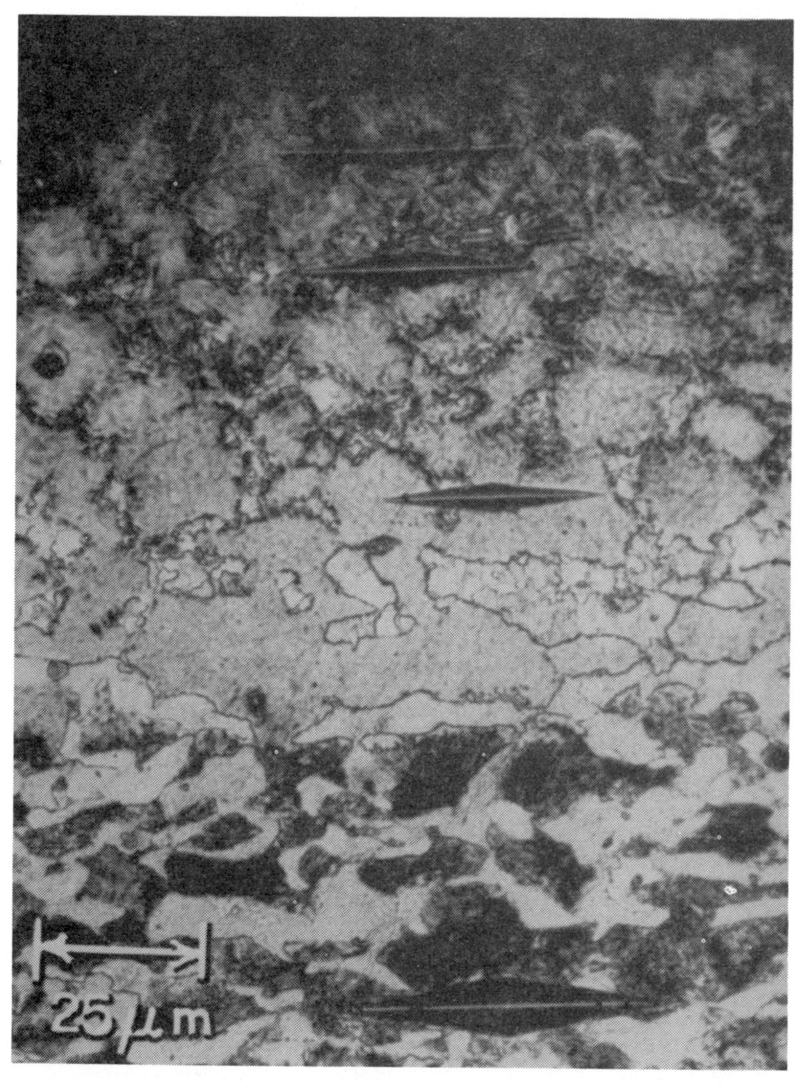

FIGURE 63. High magnification micrograph of the surface region of the laser-hardened 1045 steel shown in Figure 62. The diagonal markings are Knoop hardness indentations; the shorter the indentation, the harder the location (reprinted with permission from Singh, H. B., S. M. Copley and M. Bass. 1981. *Met. Trans.*, 12A:138, a publication of The Minerals, Metals and Materials Society, Warrendale, PA).

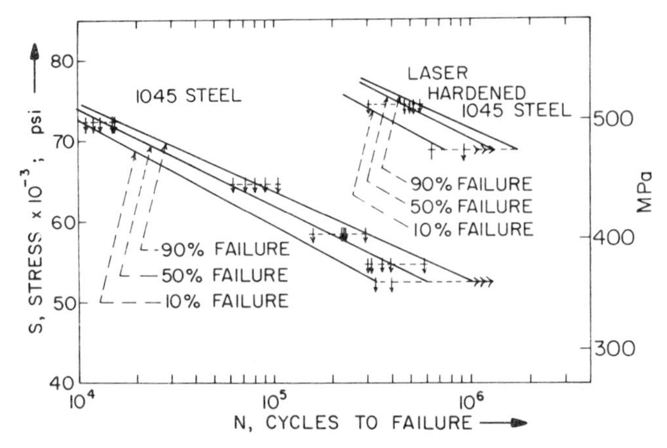

FIGURE 64. Fatigue data for a 1045 steel in the originally annealed condition and in the laser-hardened condition (reprinted with permission from Singh, H. B., S. M. Copley and M. Bass. 1981. *Met. Trans.*, 12A:138, a publication of The Minerals, Metals and Materials Society, Warrendale, PA).

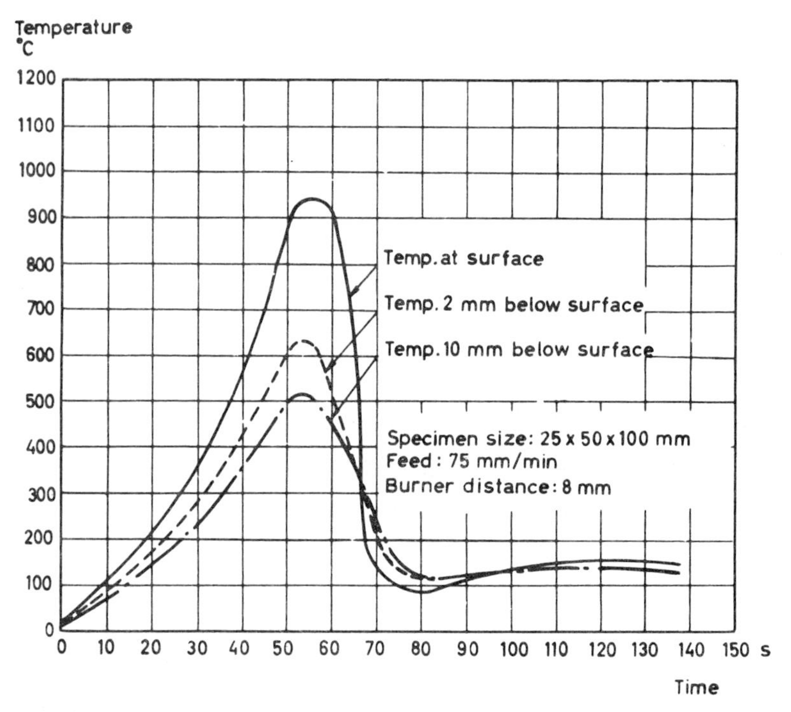

FIGURE 65. Typical temperature–time curves for flame-hardening. Cooling following heating was by water quenching (reprinted with permission from Thelning, K.-E. 1975. *Steel and Its Heat Treatment*. Boston: Butterworths).

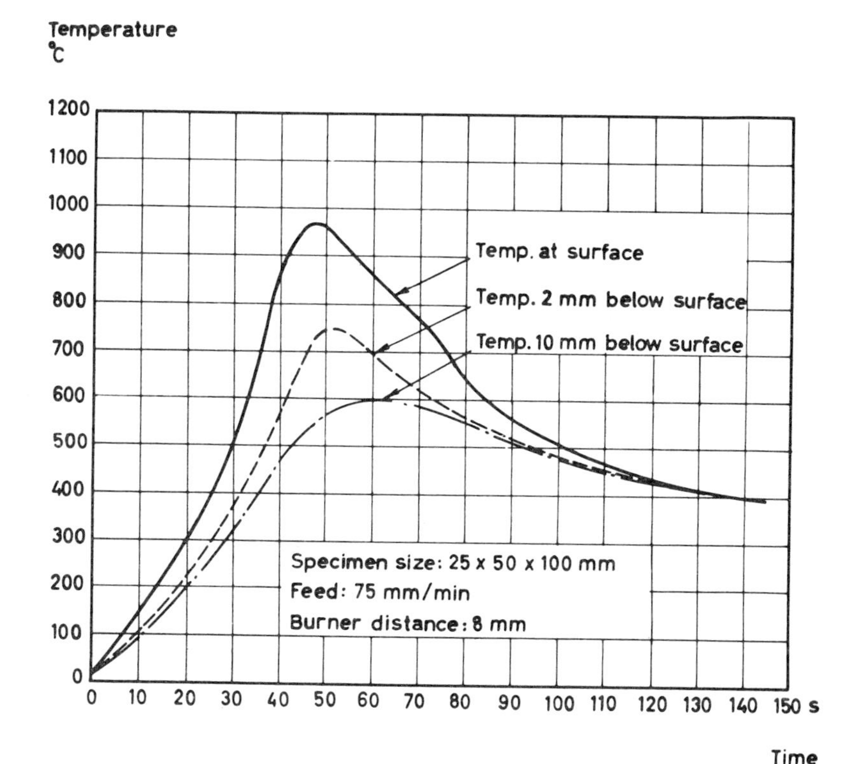

FIGURE 66. Typical temperature–time curves for flame-heating followed by air cooling (reprinted with permission from Thelning, K.-E. 1975. *Steel and Its Heat Treatment*, Boston: Butterworths).

steel part. Note the relatively slow cooling times compared to that for laser-heating. Also note, by comparison to Figure 65, that water quenching gives a considerably higher cooling rate (e.g., 10 sec from 900°C to 300°C), but still considerably less than that associated with laser-hardening.

The obvious parameters that affect the hardening process are the type of fuel being burned, the burning gas velocity, the distance between the work piece and the flame head, the relative velocity of the flame head and the work piece, and the time of flame impingement. The effects of some of these factors on the depth of hardening are shown in Figure 67. (The practical aspects of flame-hardening and equipment design are described in 1981. *Metals Handbook, 9th Edition, Vol. 4*. Metals Park, Ohio: American Society for Metals. Also see Thelning, K.-E. 1975. *Steel and Its Heat Treatment*. Boston: Butterworths.)

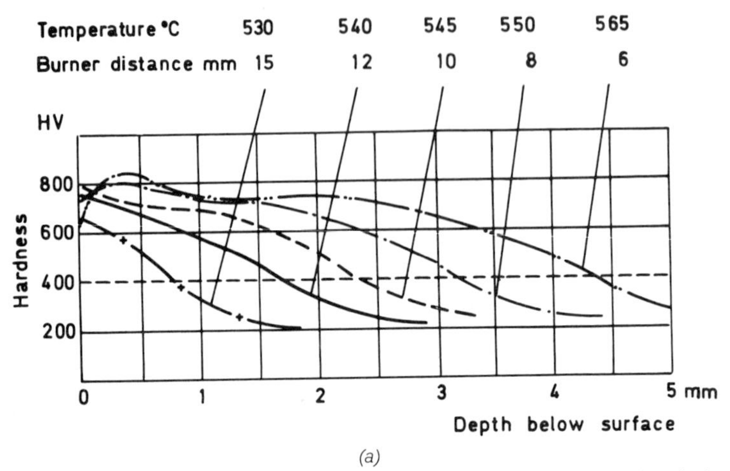

(a)

Relationship between hardness profile and burner distance during flame-hardening of a carbon steel with 0.50% C. Test specimen dimensions: 25 × 75 × 100 mm. Feed 50 mm/min Water spray quenching. Temperature measured 10 mm below the surface.

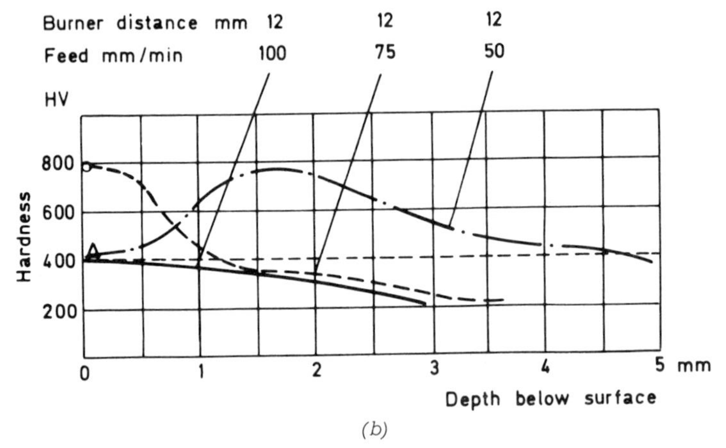

(b)

Dependence of hardness profile and microstructure on the rate of feed during flame-hardening of steel D 2 (Bofors ROP 57). Test specimen dimensions: 25 × 75 × 100 mm. Edge-hardening. Air cooling.

FIGURE 67. Hardness profiles showing the effect of several parameters in flame-hardening (reprinted with permission from Thelning, K.-E. 1975. *Steel and Its Heat Treatment*. Boston: Butterworths).

TABLE 1.
Advantages and Disadvantages of Using Electron Beams for Surface Treatments.

Advantages
Beam can be manipulated by electro-magnetic lenses, providing flexibility to the process.
Low distortion
Energy efficient; about 90% of the energy is converted to heat.
Clean operation, since the process is carried out in vacuum.
No special surface coating or treatment required.
Disadvantages
Process must be carried out in vacuum.

2g. ELECTRON BEAM-HARDENING

Surface-heating may be achieved by the use of focused electron beams. An obvious advantage of using electron beams for heating is that the beam may be easily focused and moved on the work piece by electromagnetic lenses. An important disadvantage is that the electron beam must be utilized in vacuum. The general ideas associated with rapid heating, which electron beam-heating induces, have been covered in Section 2e and will not be repeated here. Instead, Table 1 gives a list of advantages and disadvantages of using electron beam-heating for surface treatment. For further details about electron beam-heating for surface treatments, the following articles are recommended.

1. Sayegh, G. 1980. "Principles and Applications of Electron Beam Heat Treatment," *Heat Treatment of Metals*, 7:5–10.
2. Hick, A. J. 1983. "Rapid Surface Heat Treatments—A Review of Laser and Electron Beam Hardening," *Heat Treatment of Metals*, 10:3–11.
3. Zenker, R. and M. Muller. 1988. "Electron Beam Hardening Part 1: Principles, Process Technology and Prospects," *Heat Treatment of Metals*, 15:79–88.
4. Zenker, R. W. John, D. Rathjen, G. Fritsche and B. Kamfe. 1989. "Electron Beam Hardening Part 2: Influence on Microstructure and Properties," *Heat Treatment of Metals*, 16:43–51.

CHAPTER 3 **Carburizing**

3a. INTRODUCTION

In this section the process of carburizing is examined. This process involves heating into the austenite region a steel part in an environment that allows carbon to be added to the steel at the surface. The carbon then diffuses in from the surface. The process used must provide carbon at a considerably faster rate than it can diffuse in, so that a carbon gradient is produced. The steel is then quenched to form high-carbon martensite in the surface layers, which provides high hardness and high compressive residual stresses, both of which give the steel component improved properties.

There are a number of carburizing processes being used, including conventional methods, such as pack-carburizing and gas-carburizing, and several newer methods most of which are related to gas-carburizing. We will not examine the methods in detail, but will instead focus on the principles behind the methods and the microstructural aspects of the processes.

A steel of a desired carbon content can be made by placing iron or a steel of low carbon content in contact with graphite and allowing the system to come to equilibrium. For example, if the steel contains 0.2% C, and the desired final carbon content is 0.8% C, then the steel is placed in contact with the proper mass of graphite, based on the mass of the steel, to give overall 0.8% C. Then the composite is heated to the austenite region and held until equilibrium is attained. This is illustrated in Figure 68. The starting composite is a diffusion couple, and diffusion of carbon into the steel occurs. This involves the transfer of carbon atoms from the graphite lattice at the steel–graphite interface into the austenite lattice. This causes an increase in the carbon content of the austenite, and this carbon profile increases until the saturation value for the austenitizing temperature is achieved. Then as carbon diffuses into the steel, more is fed from the graphite to maintain this saturation value. As time progresses, the carbon content in the steel increases, and the size of the graphite region decreases, as shown in Figure 68. Eventually, the graphite is used up, and then finally the carbon gradient is removed. In principle, the formation of a steel by this method requires an infinite time. In actuality, to attain a distribution close

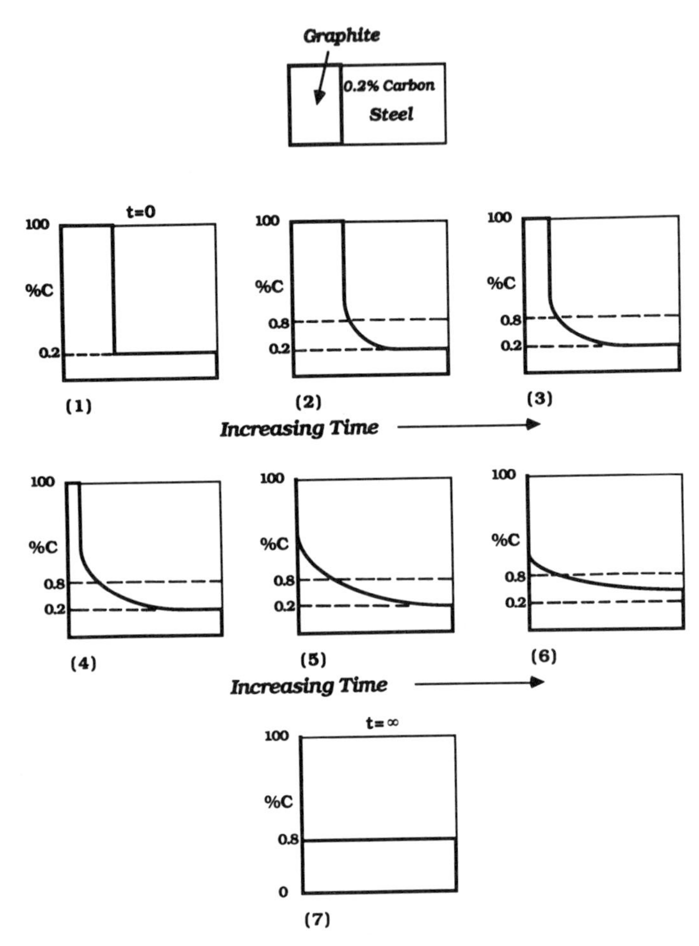

FIGURE 68. Schematic illustration of the diffusion of carbon into a Fe-0.20% C steel in contact with a piece of pure graphite. The relative amounts of the steel and the graphite are such that the overall composition of the composite is 0.8% C. Thus after equilibrium (infinite time) the composite becomes a 0.8% C steel.

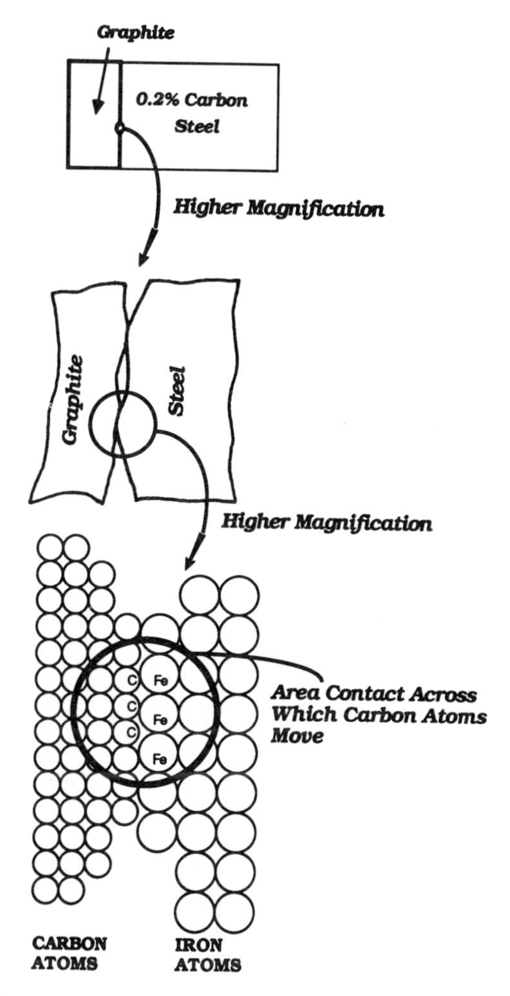

FIGURE 69. Schematic illustration of the low contact area across which carbon diffusion must occur.

to uniformity in a reasonable time (e.g., 4 hours at 900°C) requires that the sample be below a certain size (e.g., 2 mm thick). Thus, this technique is not a viable method of making steel (except in very small parts).

A more impelling reason that the above method is not a useful way of making a steel is that the technique as described above requires the steel and the graphite to be in contact, and thus the rate of carbon transfer will depend upon the contact area (see Figure 69). This will be small, unless care is taken to keep the contacting surfaces smooth and in good mechanical contact. (This is a problem in diffusion couple experiments.) Thus, adding carbon to iron or steels by relying on transfer across a steel–graphite contact surface is not feasible.

In pack-carburizing (a method still used), the steel part to be carburized is placed in a container and surrounded by graphite particles, usually in the form of charcoal. The container is then heated to the austenitizing temperature for a few hours, after which it is found that the steel has picked up carbon on the surface. But as indicated above, the carbon can not be transferred in any significant amount across the steel–graphite contact interfaces because the contact area is too small.

Then how did the steel pick up the carbon? The mechanism involves carbon transport in the gaseous form via carbon monoxide. In pack-carburizing, the space around the graphite particles and the steel part is filled with air. Upon heating to the austenitizing temperature, the oxygen in the air reacts with the carbon by the equation

$$C_{gr} + O_2 = CO_2$$

The CO_2 then is reduced by the graphite to form CO.

$$2C_{gr} + CO_2 = 2CO$$

The carbon monoxide then decomposes on the steel surface to form CO_2 and thereby to free a carbon atom.

$$2CO = \underline{C}^{\gamma} = CO_2$$

Here $\underline{C}^{\gamma}$ means carbon dissolved in austenite. The CO_2 thus formed is then reduced by graphite to form more CO, which in turn reacts at the steel surface to release carbon and to form CO_2. Thus, the source for carbon to diffuse into the steel is from the graphite, but via the gaseous reactions.

Note that this process depends upon the initial reaction with oxygen to form CO_2. The rate of the reaction can be accelerated by addition to the carburizing material of compounds such as $BaCO_3$, which decompose upon heating to release CO_2.

The chemical reactions occur at a sufficiently high rate that carbon is immediately replaced as soon as it moves into the lattice, and thus the carbon content at the surface easily increases to a value above that of the steel. This is depicted in Figure 70. With time at the carburizing temperature, the carbon content continues to build up, until the saturation value is reached; this value can be obtained from the Fe–C phase diagram, but the solubility of carbon in austenite in equilibrium with *graphite*, not with Fe_3C, must be used. It is possible from this point on that Fe_3C may form at the surface. However, it is usually found that instead, the saturation carbon content is

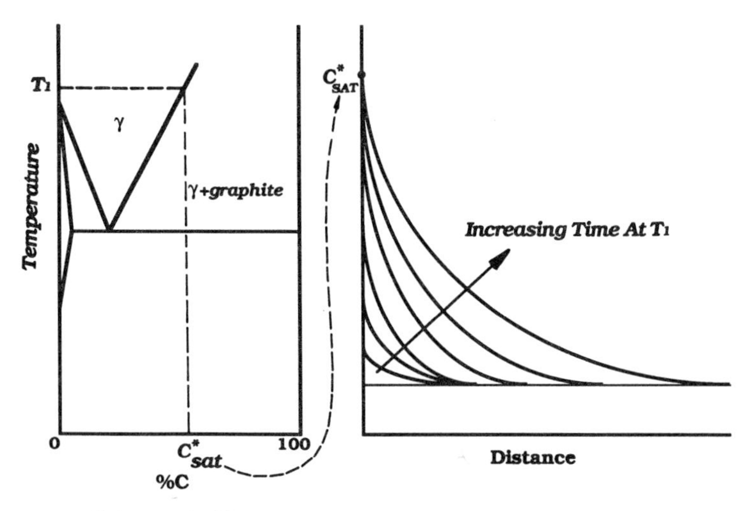

FIGURE 70. Schematic illustration of the increase in carbon content with carburizing time.

maintained at the surface, so that the carbon-depth profile just increases, as shown in Figure 70.

At infinite time, the carbon content would be uniform throughout the part, but for most part sizes, the diffusion rate of carbon is too low for this to be attained in usable times. The carburizing temperature and time are controlled to attain the desired profile, usually specified in terms of case depth. This is covered in more detail in a following section.

3b. MICROSTRUCTURE OF CARBURIZED STEELS

The carbon gradients can be examined by cooling slowly from the carburizing temperature at the end of the carburizing time. The local structure should contain, in amounts that depend upon the local carbon content, pearlite, primary ferrite, and primary iron carbide. The procedure for microstructural analysis is illustrated in Figure 71. The sample, after carburization and cooling to 25°C, is cut along the mid-plane, and the cut surface prepared metallographically. Then the microstructure is examined from the edge in towards the center. The microstructural features of carburized steels are illustrated by the following examples from *Optical Microscopy of Carbon Steels* by Samuels; a more detailed description of them is given in his book.

In Figure 72 is shown low magnification micrographs of a 0.15% C steel which has been carburized at 940°C for increasing times, then cooled

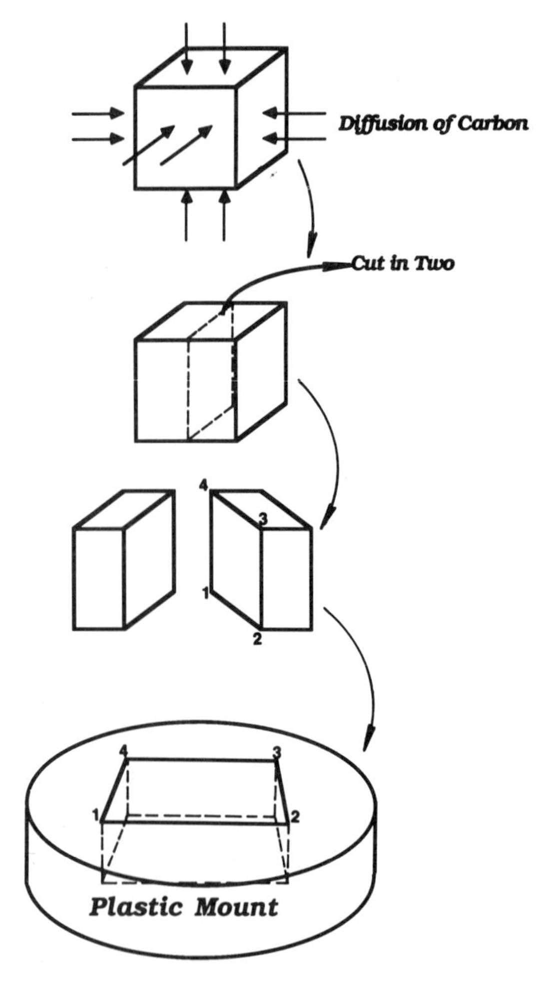

FIGURE 71. Schematic illustration of a method of preparation of a metallographic sample for observing the microstructural gradient in a carburized steel.

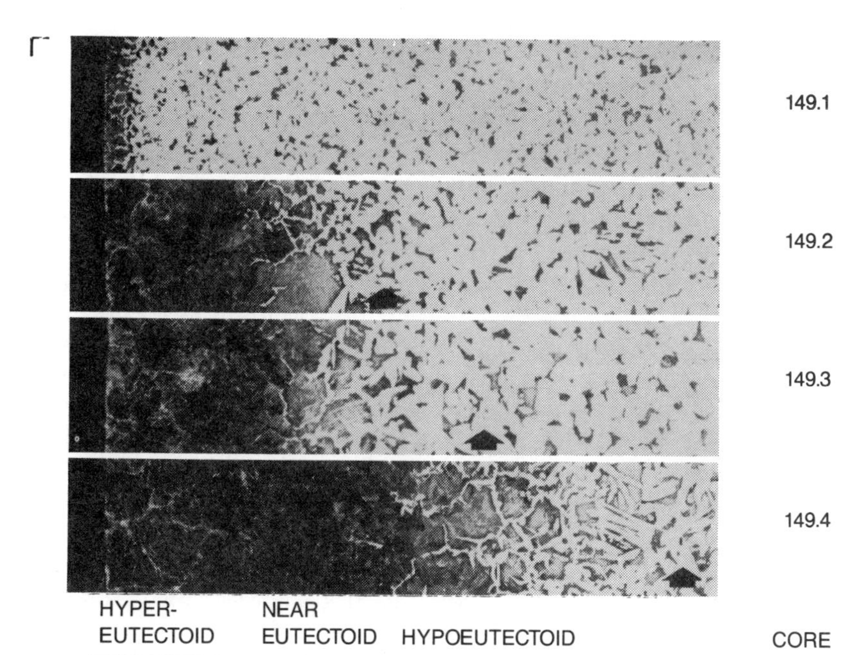

Figure No.	Material	Analysis, wt %	Condition	Hardness, HV	Etchant	Magnification
	0.15%C	0.17 C, 0.05 Si, 0.64 Mn	Pack carburized at 940°C for time indicated, cooled slowly			
149.1			0.5 h		Picral	75
149.2(a)			1 h		Picral	75
149.3(a)			2 h		Picral	75
149.4(a)			4 h		Picral	75

(a) Arrow indicates total case depth estimated from figure below.

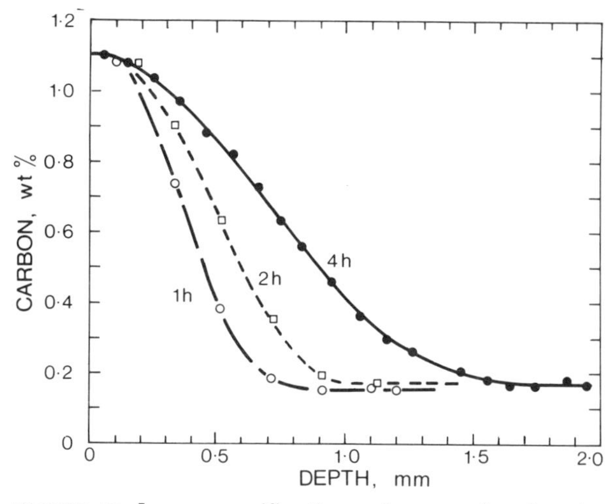

149.9. Variation of carbon content with depth in the carburized bars illustrated in Fig. 149.2 to 149.4. Turnings removed from successive layers either 0.1 or 0.2 mm thick were analyzed for carbon content.

FIGURE 72. Low magnification micrographs showing the effect of carburizing time on the microstructure of a 0.17% C steel upon pack-carburizing at 940°C. Also shown is the carbon profile (reprinted with permission from Samuels, L. E. 1980. *Optical Microscopy of Carbon Steels.* Metals Park, Ohio: American Society for Metals).

slowly to 25 °C. The core or unaffected structure consists of primary ferrite and pearlite. For the 0.15% C content, the microstructure will contain about 80% primary ferrite and 20% pearlite. Thus, at low magnification, this region appears relatively light. As the edge of the metallographic sample is approached, a position is reached where the carbon content begins to increase. Thus more and more pearlite is present, and the structure appears darker. At the location containing 0.8% C, the microstructure will be all pearlite, and at low magnification this region will appear completely dark. Closer towards the surface, if the carbon content is greater than 0.8%, primary Fe_3C is present, having formed along the austenite grain boundaries during cooling. Thus a white network of primary Fe_3C appears, visible even at low magnification, as shown in Figure 72 (see photomicrograph 149.4). Also shown in Figure 72 are the carbon profiles corresponding to the micrographs.

The microstructures at the surface for these carburized samples are shown in Figure 73. The primary ferrite is quite prominent for a carburizing time of 0.5 hrs. However, the carbon profile curves in Figure 72 show that the carbon content at the surface was about 1.1% for carburizing times of 1, 2 and 4 hours. This is reflected in Figure 73 in the identical microstructures at the surface for the carburizing times used. The white network of primary Fe_3C, which has formed on the austenite grain boundaries during slow cooling, is quite apparent.

If the steel is quenched from the carburizing temperature, the microstructure at a given location depends upon the cooling rate and the hardenability, and the latter depends upon the carbon content. At the surface both the cooling rate and the carbon content will be highest, so this location may contain the most martensite. However, it must be kept in mind that the M_f temperature decreases with carbon content, and above about 0.7% C, retained austenite will be present at 25 °C after quenching. The presence of retained austenite is illustrated by the microstructures in Figure 74. The retained austenite will cause the hardness to be low, as illustrated in Figure 75. However, the hardness value can be increased by converting the retained austenite to martensite, which can be accomplished by cooling to temperatures below 25 °C. This is illustrated in Figure 75. Methods of dealing with the problem of retained austenite are discussed in more detail in Section 3g.

3c. GAS-CARBURIZING

INTRODUCTION

In the previous section, carburizing was described in which the steel part was surrounded by graphite, and the initial reaction of the graphite with

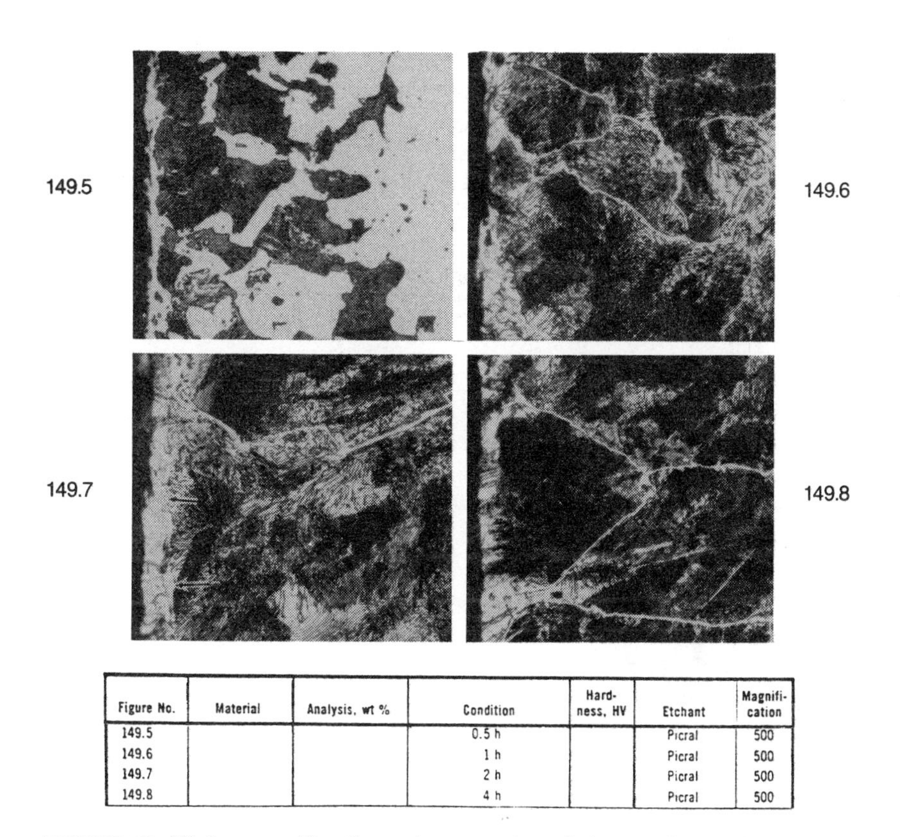

Figure No.	Material	Analysis, wt %	Condition	Hard-ness, HV	Etchant	Magnifi-cation
149.5			0.5 h		Picral	500
149.6			1 h		Picral	500
149.7			2 h		Picral	500
149.8			4 h		Picral	500

FIGURE 73. High magnification micrographs of the surface of the carburized steel of the low magnification micrographs in Figure 72 (reprinted with permission from Samuels, L. E. 1980. *Optical Microscopy of Carbon Steels*. Metals Park, Ohio: American Society for Metals).

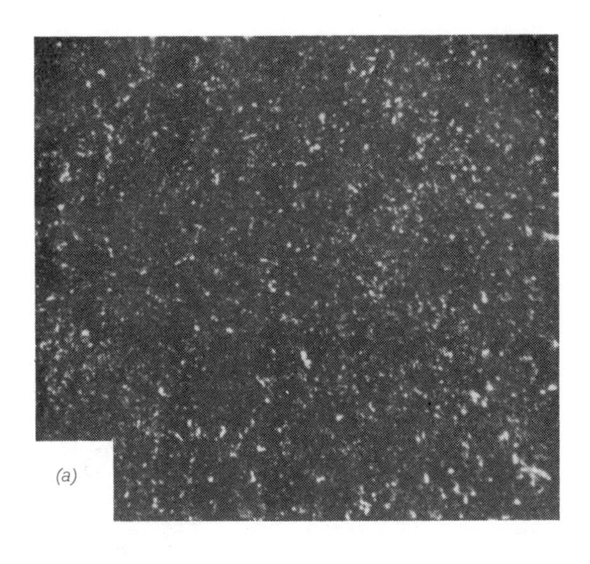
(a)

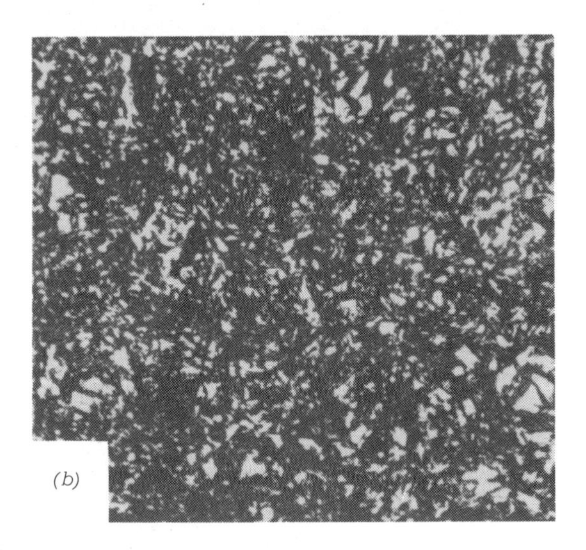
(b)

Case microstructures (nital etch; 1000×) with various amounts of retained austenite in 4620 steel that was gas-carburized to produce a surface carbon content of 0.90 to 0.95%. Microstructures are mainly tempered martensite with (a) 0%, (b) 10%, (c) 20%, (d) 35%, (e) 40% and (f) 45% retained austenite, as determined by x-ray diffraction analysis.

FIGURE 74. Microstructures illustrating the presence of retained austenite in the surface layers of carburized steels (reprinted with permission from 1977. *Carburizing and Carbonitriding.* Metals Park, Ohio: American Society for Metals).

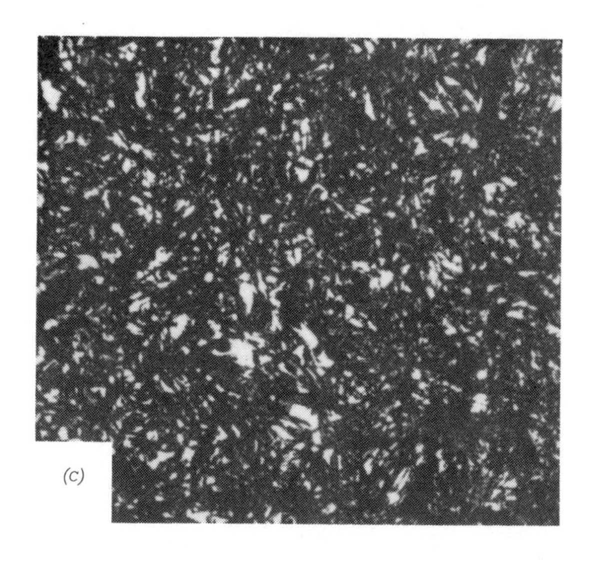

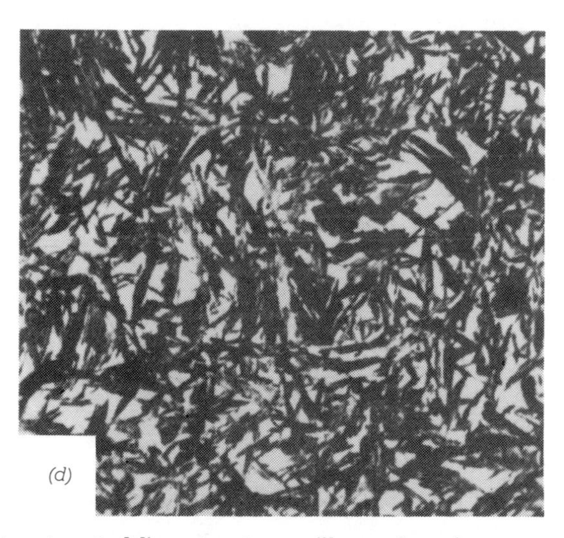

FIGURE 74 (continued). Microstructures illustrating the presence of retained austenite in the surface layers of carburized steels (reprinted with permission from 1977. *Carburizing and Carbonitriding.* Metals Park, Ohio: American Society for Metals).

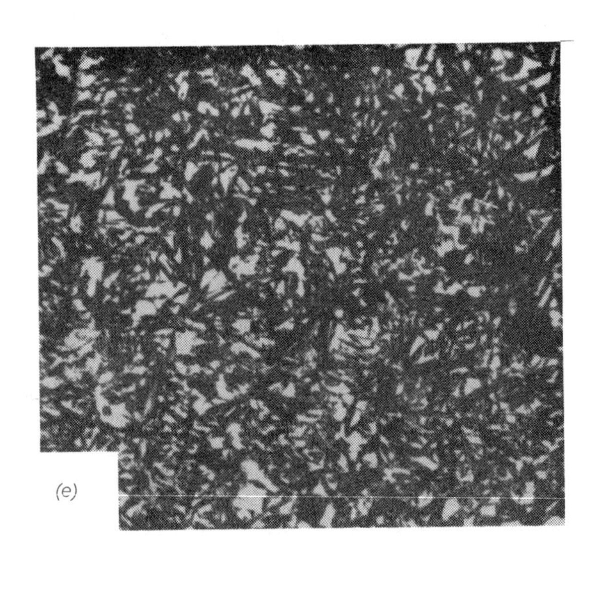

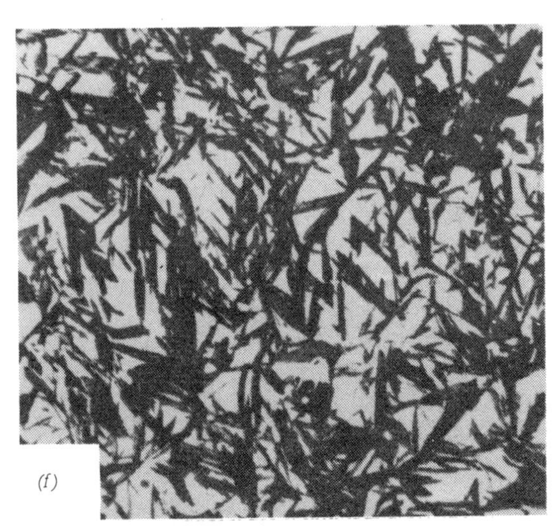

FIGURE 74 (continued). Microstructures illustrating the presence of retained austenite in the surface layers of carburized steels (reprinted with permission from 1977. *Carburizing and Carbonitriding.* Metals Park, Ohio: American Society for Metals).

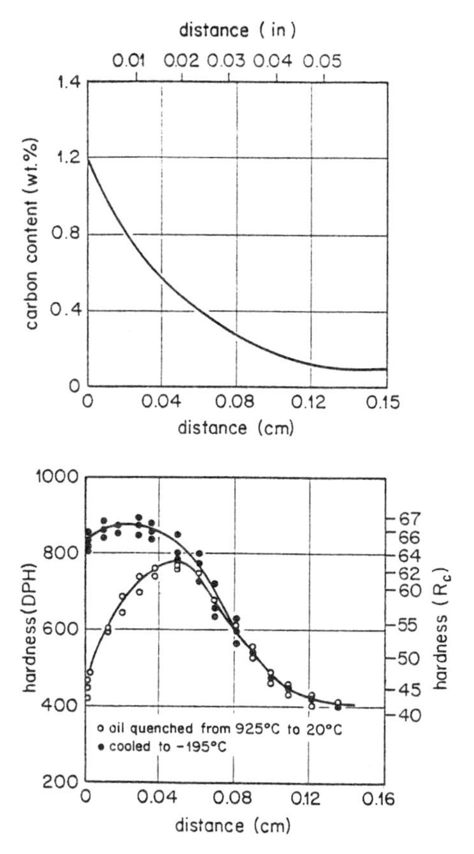

FIGURE 75. Hardness profile of a carburized 3312 steel, showing the reduction in the hardness at the surface due to retained austenite. The initial quench to 20°C did not convert all of the austenite to martensite, since the high-carbon content lowered the M_f below 20°C. Subsequent subzero cooling to −195°C converted most of the retained austenite to martensite, and the hardness increased (adapted with permission from Scott, H. and J. L. Fisher. 1942 in *Controlled Atmospheres*. Metals Park, Ohio: American Society for Metals).

oxygen in the air initiated a gaseous reaction whereby CO was formed. The decomposition of CO then occurred (at a temperature in the austenite region) on the surface of the steel to form CO_2 and release a carbon atom at the surface. This atom then moved into an interstitial site, then began a random walk process to diffuse towards the center of the steel part. If the kinetics of the reaction are sufficiently rapid, then as soon as the carbon atom moves from the surface towards the center, another carbon atom is deposited on the surface. Thus, it is assumed that the rate of movement of carbon into the steel is controlled by the diffusion of carbon in the austenite, and is not controlled by the kinetics of the chemical reaction.

An important disadvantage of pack-carburizing is the difficulty of maintaining good control of the carbon content at the surface. For example, it may be desirable to attain a carbon content at the surface of only 0.8% C (to minimize the amount of retained austenite, for example). However, in pack-carburizing, this can only be accomplished by controlling the carburizing time (for a given carburizing temperature), and the depth of carbon penetration usually desired might not be achieved with this level of carbon at the surface. Thus, the carburizing temperature or time usually used sets the carbon content at the surface higher than 0.8% C.

However, examination of the chemical reaction by which carbon is deposited on the surface suggests a method of carburization, by which the maximum carbon content at the surface can be set. From chemical thermodynamics there is a relation between the gas composition (% CO and % CO_2) and the carbon content of the austenite, from which one can calculate the amount of CO and CO_2 that must be in contact with the steel to produce a desired carbon content. Similar relations also exist for other gases, such as CH_4, which might cause carburization.

This is the approach taken in gas-carburizing. An atmosphere that has the proper chemical composition to produce the desired maximum carbon content (e.g., 0.8%) is maintained at the surface. As carburizing progresses, once the carbon content at the surface attains this equilibrium value, it does not increase; only the depth of penetration increases. The purpose of this section is to give the necessary chemical thermodynamic background in order to demonstrate how the relationships needed for the calculation are obtained, and how they are then used to set the maximum carbon content at the surface. We first examine the relations for simple binary gas mixtures of CO and CO_2 and of CH_4 and H_2. Then we examine more complex (and more realistic) gas mixtures, containing N_2, CO, CO_2, CH_4, H_2 and H_2O.

THERMODYNAMIC RELATIONS

To lay the framework for the necessary thermodynamic equation for gas-carburizing, we first consider a generalized situation. We take a system consisting of components A, B and C. (These could, for example, be CO, CO_2 and carbon dissolved in austenite.) Also we take the case where these three are connected by the chemical reaction

$$A + 2B = 3C \qquad (1)$$

We take A and C to be gases, but B is in the solid or the liquid form. The

Gibbs free energy change for this reaction at a given temperature and total pressure can be written as

$$\Delta g = 3g_C - g_A - 2g_B \tag{2}$$

where g stands for the molar Gibbs free energy. At chemical equilibrium at constant temperature and total pressure

$$\Delta g = 0$$

so

$$3g_C - g_A - 2g_B = 0 \tag{3}$$

These free energy terms must be related to measurable quantities. We can assume quite realistically that the gas mixture will be ideal, and thus we can write

$$g_A = g_A^0 + RT \ln p_A \tag{4}$$

$$g_C = g_C^0 + RT \ln p_C \tag{5}$$

where g_A^0 and g_C^0 are the Gibbs free energy for pure gases A and C at one atmosphere pressure, and p_A and p_C are the partial pressures of A and C in the gas mixture. g_A^0 and g_C^0 are referred to as the *standard free energies*, and the *standard state* to which they refer is the pure gas at one atmosphere pressure. The partial pressures are defined by the relations

$$p_A = X_A P \tag{6}$$

$$p_C = X_C P \tag{7}$$

where P is the total pressure and X_A and X_C are the mole fractions of A and C in the gas mixture.

For the component B, which is not gaseous, the relation is

$$\mu_B = g_B^0 + RT \ln a_B \tag{8}$$

μ_B, called the chemical potential of B in the solid or liquid phase, is the partial derivative of the free energy of the phase with respect to a change in the amount of B present, all other variables (temperature, pressure, amount of other components) being constant. B could be an element in a solid or liquid solution (of a certain composition), it could be an element in a solid compound, or it could be a pure liquid or solid element. g_B^0 is called the standard free energy, and for our purposes we will take this to be the molar Gibbs free energy for pure B. We do not need to specify the total pressure since g is rather insensitive to pressure for condensed phases. [For the free energy of the gases A and C, g_A and g_C, we did have to specify the total pressure (to be one atmosphere) since the free energy for a gas is quite sensitive to the pressure.] However, we do have to specify the state of pure B. It is either liquid or solid, and if solid, then we must state the crystal structure. For example, if B is carbon, we must indicate whether the standard state is graphite or diamond.

If the chemical reaction involves pure B, then the chemical potential is identical to the molar free energy. In Equation (8), the term a is called the *activity*. Note that if B is present as pure B, then $a_B = 1$. However, if B is present in a solution (solid or liquid), then $a_B \neq 1$.

Note that the definition taken for the standard state of B, namely pure B in a specified crystalline form or as a liquid, requires that when B is present as pure B in this form, the activity is unity. The activity of B is a function of the composition of the phase of which B is a part, and this dependence must be determined experimentally. When the activity of B in the solution (liquid or solid) is equal to the mole fraction in the solution, then the solution is called ideal, and the activity–composition relation follows Raoult's law.

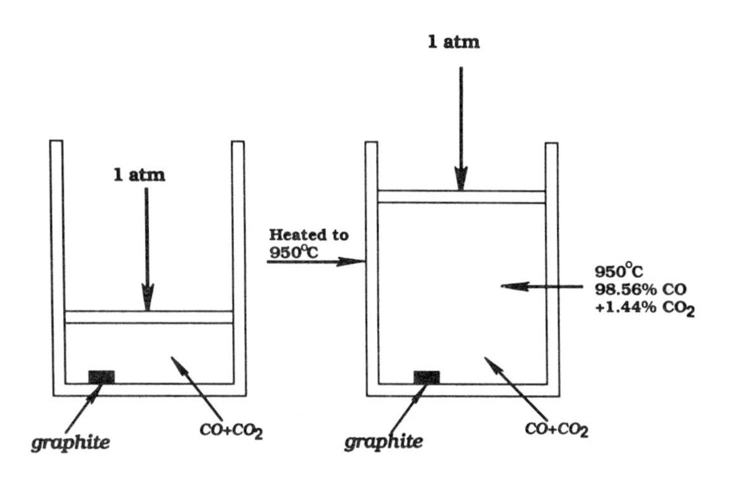

FIGURE 76. Schematic illustration of an experiment to determine the standard free energy change of the reaction $2CO \rightarrow CO_2 + C_{gr}$.

Equations (4), (5) and (8) are substituted into Equation (3) to obtain

$$3(g_C^0 + RT \ln p_C) - (g_A^0 + RT \ln p_A) - 2(g_B^0 + 2RT \ln a_B) = 0 \quad (9)$$

which can be rearranged to obtain

$$3g_C^0 - g_A^0 - 2g_B^0 = -RT \ln [(p_A/p_C^3)(1/a_B^2)] \quad (10)$$

which can be written as

$$\Delta g^0 = -RT \ln [(p_A/p_C^3)(1/a_B^2)] \quad (11)$$

where Δg^0 is called the *standard free energy change*. It carries with it the standard states used for the components in the chemical reaction. The equilibrium constant K_p is defined by

$$K_p = [(p_A/p_C^3)(1/a_B^2)] \quad (12)$$

CO–CO$_2$ EQUILIBRIUM WITH CARBON IN AUSTENITE

The above equations are now applied to the reaction

$$CO_2 + C = 2CO \quad (13)$$

where CO and CO_2 are understood to be gases. Thus, Equation (11) becomes

$$\Delta g^0 = -RT \ln [(p_{CO}^2/p_{CO_2})(1/a_C)] \quad (14)$$

We examine the case where carbon is present as pure graphite. In this case Δg^0 is the free energy change at a given temperature when one mole of pure CO_2 at one atmosphere reacts with one mole of pure graphite to form two moles of pure CO at one atmosphere. Δg^0 is temperature-dependent, and this dependence must be determined experimentally. To illustrate how this may be done, consider placing graphite, CO and CO_2 in a cylinder, with nothing else present, then heating the cylinder to a certain temperature (see Figure 76). During heating, the chemical reaction given in Equation (13) will occur (in one direction or the other), and when the system reaches the temperature of interest, eventually this reaction will come to equilibrium. Then a gas sample is taken and analyzed to obtain the values of X_{CO} and X_{CO_2}. Assuming that the system is at a total pressure of one atmosphere, the partial pressures of CO and CO_2 are equal to their mole fractions. If this experiment is carried out at 950°C (1223 K), the chemical analysis of the gas will be found to be 98.56% CO and 1.44% CO_2.

Thus the ratio $p^2_{CO}/P_{CO_2} = 66.5$, and substitution into Equation (14) yields a value of $-42{,}670$ joules for Δg^0. This process is repeated for different temperatures to obtain the temperature dependence of Δg^0.

Figure 77 shows the temperature dependence of the equilibrium gas composition for CO, CO_2 and graphite as described above. Note that at equilibrium at low temperatures almost no CO is present, and above about 1000°C almost no CO_2 is present.

From experiments equivalent to that described above, the temperature dependence of Δg^0 for the chemical reaction is obtained. One relation is

$$\Delta g^0 = +170{,}700 - 174.5T \qquad (16)$$

where Δg^0 is in joules and T is in K. Since Δg^0 is determined from experimental data, the expression found in the literature will depend upon the experimental data used.

At constant temperature and pressure, the Gibbs free energy change is

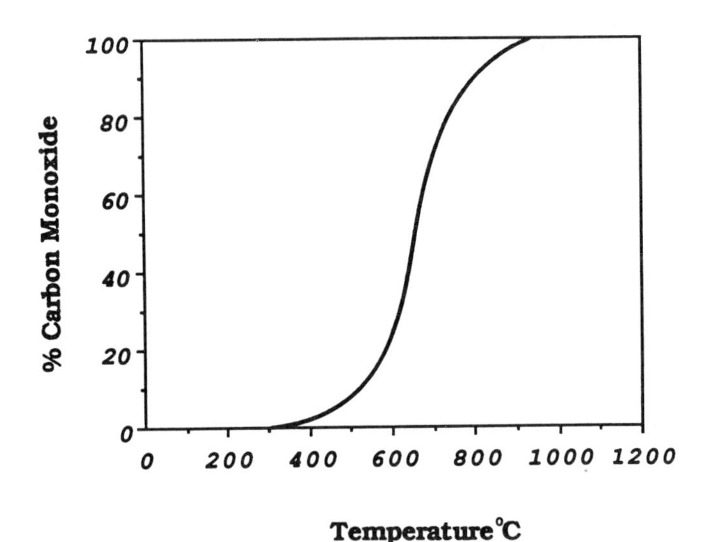

FIGURE 77. The variation with temperature of the amount of CO (and CO_2) in equilibrium with graphite at a total pressure of 1 atmosphere (adapted with permission from Gaskell, D. R. 1973. *Introduction to Metallurgical Thermodynamics*. New York: McGraw-Hill Book Company).

related to the enthalpy change Δh^0 and the entropy change Δs^0 for the chemical reaction by

$$\Delta g^0 = \Delta h^0 - T\Delta s^0 \tag{17}$$

It is usually found that Δh^0 and Δs^0 are almost temperature-independent, so that Equation (17) is a straight line. The enthalpy change, when one mole of graphite is reacted with one mole of CO_2 to form two moles of CO, is thus +170,700 joules. Since this is positive, the reaction is endothermic. The entropy change is +174.5 joules/K. This positive value is expected since a condensed phase (graphite) is reacting with one mole of gas (CO_2) to form *two* moles of another gas (CO), which should cause an increase in entropy.

In Figure 78 is plotted p^2_{CO}/p_{CO_2} on a logarithmic scale versus temperature. We will use this type of plot in developing Figure 81.

Now the above approach is applied to the case where the carbon is dissolved in pure iron in the austenite phase. In this case we must use Equation (14), involving the activity of carbon in the austenite. What we want to determine is the dependence of the activity of carbon in austenite on the composition of the austenite. This must be done experimentally. We shall conduct an experiment similar to that in Figure 76. Here, though, we use a small piece of iron (or of Fe–C alloy) (Figure 79). A gas mixture of CO and CO_2 is allowed to flow across the piece of iron until it attains a uniform carbon content, after which it is analyzed to determine its value. If the temperature is 950°C and the gas composition 97.8% CO and 2.20% CO_2, the carbon content will be found to be 0.98% C. Using Equation (14) and the value of Δg^0 from Equation (16), the activity of carbon in the austenite is calculated to be 0.65. (It is important to realize that this activity value is relative to pure graphite as the standard state, because the value of Δg^0 which was used in the calculation of Equation (14) is based on pure graphite.) Figure 80 shows plots of the activity of carbon in austenite as a function of carbon content for different temperatures.

It is useful to change the standard state of carbon to that of austenite containing 1 wt.% C. This is done by using the relation below, for a fixed composition of austenite.

$$\mu_C = \mu_C \tag{18}$$

Equation (8), a relation involving activity for which the standard state is pure graphite, is used for the left-hand side of Equation (18). For the right-

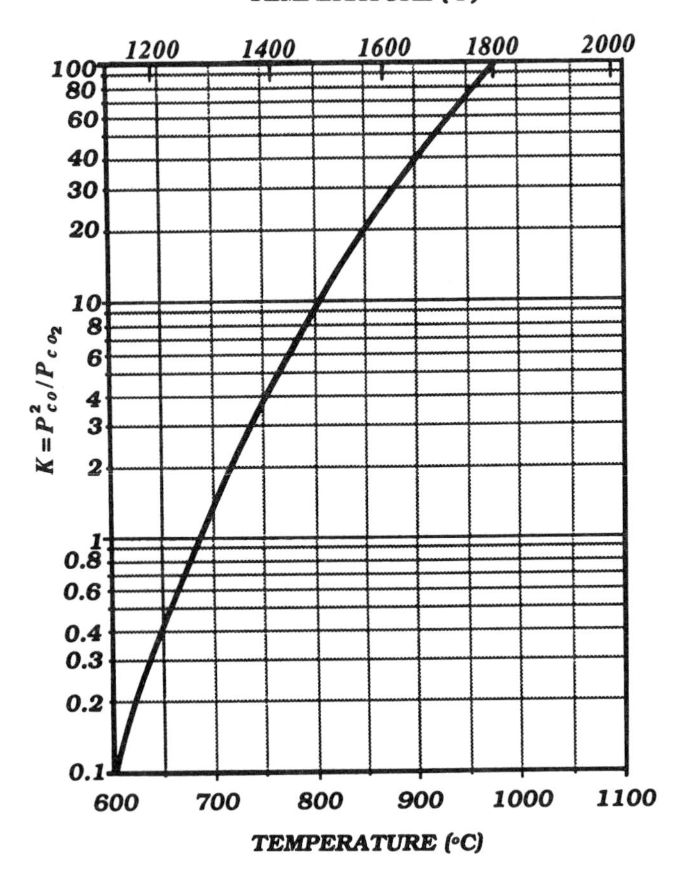

FIGURE 78. The temperature dependence of the ratio p^2_{CO}/p_{CO_2} for equilibrium with graphite.

hand side, the standard state is taken as 1 wt.% C in austenite, and the activity is taken to be the weight percent carbon, W_C. Thus

$$g^0_C + RT \ln a_C = g^*_C + RT \ln W_C \tag{19}$$

where a_C denotes the activity of carbon in austenite with pure graphite as the standard state. The other standard state is austenite containing 1 wt.% C, g^*_C is the free energy for this state, and the activity is W_C. Rearranging Equation (19)

$$a_C = W_C \{\exp[(g^*_C - g^0_C)/RT]\} \tag{20}$$

Thus, the activity of carbon in austenite can be converted from one standard state to the other if a value for the exponential term in Equation (20) is known. A method of obtaining this conversion is now described.

From chemical thermodynamics, it is established that for multiphase equilibrium the chemical potential of any given component is identical in all phases. Thus if austenite is in equilibrium with graphite

$$\mu_C^{gr} = \mu_C^{\gamma} \tag{21}$$

In this case, the austenite is saturated with carbon, so that $a_C = 1$. Also, for pure graphite as the standard state, $\mu_C = g_C^0$. Since the austenite is saturated, $W_C = W_C^*$. Then

$$g_C^0 + RT \ln a_C = g_C^0 = g_C^* + RT \ln W_C^* \tag{22}$$

W_C^* is the saturation value, obtained from the Fe–C phase diagram (using

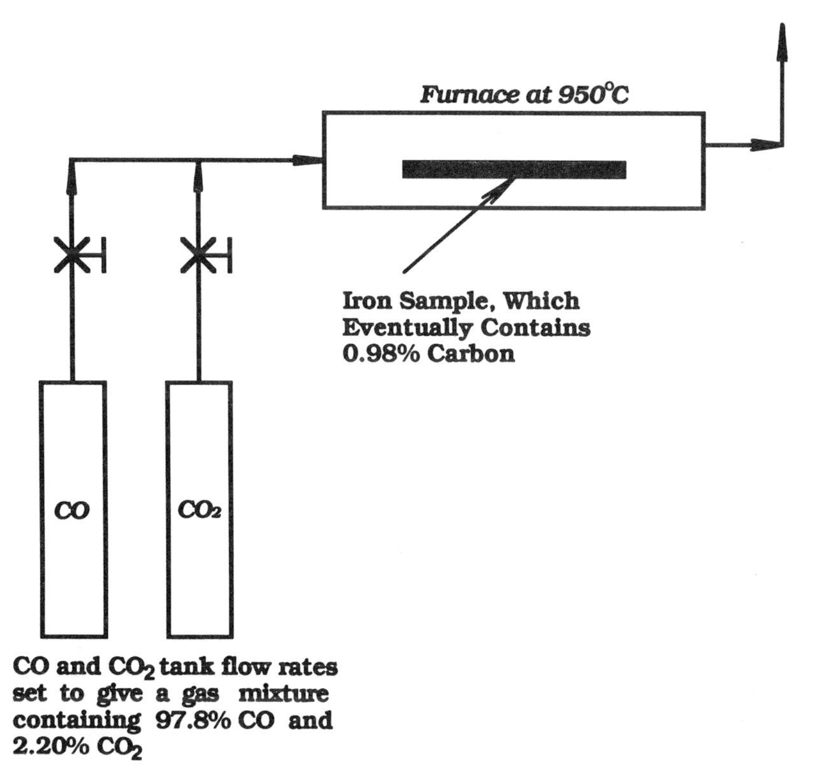

FIGURE 79. Schematic illustration of an experiment to determine the activity of carbon in austenite as a function of the carbon content of the austenite.

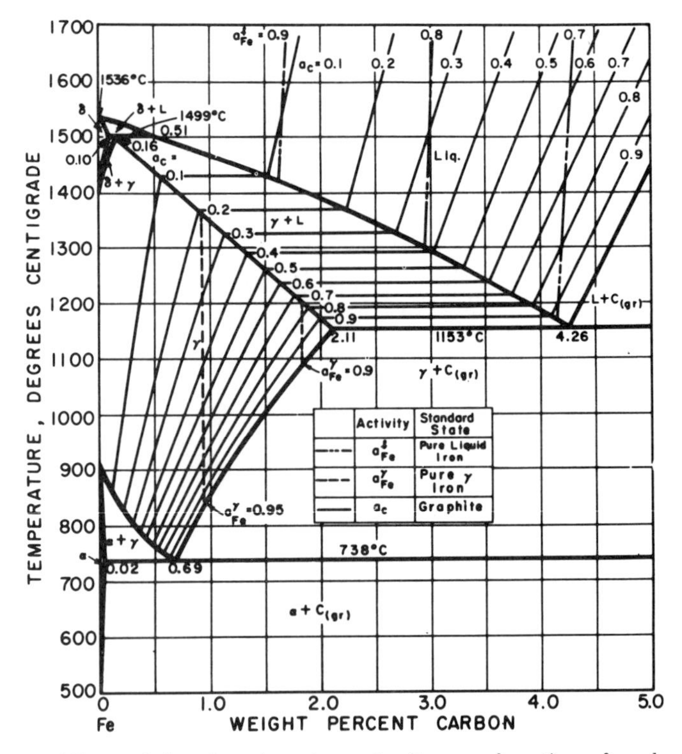

FIGURE 80. The activity of carbon in austenite as a function of carbon content and temperature (reprinted with permission, and adapted from Benz, M. G. and J. F. Elliott. 1961. *Trans. AIME*, 221:323, a publication of The Minerals, Metals and Materials Society, Warrendale, PA).

the solubility curve of carbon in austenite in equilibrium with *graphite*, not Fe_3C). Thus

$$\exp[(g_C^* - g_C^0)/RT] = W_C^* \tag{23}$$

This can be substituted into Equation (20) to obtain

$$a_C = W_C/W_C^* \tag{24}$$

For the reaction

$$\underline{C}^{\gamma} + CO_2 = 2CO$$

where $\underline{C}^{\gamma}$ refers to carbon dissolved in austenite, we get

$$+ 170{,}700 - 174.5T = -RT \ln [(p^2_{CO}/p_{CO_2})(1/a^{\gamma}_{C})] \tag{25}$$

and using Equation (24) we get

$$+ 170{,}700 - 174.5T = -RT \ln [(p^2_{CO}/p_{CO_2})(W^*_{C}/W_{C})] \tag{26}$$

where T is in K and R = 8.314 joules/mole-K. This equation can be used to develop a graphical representation of the relation between the ratio p^2_{CO}/p_{CO_2} required to achieve chemical equilibrium with carbon in austenite of a given carbon content. This can be done as a function of temperature, and the result is shown in Figure 81. (Note that the curve for

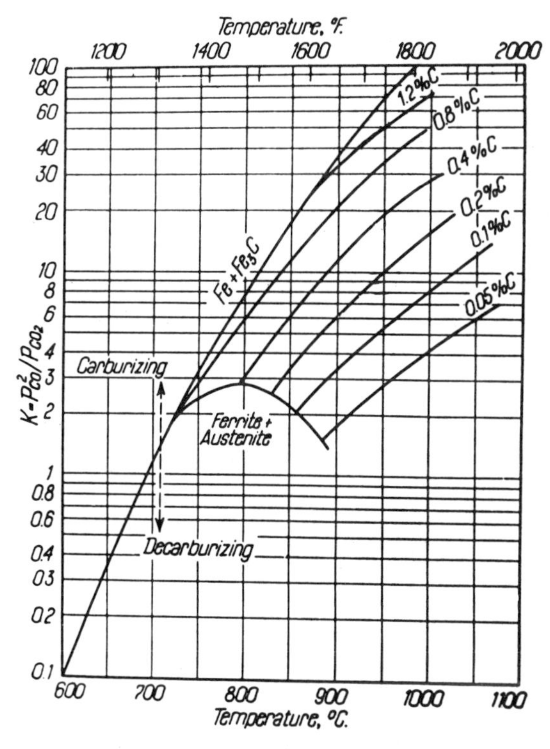

FIGURE 81. The temperature dependence of the ratio p^2_{CO}/p_{CO_2} for equilibrium with carbon in austenite of different carbon contents (reprinted with permission from Austin, J. B. and M. J. Dary. 1942. In *Controlled Atmospheres*. Metals Park, Ohio: American Society for Metals).

equilibrium with austenite saturated with carbon has been calculated, as this corresponds to equilibrium with graphite, and was plotted in Figure 78.) This format is convenient, as one can bypass the calculations to determine the equilibrium condition by using the graphical depiction.

Thus, if one desires to carburize at 950°C steel containing 0.2% C so that the surface carbon content does not exceed 0.80% C, then a mixture of CO and CO_2 can be used if the ratio of the partial pressures $p^2_{CO}/p_{CO_2} = 34$. To determine the gas composition required, the equation $p_{CO} + p_{CO_2} = P$ must be solved, where P is the total pressure, which is usually 1 atmosphere.

In many gases, nitrogen may be present because CO–CO_2 mixtures may be generated by burning fuel with air. Thus, equilibrium data are sometimes presented as shown in Figure 82. These data are somewhat misleading, because the amount of nitrogen present is not fixed. For example, at 927°C (1700°F), the CO and CO_2 content can be calculated for equilibrium with austenite containing 0.2% C. If it is assumed that the total pressure is one atmosphere, *and* that the amount of nitrogen (or any inert gas) is 79%, then the $P_{CO} + P_{CO_2} = 0.21$. This relation and the relation from the thermodynamic equations are solved simultaneously to obtain 20.46% CO and 0.54% CO_2. Note in Figure 83 that from these values, a point A is found on the curve for 0.2% C. If the nitrogen content is 85%, then $p_{CO} + p_{CO_2} = 0.15$, and a calculation gives a gas containing 14.72% CO and 0.28% CO_2 to be in equilibrium with austenite containing 0.2% C. This is point B on the curve in Figure 83. Thus, for these curves, the amount of inert gas is found by subtracting the sum of the amount (%) of CO and CO_2 present from 100.

CH_4–H_2 EQUILIBRIUM WITH CARBON IN AUSTENITE

By a similar procedure to that described above, the equilibrium relations for the reaction of the gas mixture of CH_4 and H_2 with austenite can be determined. The chemical reaction is

$$CH_4 = 2H_2 + \underline{C}^{\gamma} \tag{27}$$

The data for gases containing only CH_4 and H_2 are shown in Figure 84. Note that the gas ratio is considerably higher than that for CO–CO_2 (Figure 82). This translates into a relatively high H_2 content compared to CH_4. For example, at 950°C, the ratio necessary to be in equilibrium with austenite containing 0.8% C is $p^2_{H_2}/p_{CH_4} = 100$. Using $p_{CH_4} + p_{H_2} = 1.00$ atm, the equilibrium gas content is 99.00% H_2 and 1.00% CH_4. Thus only a small amount of CH_4 must be present to attain equilibrium.

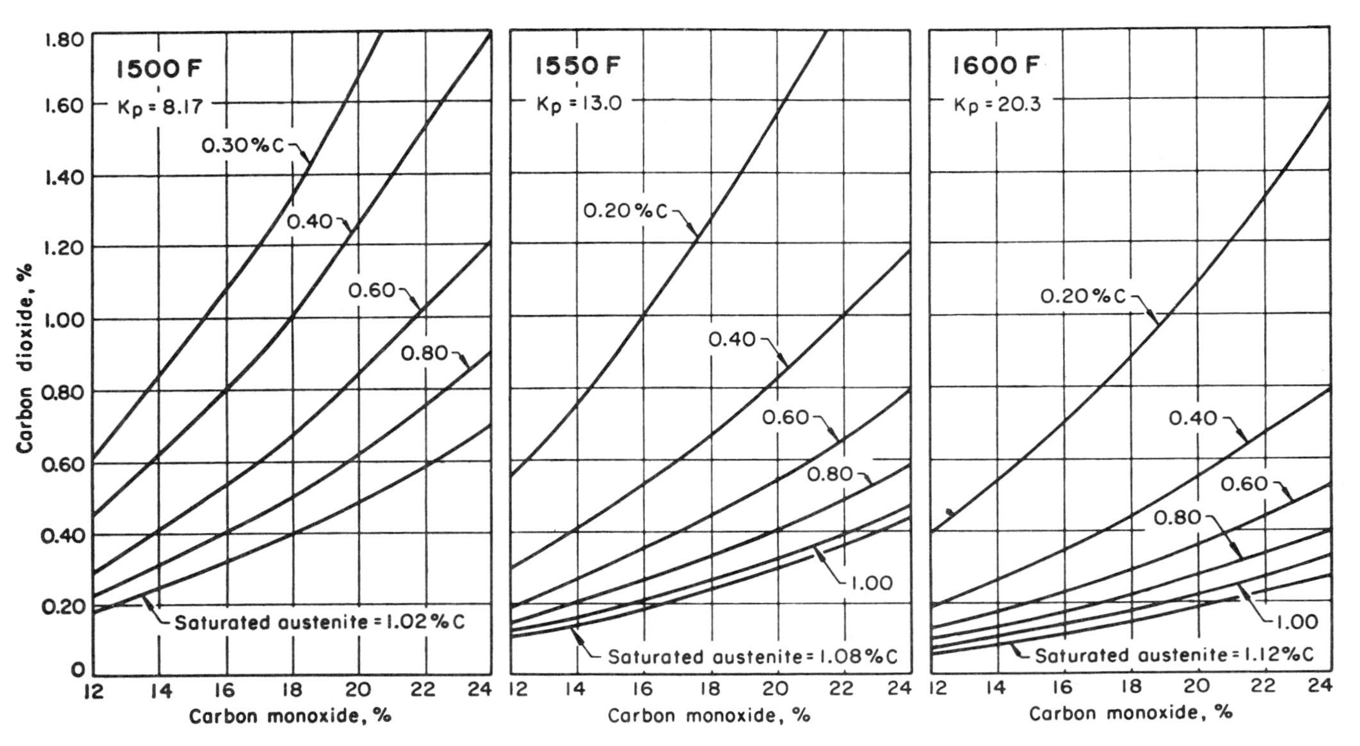

FIGURE 82. Percentages of CO and CO_2 in equilibrium with austenite of various carbon contents at several temperatures. The remainder of the gas is inert nitrogen (reprinted with permission from *Metals Handbook, Vol. 2.* 1964. Metals Park, Ohio: American Society for Metals).

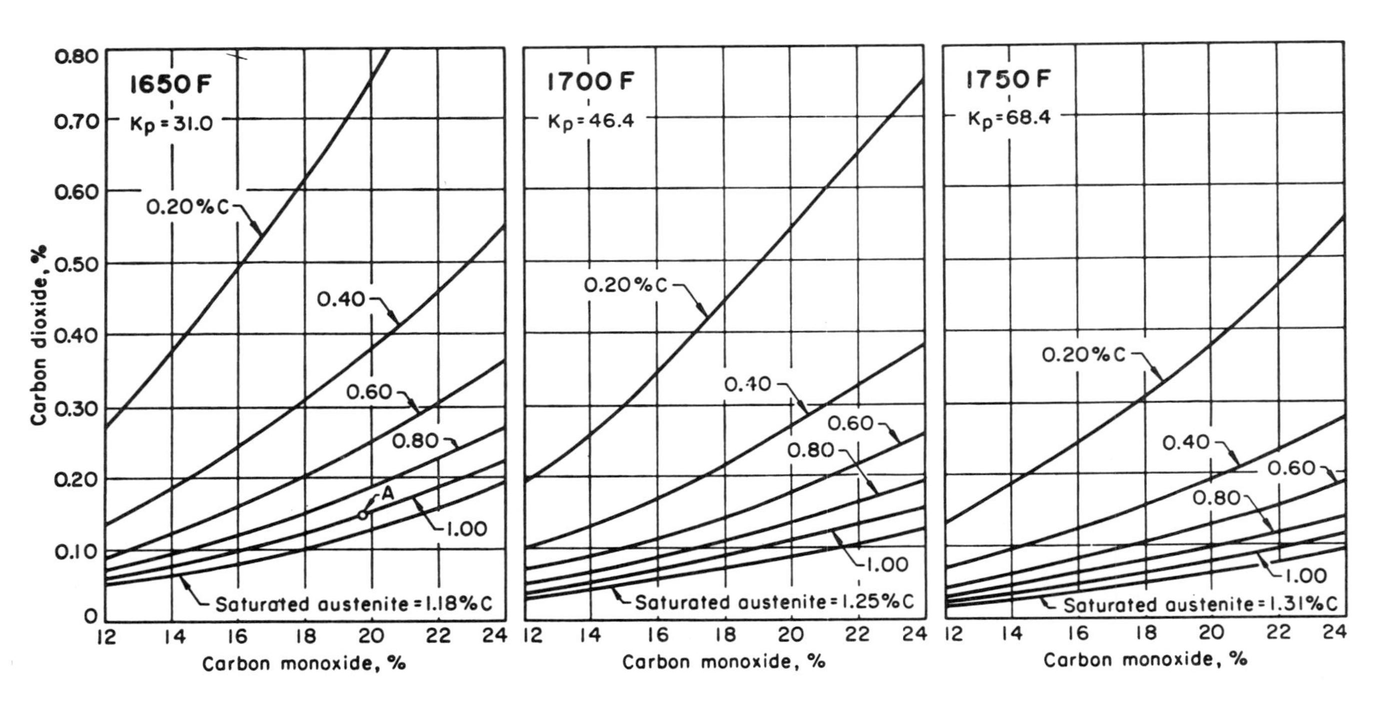

FIGURE 82 (continued). Percentages of CO and CO_2 in equilibrium with austenite of various carbon contents at several temperatures. The remainder of the gas is inert nitrogen (reprinted with permission from *Metals Handbook, Vol. 2.* 1964. Metals Park, Ohio: American Society for Metals).

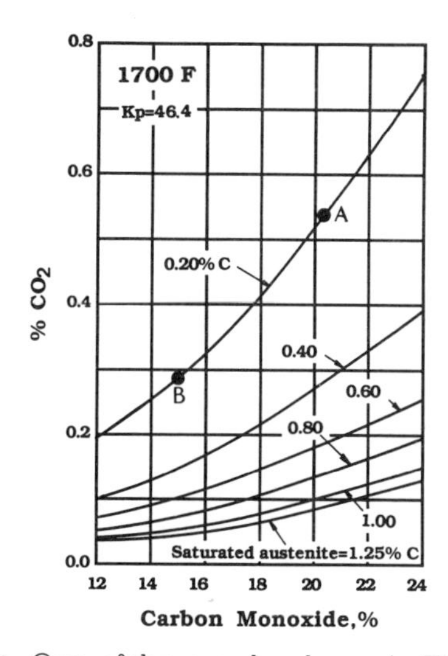

FIGURE 83. One of the graphs shown in Figure 82.

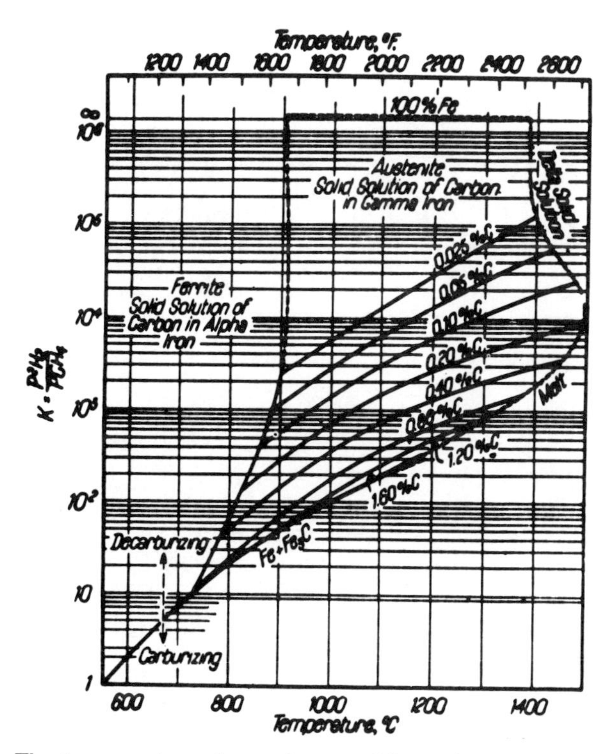

FIGURE 84. The temperature dependence of the ratio $p^2_{H_2}/p_{CH_4}$ for equilibrium with carbon in austenite of different carbon contents (reprinted with permission from *Metals Handbook, Vol. 2.* 1964. Metals Park, Ohio: American Society for Metals).

CO-CO_2-H_2-H_2O EQUILIBRIUM WITH CARBON IN AUSTENITE

In most carburizing gases, not only are CO and CO_2 present, but H_2 and H_2O are as well. [Since H_2 is present, then CH_4 will be present, via Equation (27). However, we will neglect its presence, and justify this at the end of this section.] Thus, there are two independent chemical reactions that must be satisfied at equilibrium.

$$\underline{C}^{\gamma} + CO_2 = 2CO \tag{13}$$

$$CO_2 + H_2 = CO + H_2O \tag{28}$$

The standard free energy change for this latter reaction has also been determined, and thus the relation between the composition of these gases, via the equilibrium constant, can be determined.

$$K_1 = [(X_{CO} X_{H_2O})/(X_{CO_2} X_{H_2})] \tag{29}$$

At equilibrium at a given temperature this equation must be satisfied, along with the equilibrium constant for the reaction in Equation (13)

$$K_2 = [(p^2_{CO}/p_{CO_2})(1/a^{\gamma}_{c})] \tag{30}$$

and the relation

$$p_{CO} + p_{CO_2} + p_{H_2} + p_{H_2O} = P \tag{31}$$

where P is the total pressure (usually 1 atmosphere). To determine the required gas composition for a given carbon content (activity of carbon) in austenite, the relations in Equations (29), (30) and (31) must be solved simultaneously.

At 927°C, K_1 = 1.43. If a gas containing 40% H_2, 18.67% CO, 1% H_2O and 40% nitrogen is used, then the calculated CO_2 content is 0.33%. Then using Equation (26) applied at 1200 K (917°C), the calculated carbon content of the austenite is 0.30%. From such a procedure, the relation between the water content of the gas and the carbon content of the austenite, for a gas containing 20% CO, 40% H_2 and 40% N_2, can be obtained. Also this can be put in terms of the CO_2 content and the carbon content of the austenite. These relationships are shown in Figure 85. Note the agreement between the values calculated here and the graphical data. These considerations show that the carbon content of the austenite can be set by controlling the water vapor content instead of by controlling the CO_2 content.

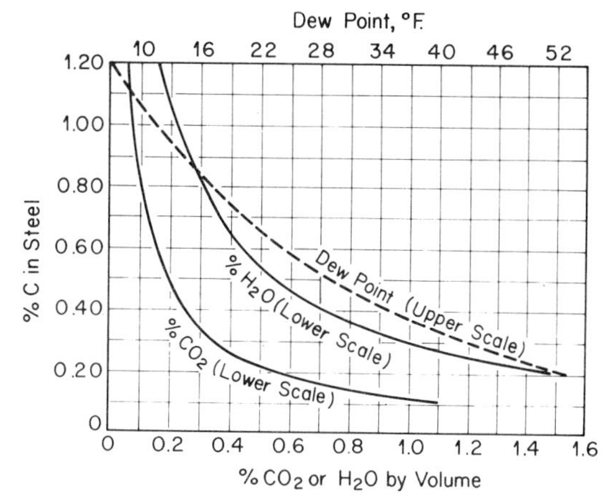

FIGURE 85. The equilibrium carbon content of austenite at 927°C as a function of the CO_2 and the H_2O content of a gas containing 20% CO, 40% H_2 and 40% N_2 (adapted from Cullen, O. E. 1954. *Metals Progress*, 66:114).

The water content of gases is frequently reported as a dew point temperature, instead of the % H_2O present. The experimental relation between the dew point and the water content is shown in Figure 86. Thus, graphical data of the equilibrium carbon content of austenite can be given in terms of the dew point, such as in Figure 87.

We now return to the question of neglecting the methane reaction. From

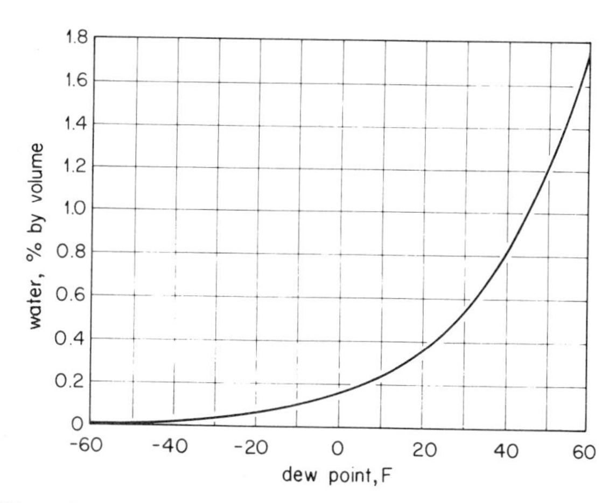

FIGURE 86. The dew point curve for water. The total pressure is 1 atmosphere (reprinted with permission from Brooks, C. R. 1979. *Heat Treatment of Ferrous Alloys*. New York: Hemisphere Publishing Corporation/McGraw-Hill Book Company).

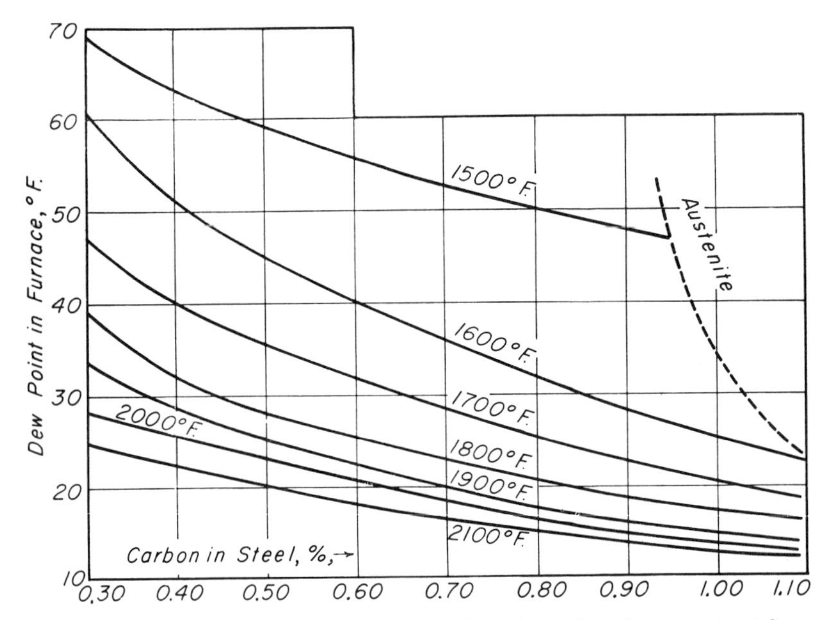

FIGURE 87. The dew point of water as a function of carbon content for a gas containing 40% H_2 and 20% ($CO + CO_2$) (adapted from *Metal Progress Data Sheets, No. 79*. 1954. Metals Park, Ohio: American Society for Metals).

Figure 84 we have determined that at 927°C a 0.8% C steel will be in equilibrium with a gas with a ratio $p_{H_2}^2/p_{CH_4} = 90$. Using a gas containing 40% H_2, the methane content will be 0.16%. Thus, only a small amount of methane has to form by the reaction of H_2 with the carbon in the austenite to achieve equilibrium by this reaction. (It also turns out that the rate of the carburizing reaction here is relatively low, a point that will be discussed in the next section.)

Many carburizing gases are generated by burning fuel and air under conditions that generate a gas mixture containing carburizing species. The gas compositions typical of those formed by burning natural gas and air are shown in Figure 88. This is an exothermic reaction, not requiring any additional heat in the burner to sustain the reaction, so it is referred to as an exothermic composition. As the amount of oxygen, via the air, is increased, the amount of H_2 and CO present decrease, and the amount of CO_2 increases. When the amount of oxygen used is more than sufficient to cause complete combustion, oxygen is leftover, and the gas consists of CO_2 and O_2. Note that for the gas mixture in Figure 88 the remainder of the gas is inert nitrogen. It is seen that for air/gas ratios of greater than

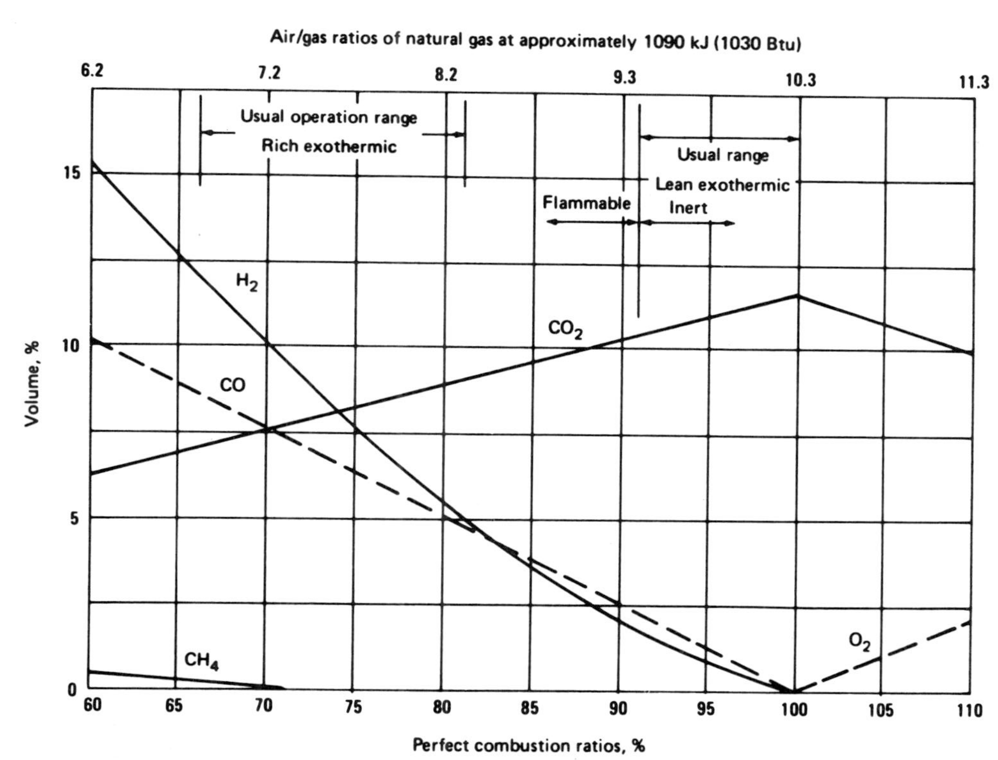

FIGURE 88. Typical chemical analysis of an exothermic gas mixture formed by burning natural gas with air (reprinted with permission from *Metals Handbook, Vol. 4, Heat Treating.* 1981. Metals Park, Ohio: American Society for Metals).

about 7.2, the amount of CH_4 present is negligible. Thus, it will not enter into the consideration of carburizing.

The gas composition that is generated depends upon the gas burned. Figure 89 shows data using propane; note that this gas mixture is significantly different from that in Figure 88 for burning methane.

Figure 90 shows some data for a commercial carburizing process. The details of the operation are given in the caption of the figure. Note that the dew point is followed to monitor the carburizing process. It is also seen that the curvature of the carbon gradient reflects the changes in the carburizing process. It does not have the shape for carburizing with a constant surface carbon content (discussed in Section 3d).

EFFECT OF ALLOYING ELEMENTS ON CARBON ACTIVITY

Figure 80 shows the effect of carbon content on the activity of carbon in austenite for binary Fe–C alloys. However, the presence of alloying elements (e.g., Cr, Ni, Mo) will affect the activity of carbon, and hence will modify the carburizing process from that expected for plain carbon steels. Thus, significant differences in the surface carbon content can be found in alloy steels compared to plain carbon steels for the same gas atmosphere.

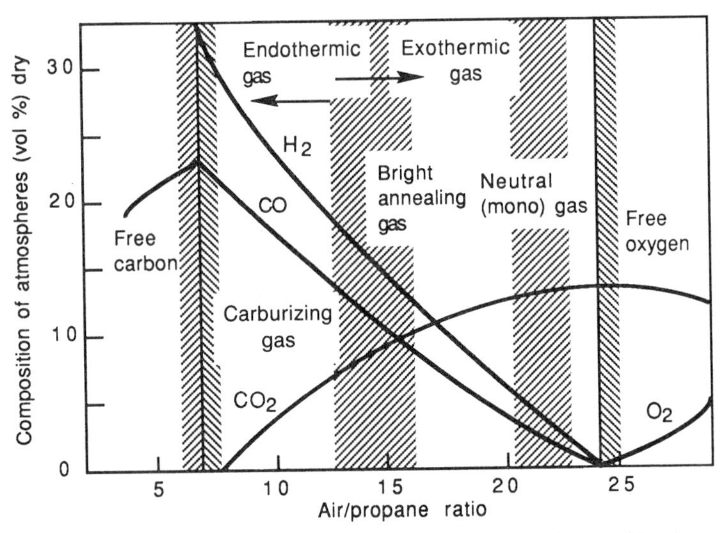

FIGURE 89. Typical chemical analysis of a gas mixture formed by burning propane gas with air (reprinted with permission from Nemenyi, R. and G. H. J. Bennett. *Controlled Atmospheres for Heat Treatment*. © 1984, Pergamon Press PLC).

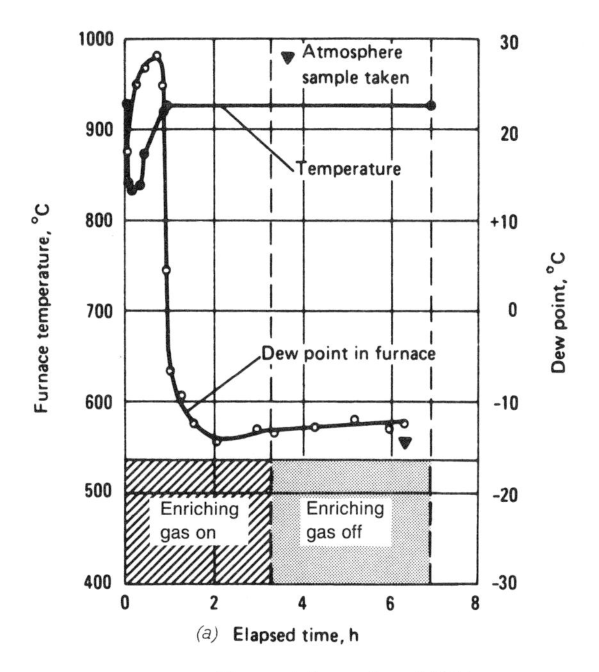

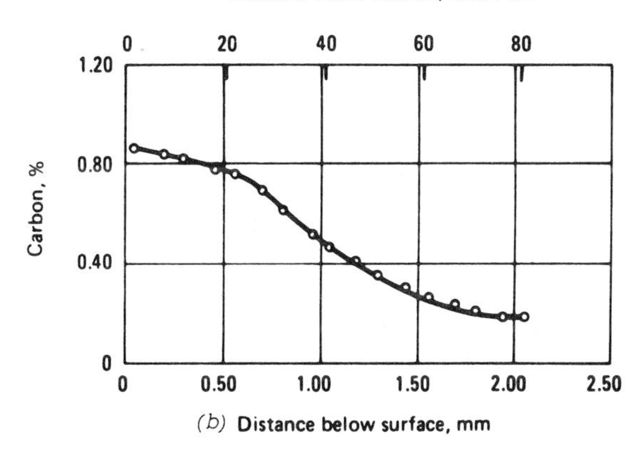

Load: ring gears, 231 kg (510 lb) net, 390 kg (860 lb) gross. **Furnace:** metallic-retort pit furnace, electrically heated. **Carrier gas:** 2.8 m^3/h (100 ft^3/h) of endothermic gas throughout the cycle, including 5 h and 59 min at temperature. **Enriching gas:** 0.3 m^3/h (12 ft^3/h) of natural gas; addition started when load was placed in furnace; flow of natural gas stopped after 2½ h at temperature, or 42% of at-temperature cycle. **Generator dew point:** −6 to −4 °C (+22 to +25 °F) throughout cycle. **Heating chamber pressure:** 5 to 8 mm (0.20 to 0.31 in.) water column. **Carburizing temperature:** 925 °C (1700 °F). **Cooling method:** Slow cooled from 925 °C (1700 °F) in cooling pit. **Analysis of atmosphere near end of cycle:** 20.8% CO, 0.4% CO_2, 34.0% H_2, 0% O_2, 0.8% CH_4, rem N_2

(a) Furnace temperature and atmosphere conditions for carburizing. (b) Carbon gradient produced by the cycle shown in (a)

FIGURE 90. A carburizing process for heat treatment of gears (reprinted with permission from *Metals Handbook, Vol. 4, Heat Treating*. 1981. Metals Park, Ohio: American Society for Metals).

One relation that has been determined to account for the effect of Cr and Ni is that of Natesan and Kassner (1973. *Metallurgical Transactions*, 4:2557). The carbon activity (relative to pure carbon as graphite) is given by (composition in wt.%).

$$\ln a_{\mathrm{C}}^{(\mathrm{FeCrNiC})} = \ln (0.048 \text{ pct C}) + \left(0.525 - \frac{300}{T}\right) \text{pct C}$$

$$- 1.845 + \frac{5100}{T} - \left(0.021 - \frac{72.4}{T}\right) \text{pct Ni}$$

$$+ \left(0.248 - \frac{404}{T}\right) \text{pct Cr}$$

$$- \left(0.0102 - \frac{9.422}{T}\right) \text{pct Cr}^2 \qquad (32)$$

Figure 91 shows some of their data for 8% Ni steels with various amounts of C and Cr. To illustrate the effect of alloy content on the activity, consider the example on page 94, where the gas composition was determined to attain an equilibrium carbon content of 0.33 wt.% at 927°C (1200 K). This value was for a binary Fe–C alloy, and Figure 80 can be used to determine the activity of carbon at this temperature to be about 0.22. This same gas will be in equilibrium with carbon in austenite at this same activity. However, the corresponding carbon content depends upon the alloy content. Using this activity value of 0.22 in Equation (32) for a temperature of 1200 K, a Ni content of 8% and a Cr content of 4%, the calculated carbon content is about 0.4%. Thus, this gas would equilibrate with the alloy steel at a carbon content of 0.4%, not 0.33%.

KINETICS OF THE GAS REACTIONS

In the previous discussions, it has been implied that the carbon gradient will be controlled by the diffusion of carbon into the steel. This implies that carbon is deposited by the chemical reactions more rapidly than it can be absorbed. However, this is not entirely supported by a critical examination of the chemical kinetics of the carburizing gas reactions.

The rate of gaseous chemical reactions is found to be related to the "concentration" of the reacting species, to a power that appears as the coefficient in the chemical reaction. The chemical reactions of interest in gas-carburizing are those in Figure 92. The corresponding equations for the rate of deposition of carbon on the surface are given in Figure 93. Note that the negative sign denotes removal of carbon. The carburizing reaction

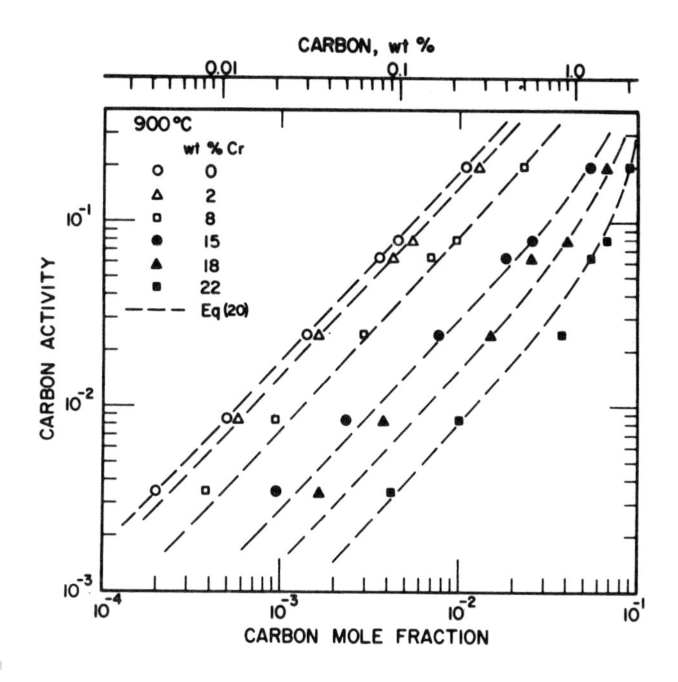

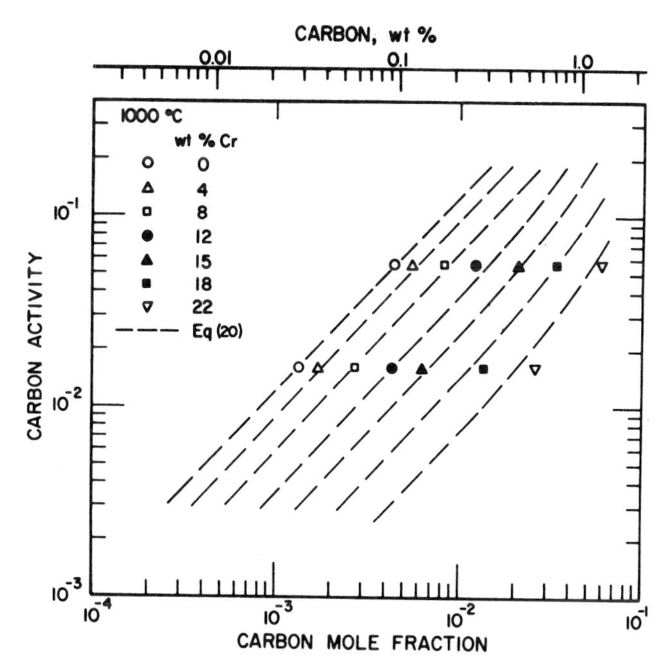

FIGURE 91. Effect of chromium and nickel content on the activity of carbon in austenite in Fe–Cr–8% Ni alloys (reprinted with permission from Natesan, K. and T. F. Kassner. 1973. *Met. Trans.*, 4:2557, a publication of The Minerals, Metals and Materials Society, Warrendale, PA).

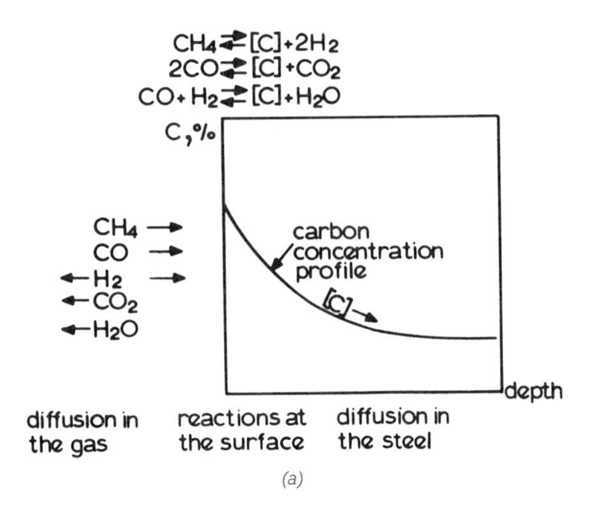

(a)

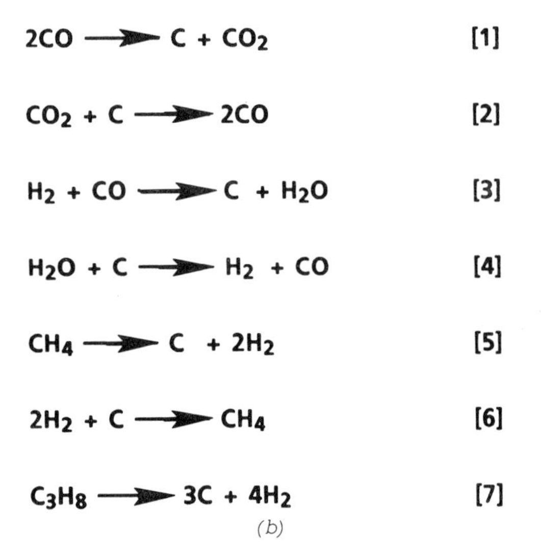

(b)

FIGURE 92. (a) Schematic diagram showing the chemical reactions involved in gas-carburizing (adapted from Collin, R., S. Gunnarson and D. Thulin. 1972. *J. Iron and Steel Inst.*, 210:785; by permission of The Institute of Metals). (b) Chemical reactions involved in gas-carburizing (adapted from Kaspersma, J. H. and R. H. Shay. 1980. *J. Heat Treating*, 1:21).

Experimental and Derived Rate Constants for Carburizing and Decarburizing Reactions

Rate Constant, k	Value (mol/cm2 min atm or mol/cm2 min atm2)	
	1700°F (927°C)	1550°F (843°C)
k_1	1.4×10^{-5}	1.3×10^{-5}
k_2	7.0×10^{-4}	1.8×10^{-4}
k_3	2.2×10^{-4}	1.4×10^{-4}
k_4	8.3×10^{-3}	1.9×10^{-3}
k_5	1.0×10^{-5}	3.0×10^{-6}
k_6	3.0×10^{-7}	9.0×10^{-7}
k_7	1.3×10^{-3}	4.5×10^{-4}

$$V \frac{dc_c}{dt} = Ak_1 P^2_{CO}$$

$$V \frac{dc_c}{dt} = -Ak_2 P_{CO_2} a_c$$

$$V \frac{dc_c}{dt} = Ak_3 P_{H_2} P_{CO}$$

$$V \frac{dc_c}{dt} = -Ak_4 P_{H_2O} a_c$$

$$V \frac{dc_c}{dt} = Ak_5 P_{CH_4}$$

$$V \frac{dc_c}{dt} = -Ak_6 P^2_{H_2} a_c$$

$$V \frac{dc_c}{dt} = Ak_7 P_{C_3H_8}$$

where:

- V = Volume of shimstock in cm^3
- A = Surface area of shimstock in cm^2
- c_c = Concentration of carbon in steel shimstock as mol/cm^3
- k = Rate constant for Reactions [1] through [7], with dimensions of either mol/cm^2 min atm or mol/cm^2 min atm^2
- t = Time in minutes
- P = Component partial pressure in atmospheres
- a_c = Carbon activity

FIGURE 93. Reaction rate constants at 927 and 843°C for the chemical reactions shown (adapted from Kaspersma, J. H. and R. H. Shay. 1980. *J. Heat Treating*, 1:21).

for the decomposition of methane [Equation (5) in Figure 92] will not have the same rate as the decarburizing reaction using hydrogen [Equation (6) in Figure 92]. The sum of these two reactions give the net effect for the general reaction of methane decomposing to release carbon and form hydrogen.

Experimentally determined rate constants for 927°C and 843°C are listed in Figure 93. Note that the constant for the methane reaction [Reaction (5)] is low, and that the reaction with the highest rate constant is Reaction (4). Although the reaction rate constant is a factor in setting the reaction rate, the partial pressures and the activity of carbon are also involved, as is shown by the equations in Figure 93. The results of calculations of the reaction rates at 927°C are shown in Figure 94 for the reactions listed. The rate constants used were those in Figure 93. The range of partial pressures typical of the gaseous components found in gas-carburizing are listed in Figure 94. A mean value is listed for each component, and these were used to calculate the reaction rates shown in Figure 94. For these calculations, the austenite was assumed to be saturated with carbon, so that $a_{\underline{C}}^{\gamma} = 1$. Note that the methane reaction has a low reaction rate, which justifies neglecting its involvement in carburizing in these types of gas mixtures. Also note that the main carburizing reaction is Reaction (3), which has a rate that is an order of magnitude higher than any of the others. Figure 95 gives equations for the reaction rate constants that incorporate temperature and partial pressures.

3d. DIFFUSION OF CARBON AND CASE DEPTH

We now examine the diffusion of carbon into austenite, and the carbon profile that forms. We first consider the case where the gradient is controlled only by the diffusion of carbon, which means that carbon is supplied sufficiently rapidly by the chemical reactions that these reactions do not limit the movement of carbon into the surface. We assume that during carburizing the carbon content of the surface increases with time until a value is reached that is the equilibrium value dictated by the gas composition; this may be the saturation value. Limiting our treatment to one-dimensional diffusion, we can obtain a solution to Fick's second law for the proper boundary conditions. These include the assumption that at $x = 0$ (at the surface), the carbon content is fixed at C_s, and at $x = \infty$ (the core), the carbon content is C_0; both of these values are independent of time. The solution then is

$$C(x,t) = C_s - (C_s - C_0) \operatorname{erf} (x/2\sqrt{Dt}) \tag{33}$$

Carburizing (+) and Decarburizing (-) Rates for Reactions 1 Through 7

Equation Number	Reaction Rate (mol/cm2 min)
$2CO \xrightarrow{k_1} C + CO_2$	$+2.0 \times 10^{-7}$
$CO_2 + C \xrightarrow{k_2} 2CO$	-1.3×10^{-6}
$H_2 + CO \xrightarrow{k_3} C + H_2O$	$+7.9 \times 10^{-6}$
$H_2O + C \xrightarrow{k_4} H_2 + CO$	-1.5×10^{-5}
$CH_4 \xrightarrow{k_5} C + 2H_2$	$+6.0 \times 10^{-7}$
$2H_2 + C \xrightarrow{k_6} CH_4$	-2.7×10^{-8}
$C_3H_8 \xrightarrow{k_7} 3C + 4H_2$	$+4.0 \times 10^{-7}$

T = 1700°F (927°C) $a_c = 1$

Typical Partial Pressures of Furnace Atmosphere Components

Component	Partial Pressure, P_i (atmospheres)	
	Range	Mean
CO	$5\text{-}20 \times 10^{-2}$	12×10^{-2}
CO_2	$5\text{-}30 \times 10^{-4}$	18×10^{-4}
H_2	1.5×10^{-1}	3×10^{-1}
H_2O	$5\text{-}30 \times 10^{-4}$	18×10^{-4}
CH_4	$2\text{-}10 \times 10^{-2}$	6×10^{-2}
C_3H_8	$1\text{-}5 \times 10^{-4}$	3×10^{-4}

FIGURE 94. Calculated rates of reaction at 927°C for the chemical reactions shown. The mean partial pressures shown were used in the calculations (adapted from Kaspersma, J. H. and R. H. Shay. 1980. *J. Heat Treating*, 1:21).

Carburizing reactions:

$CH_4 \rightleftharpoons [C] + 2H_2$ .... (1)

$2CO \rightleftharpoons [C] + CO_2$ .... (2)

$CO + H_2 \rightleftharpoons [C] + H_2O$ .... (3)

Reaction rate coefficients

$$k_1 = 1{\cdot}96 \,.\, 10^{-2} \,.\, pH_2^{3/2} \exp\left(-\frac{17600}{T}\right) \qquad (1)$$

(calculated from Grabke[7])

$$k_2 = 184 \left(\frac{pCO_2}{pCO}\right)^{-0{\cdot}3} . \, pCO_2 \,.\, \exp\left(-\frac{22400}{T}\right) \qquad (2)$$

(calculated from Grabke[8])

$$k_3 = \frac{4{\cdot}75 \,.\, 10^5 \exp\left(-\frac{27150}{T}\right) pH_2O/\sqrt{pH_2}}{1 + 5{\cdot}6 \,.\, 10^6 \exp\left(-\frac{12900}{T}\right) . \, pH_2O/pH_2} \qquad (3)$$

(calculated from Collin *et al.*[2])

FIGURE 95. The carburizing reactions, and the temperature and composition dependence of their reaction rate constants (adapted from Kaspersma, J. H. and R. H. Shay. 1980. *J. Heat Treating*, 1:21).

where erf is the error function, D is the diffusivity of carbon in austenite and t is time. The temperature dependence of the diffusivity is given by

$$D = D_0 \exp[-Q/RT] \qquad (34)$$

where D_0 is a constant, Q is the activation energy for diffusion (a constant), R is the ideal gas constant and T is the absolute temperature.

The terms in Equation (33) are illustrated in Figure 96. Also shown there is the definition of $\sqrt{Dt}$, a term that can be used as an approximation for the depth of carbon penetration. In fact, the depth x and time t can be combined in a normalized depth, as shown in Figure 97.

A useful equation for approximate calculations is $x \cong \sqrt{Dt}$, which is based on the time to attain one-half of the concentration change (see Figure 96). Thus, if the case depth noted by the arrows in Figure 72 is plotted against the time t, an approximately straight line is obtained.

The diffusion equation can be solved for other shapes, and Figure 98 illustrates results for the situation in which the case depth is taken to be that point where the carbon content decreases to 25% of the surface value minus the core value. For example, a flat slab of half-thickness of 10 mm

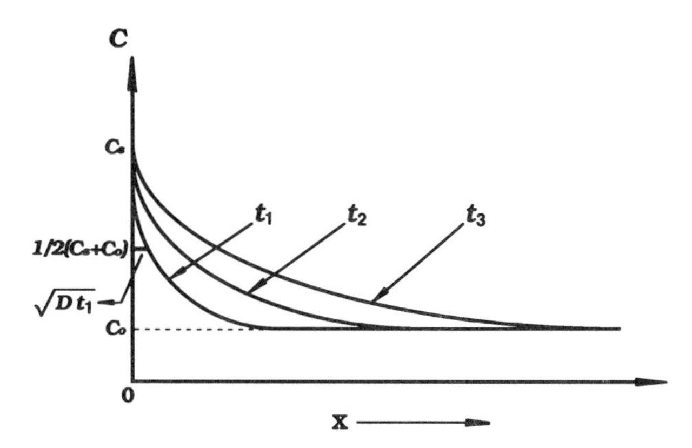

FIGURE 96. Concentration profile defining terms in Equation (33).

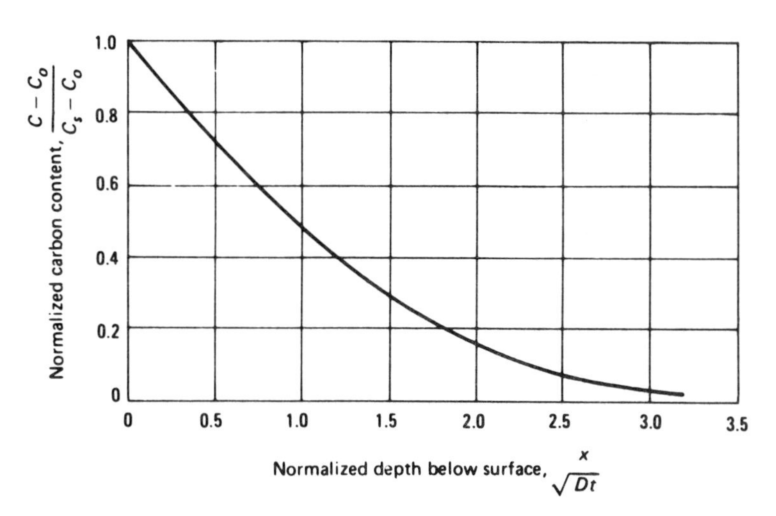

FIGURE 97. Normalized carbon profile (reprinted with permission from *Metals Handbook, Vol. 4, Heat Treating.* 1981. Metals Park, Ohio: American Society for Metals).

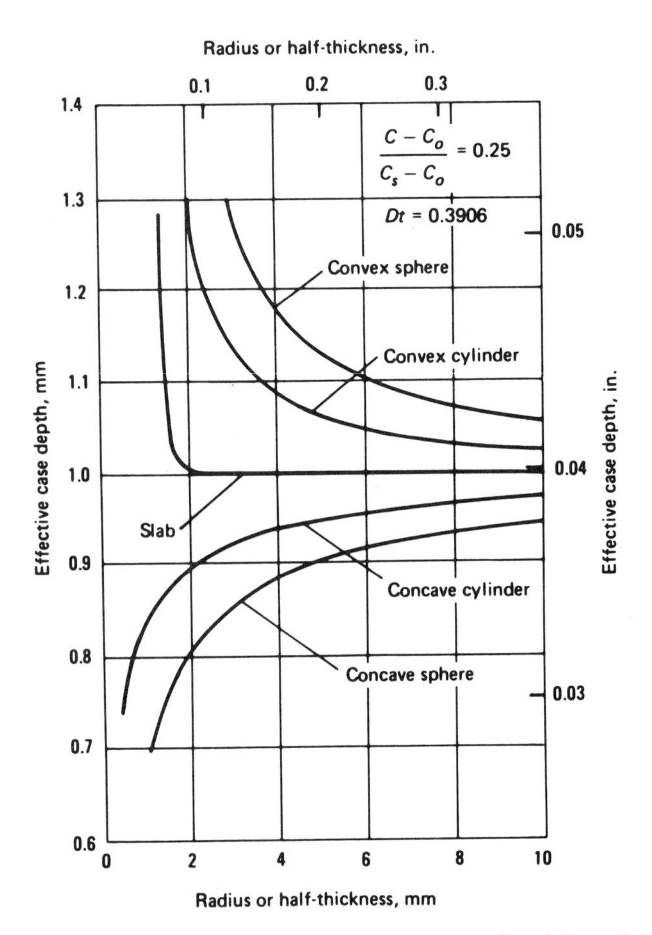

FIGURE 98. Influence of surface curvature on case depth (reprinted with permission from *Metals Handbook, Vol. 4, Heat Treating.* 1981. Metals Park, Ohio: American Society for Metals).

will have a case depth of 1.0 mm. A cylinder (convex cylinder, diffusing into the surface of the cylinder) of radius 10 mm will have a case depth higher than this, about 1.05 mm, an increase of about 5%.

An expression frequently quoted and used to estimate the case depth is the one given by Harris. It is

$$d = \{(31.6\sqrt{t})/[10^{6700/T}]\} \quad (35)$$

where d is in inches, t is time in hours and T is temperature in R. Some calculated values of case depth as a function of carburizing time and temperature are listed in Figure 99. Note that to obtain case depths of 0.02 to 0.1 inches in a few hours requires temperatures in the range of 900–1000°C.

The above description does not consider that the chemical reactions may be controlling the carbon content at the surface. This can be taken into account by noting that the flux at the surface from the diffusion equation is equal to that from the chemical reactions. Thus we can write

$$J\text{ (diffusion)} = J\text{ (chemical reaction)}$$

The diffusion flux can be obtained by applying Fick's first law at the surface ($x = 0$) to Equation (33). The flux from the chemical reactions can be obtained by applying a model to the chemical reactions, and summing

Time, t, hr	Case depth, in., after carburizing at 1600 F	1650 F	1700 F
2	0.025	0.030	0.035
4	0.035	0.042	0.050
8	0.050	0.060	0.071
12	0.061	0.073	0.087
16	0.071	0.084	0.100
20	0.079	0.094	0.112
24	0.086	0.103	0.122
30	0.097	0.116	0.137
36	0.108	0.126	0.150

Case depth = $0.025\sqrt{t}$ for 1700 F; $0.021\sqrt{t}$ for 1650 F; $0.018\sqrt{t}$ for 1600 F.

For normal carburizing (saturated austenite at the steel surface while at temperature).

FIGURE 99. Values of case depth as a function of carburizing temperature and time, calculated using Equation (35) (reprinted with permission from *Gas Carburizing*. 1964. Metals Park, Ohio: American Society for Metals).

them. However, if we assume that Reaction (3) is rate-controlling, then we can use

$$J = -(\varrho k_3)[(a_{C_{gr}} - a_{\underline{C}}^{\gamma})/(a_{\underline{C}}^{\gamma})] \tag{36}$$

where k_3 is the reaction rate constant, ϱ is the density of the steel, and $a_{C_{gr}}$ is the carbon activity corresponding to the gas used [e.g., the activity of carbon for Reaction (3), Figure 94] and $a_{\underline{C}}^{\gamma}$ is the activity of carbon at the surface (which is time-dependent). These activities of carbon can be converted to weight% via relations already cited.

These equations can be solved to obtain the carbon profile. Some experimental carburizing data are compared to calculated data in Figure 100. Note that the agreement is good.

The approach can be used to determine the gradient when the gas composition varies, including decarburizing following carburization. Figure 101 shows some calculated curves for different conditions. The gas contained 31.5% H_2, 23.5% N_2 and 0.60% CH_4 (corresponding to a normal propane-enriched endogas).

A treatment similar to the one above has been described by Goldstein and Moren, except that they made their calculations for assumed surface carbon contents, and thus did not take into account the chemical kinetics of the carburizing reactions. They also did take into account the dependency of the diffusion coefficient on carbon content. This effect is small, as shown by the calculated carbon profiles in Figure 102. They applied their method to complex carburizing cycles for which experimental carbon profiles were available. Figure 103 shows the results, and it can be seen that the agreement of the calculated profiles with the experimental profiles is good.

Goldstein and Moren also took into account the effect of third elements (e.g., Cr, Mn) on the diffusivity of carbon. Their calculations were made assuming the surface carbon content to be that of saturation, but they used the proper saturation value for each ternary alloying element for which a calculation was made. The situation is complicated by the oxidation that may occur when using gases with, for example, high CO_2 contents (see next section). For example, if the steel contains Cr, and during carburization oxidation occurs, a Cr-oxide will form on the surface, and the underlying layers will be low in Cr. This causes the effect of the Cr content on the diffusion coefficient of carbon to vary with depth, and may lead to some unusual carbon gradients.

At this point, we will consider the definition of case depth. The term denotes a measure of the depth of carbon penetration, and it is necessary to establish a definition to use in describing required carburizing treatments and in correlating the carbon profiles with properties (e.g., fatigue

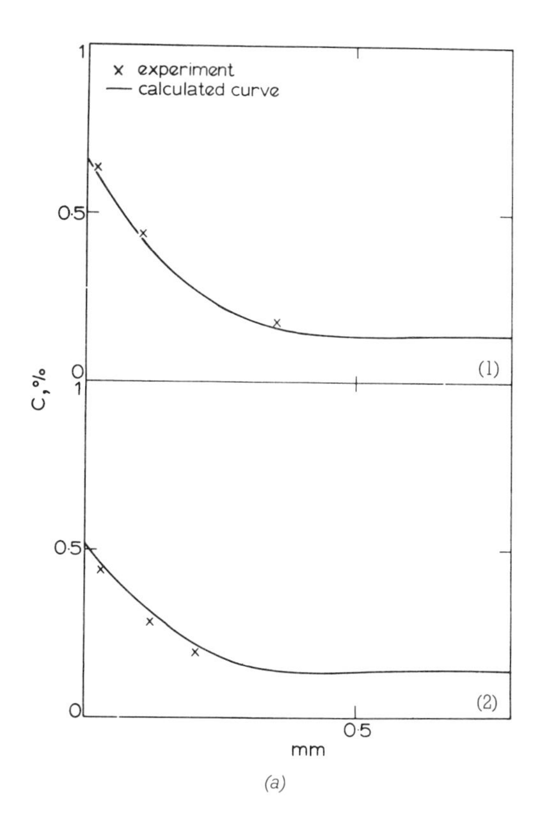

(*a*)

(1) 46 min from start: furnace temperature 880 °C, 27.7% H_2, 20.5% CO, 0.24% CO_2 and 0.39% CH_4; (2) 40 min from start: furnace temperature 890 °C, 12.5% H_2, 13% CO and 0.11% CO_2

Carburizing in a continuous furnace, comparison between experimental and calculated results

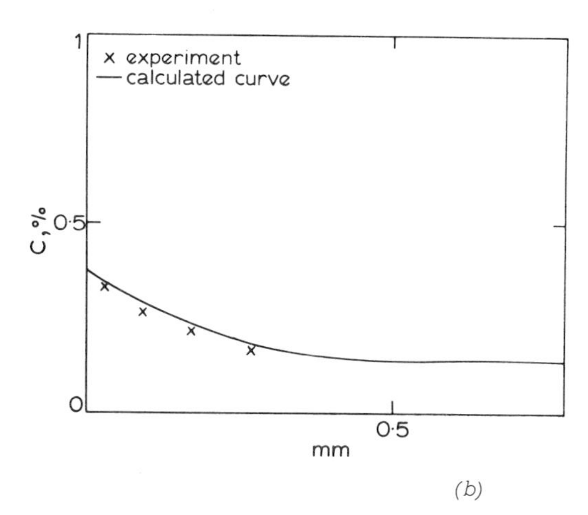

(*b*)

40 min from start: furnace temperature 940 °C, 11.5% H_2, 13.8% CO and 0.11% CO_2

Carburizing in a pit furnace, comparison between experimental and calculated results

FIGURE 100. Calculated carburizing carbon profiles compared to experimental data (adapted from Collin, R., S. Gunnarson and D. Thulin 1972. *J. Iron and Steel Inst.*, 210:785; by permission of The Institute of Metals).

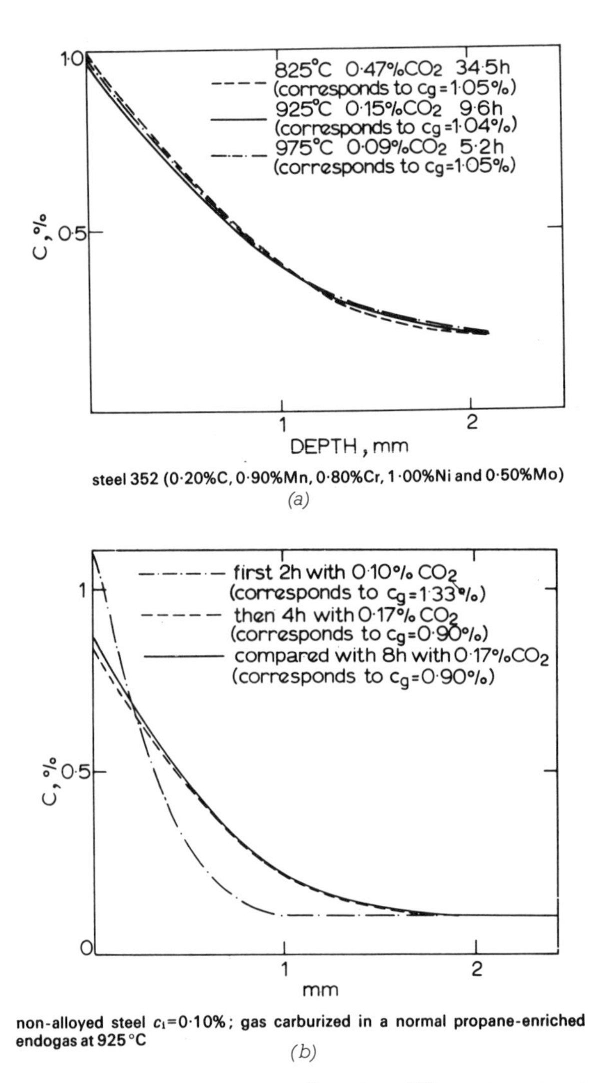

FIGURE 101. Calculated carbon profiles for different carburizing conditions (adapted from Collin, R., S. Gunnarson and D. Thulin. 1972. *J. Iron and Steel Inst.*, 210:785; by permission of The Institute of Metals).

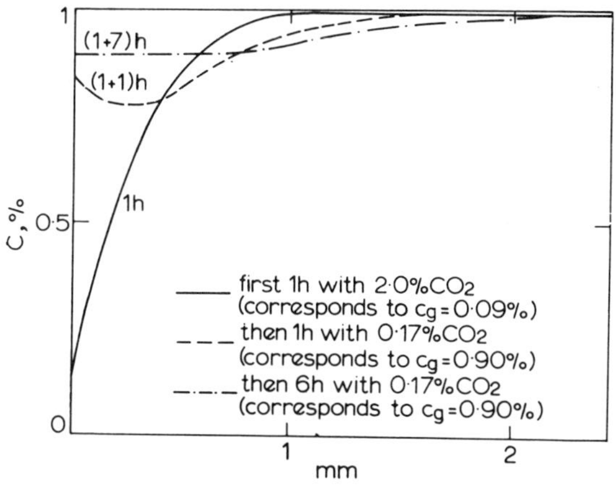

non-alloyed steel c_1=1·00%; gas decarburized and then recarburized in a normal propane-enriched endogas at 925 °C

(c)

FIGURE 101 (continued). Calculated carbon profiles for different carburizing conditions (adapted from Collin, R., S. Gunnarson and D. Thulin. 1972. *J. Iron and Steel Inst.*, 210:785; by permission of The Institute of Metals).

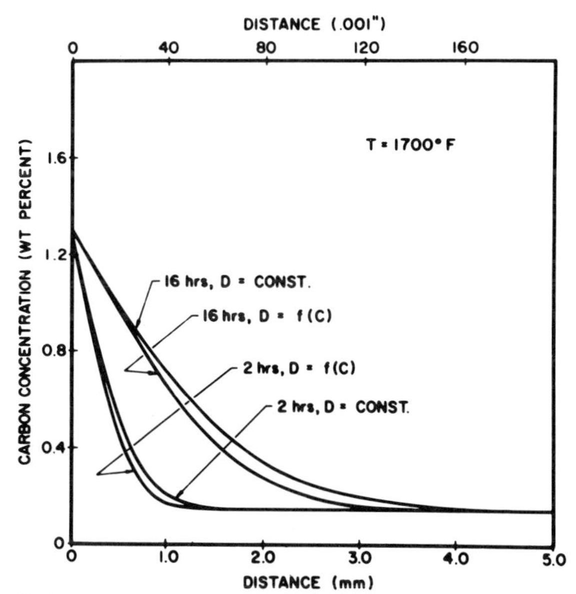

Carbon concentration vs distance curves calculated for 2 and 16 h treatments at 1700 °F, $D = f(C)$ and $D \neq f(C)$, Binary Fe-C.

FIGURE 102. Calculated carbon concentration profiles using a constant diffusion coefficient and a concentration-dependent coefficient (reprinted with permission from Goldstein, J. I. and A. E. Moron. 1978. *Met. Trans.*, 9A:1515, a publication of The Minerals, Metals and Materials Society, Warrendale, PA).

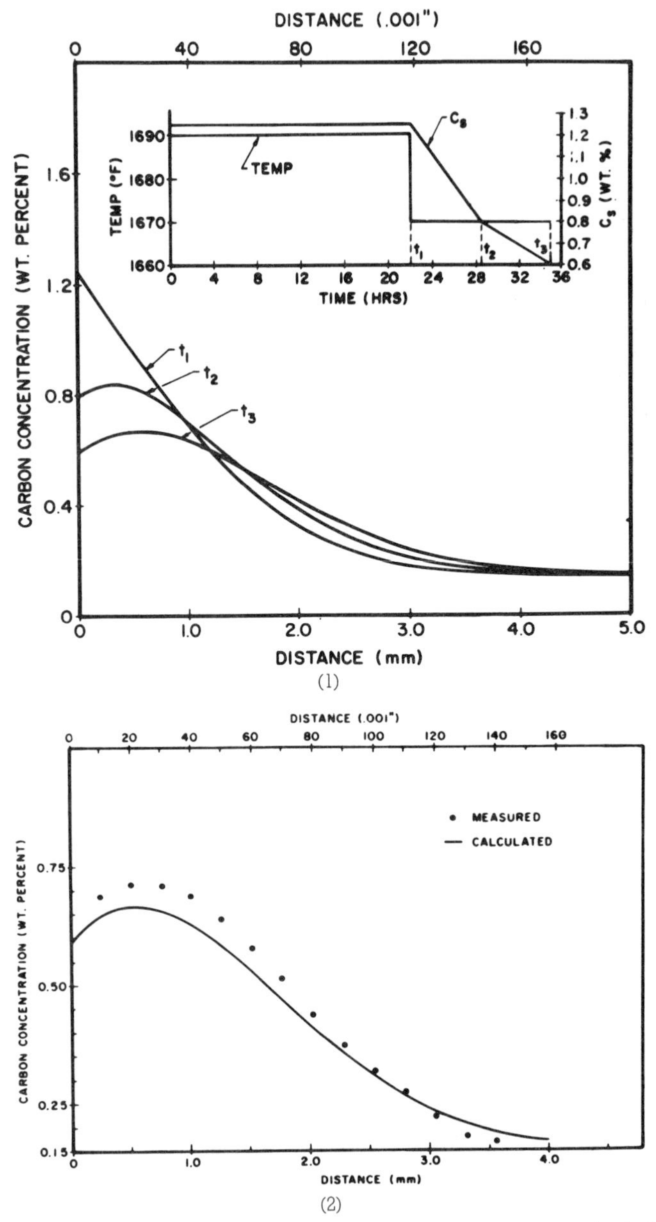

(1) Carbon concentration vs distance curves calculated for a two step carburization treatment $D = f(C)$. (2) Comparison of calculated and measured carbon profiles for the two step carburization treatment, (1).

FIGURE 103a. Calculated carbon concentration profiles for complex carburization treatments (reprinted with permission from Goldstein, J. I. and A. E. Moren. 1978. *Met. Trans.*, 9A:1515, a publication of The Minerals, Metals and Materials Society, Warrendale, PA).

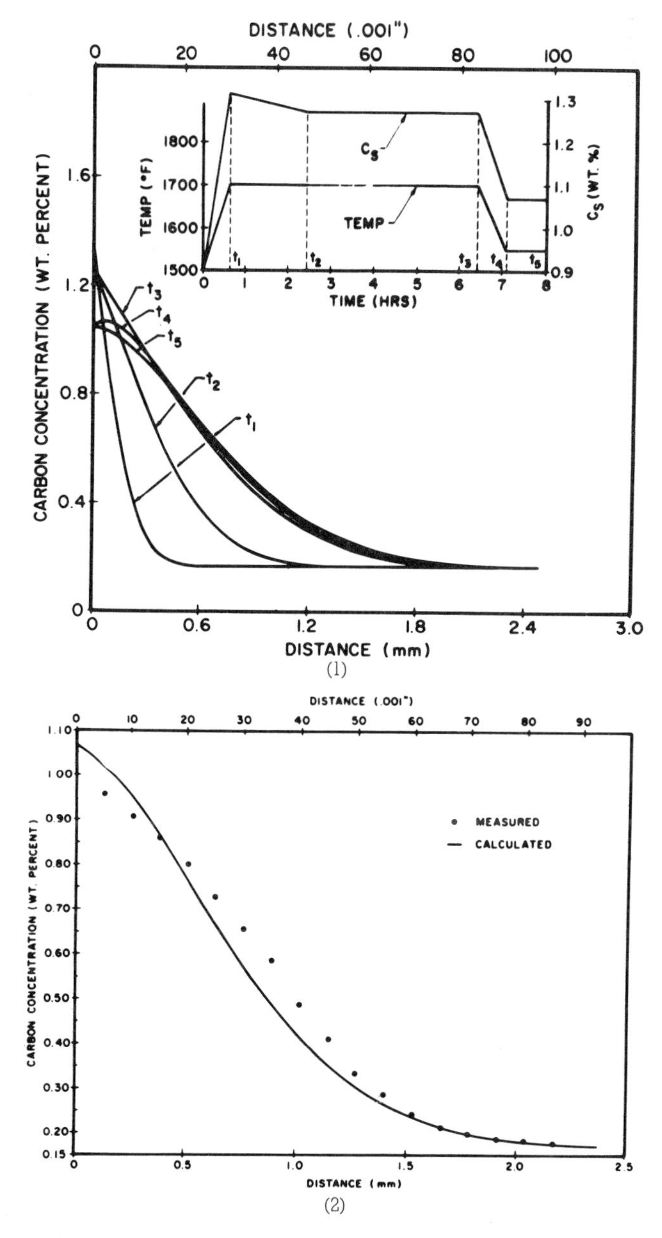

(1) Carbon concentration vs distance curves calculated for a complex carburization-diffusion treatment in a vestibule type furnace, $D = f(C)$. (2) Comparison of calculated and measured carbon profiles for the carburization treatment, (1).

FIGURE 103b. Calculated carbon concentration profiles for complex carburization treatments (reprinted with permission from Goldstein, J. I. and A. E. Moren. 1978. *Met. Trans.*, 9A:1515, a publication of The Minerals, Metals and Materials Society, Warrendale, PA).

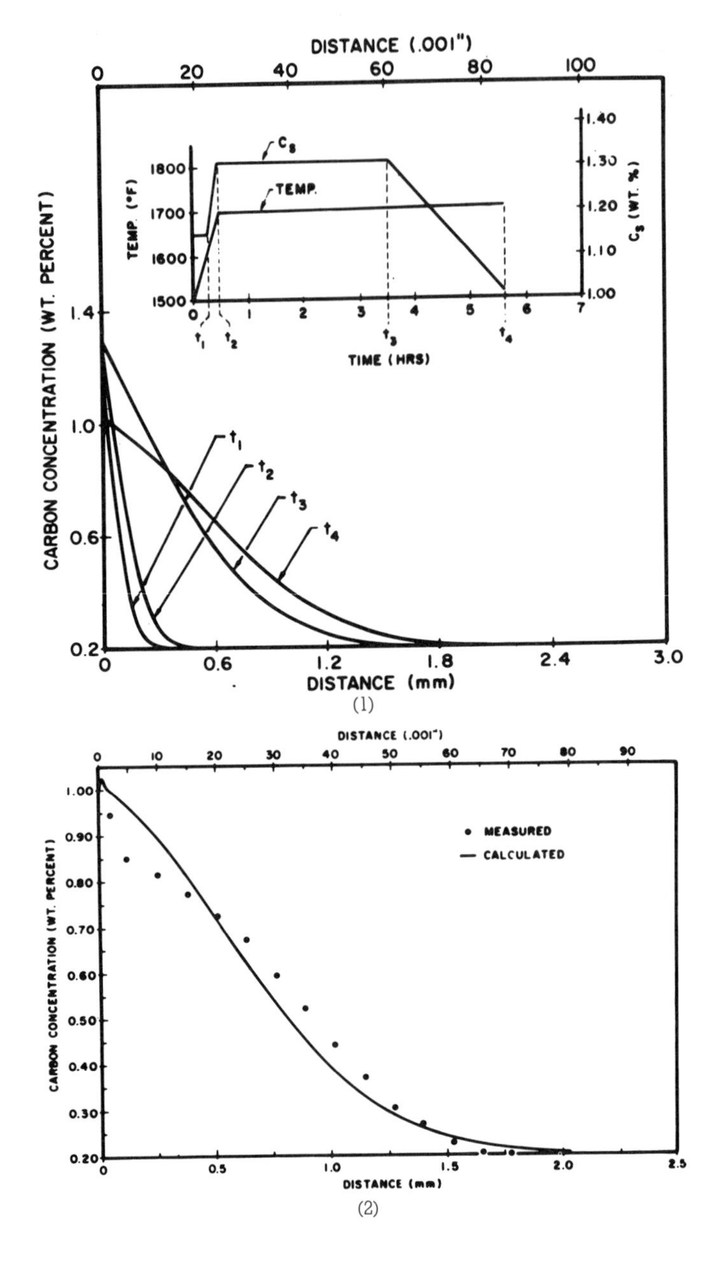

(1) Carbon concentration vs distance curves calculated for a complex carburization-diffusion treatment in a pit furnace, $D = f(C)$. (2) Comparison of calculated and measured carbon profiles for the carburization treatment, Figure (1).

FIGURE 103c. Calculated carbon concentration profiles for complex carburization treatments (reprinted with permission from Goldstein, J. I. and A. E. Moren. 1978. *Met. Trans.*, 9A:1515, a publication of The Minerals, Metals and Materials Society, Warrendale, PA).

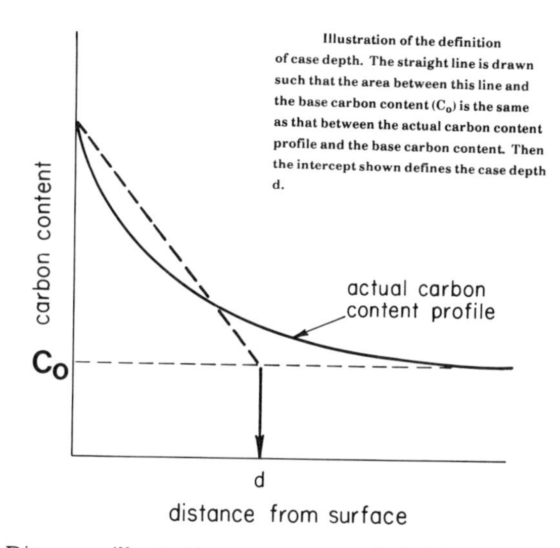

FIGURE 104. Diagram illustrating a common definition of case depth (reprinted with permission from Brooks, C. R. 1979. *Heat Treatment of Ferrous Alloys*. New York: Hemisphere Publishing Corporation/McGraw-Hill Book Company).

strength). Case depth can be defined in several ways; for example, it can be defined as the depth to a given carbon content, or the depth at which the carbon content is a certain fraction of the core value. Figure 104 shows one method that is commonly used, and on which the equation by Harris cited above is based.

3e. OXIDATION DURING CARBURIZING

Since oxygen-containing gases are used in carburizing, there is concern as to whether the steel will oxidize during carburizing, and what the effect of this will be on the carburizing process.

For the oxidation of iron, we are concerned with the reactions

$$Fe + CO_2 = FeO + CO \tag{37}$$

and

$$Fe + H_2O = FeO + H_2 \tag{38}$$

These two reactions can be dealt with in terms of thermodynamic relations as was done for the carburizing reactions, and relations between the

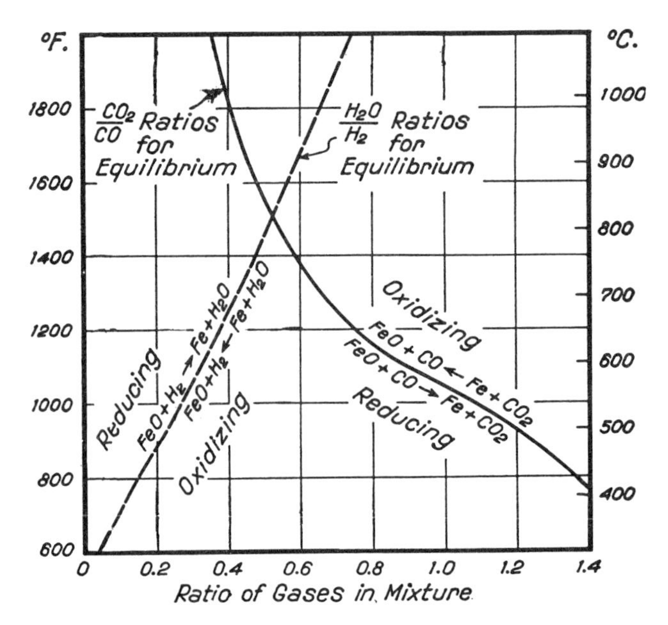

FIGURE 105. The dependence of the ratio p_{CO_2}/p_{CO} and p_{H_2O}/p_{H_2} on temperature for oxidation-reduction equilibrium of iron (adapted from Hotchkiss, A. G. 1937. *Metal Progress*, 31:375).

temperature and the required gas composition (the neutral gas) which neither oxidizes nor deoxidizes the iron can be obtained. The results are plotted in Figure 105. For example, to carburize a steel at 927°C in a mixture of CO and CO_2 at one atmosphere total pressure so that the surface is 0.8% C, requires a gas composition 95.66% CO and 4.34% CO_2. The ratio of $CO_2/CO = 0.045$, and Figure 105 shows that this atmosphere will be reducing.

If the steel contains alloying elements (which is usually the case), then the oxidation of these elements must be considered. For example, it might be expected that the highly oxidizable Cr present would form Cr-oxide, or some Cr-containing oxide. In this case the chemical reaction is

$$2\underline{Cr} + 3CO_2 = Cr_2O_3 + 3CO \tag{39}$$

where $\underline{Cr}$ refers to chromium dissolved in austenite. If the oxide is pure Cr_2O_3, the thermodynamic relation to be used is

$$\Delta g^0 = -RT \ln [(p^3_{CO}/p^3_{CO_2})(1/a^2_{\underline{Cr}})] \tag{40}$$

where $a_{\underline{Cr}}$ is the activity of chromium in austenite, which depends upon the Cr content of the austenite, and which must be determined experimentally. If the oxide is not the pure species (e.g., the Cr_2O_3 contains some dissolved Fe), a relation similar to Equation (40) is used, but an activity value for Cr_2O_3 different than unity will appear in the equation. Also, the activity of the element of interest in the austenite will depend upon the amount of other elements present (e.g., Mn, Mo), so that this effect should be taken into account. Some of these aspects have been treated by Harvey (Harvey, F. J. 1978. *Met. Trans.*, 9A:1507).

3f. DECARBURIZATION

The concepts applied in the discussion of carburization are applicable to decarburization. (Also, the thermodynamics relations used or presented graphically can be used to determine the gas composition required to be neutral, so that neither carburization nor decarburization occurs.) The equations to calculate the effect of decarburization on the carbon profile will not be analyzed. However, care must be taken to ensure that the proper equation is used, based on the proper boundary conditions.

In this section, microstructures will be presented that illustrate the effect of decarburization. (These are taken from *Optical Microscopy of Carbon Steels* by Samuels and are described in more detail in his book.) The microstructure of a hot rolled bar of a 0.8% C steel that had been air-cooled from the austenite region is shown in Figure 106. The steel was austenitized in air, and the carbon was removed by oxidation with the oxygen in the air. This occurred during the hot rolling, and probably also in hot deformation treatments prior to this hot rolling operation. Increasing amounts of primary ferrite are apparent as the surface is approached. The microstructure upon quenching this steel from the austenite region is also shown in Figure 106. Here the degree of decarburization is more difficult to see, although etching with nital does show that the surface region has been affected.

Decarburization is not so apparent if the carbon content is low. This is seen by comparing the microstructure in Figure 107 (micrograph 143.1) to those in Figure 106. Figure 108 also shows examples of this difficulty. However, microhardness measurements on quenched samples are useful in detecting decarburization, as shown in Figure 107 (graph in 143.4). This is also illustrated for a high-carbon steel in Figure 109.

It may also be difficult to detect decarburization in a spheroidized microstructure, as illustrated by the microstructures in Figure 110. Decarburization is more obvious if the carbon content is high, as illustrated by the microstructures in Figure 111.

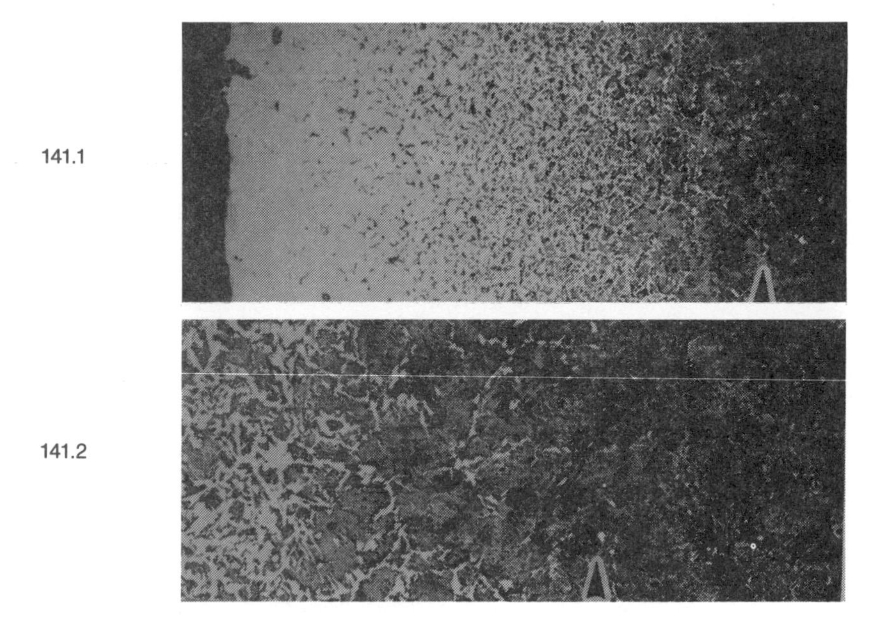

Figure No.	Material	Analysis, wt %	Condition	Hard-ness, HV	Etchant	Magnifi-cation
	0.8%C	0.78 C, 0.30 Mn	Transverse section of a hot rolled bar; normalized			
141.1					Picral	50
141.2					Picral	100
			Austenitized at 850°C, quenched in water			
141.3(a)					Nital	50
141.4(a)					Picral	50

(a) Arrow indicates total depth of decarburization, as judged from Fig. 141.1.

FIGURE 106. Micrographs illustrating decarburization in a 0.78% C steel (reprinted with permission from Samuels, L. E. 1980. *Optical Microscopy of Carbon Steels*. Metals Park, Ohio: American Society for Metals).

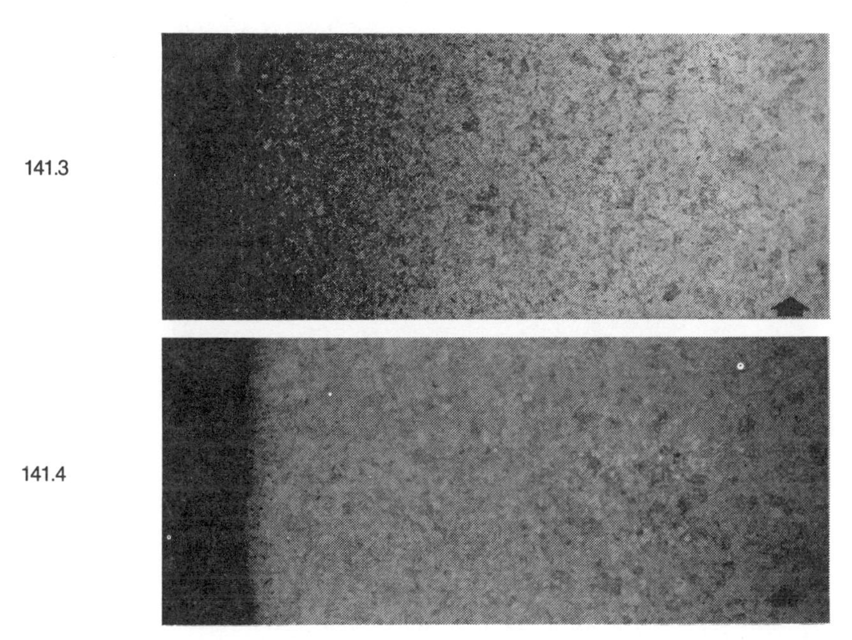

FIGURE 106 (continued). Micrographs illustrating decarburization in a 0.78% C steel (reprinted with permission from Samuels, L. E. 1980. *Optical Microscopy of Carbon Steels*. Metals Park, Ohio: American Society for Metals).

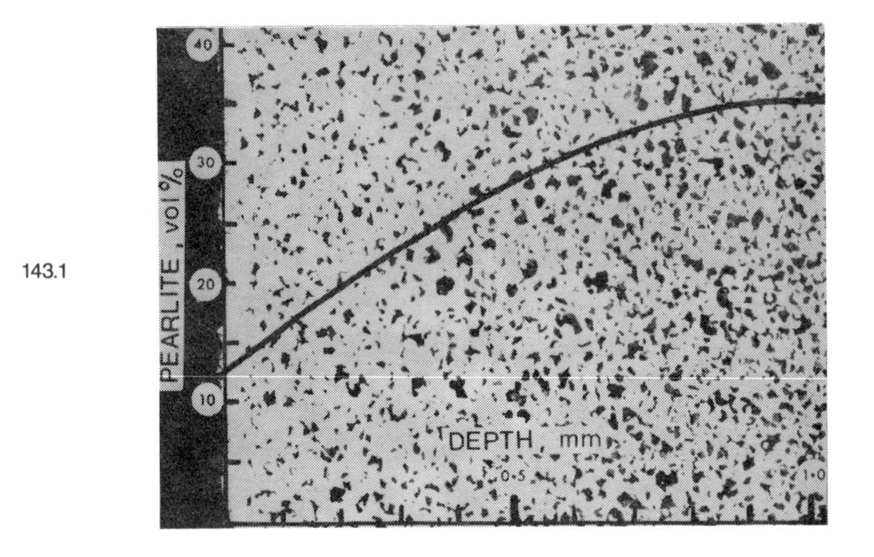

Figure No.	Material	Analysis, wt%	Condition	Hard-ness, HV	Etchant	Magnifi-cation
	0.4%C	0.41 C, 0.24 Si, 0.70 Mn				
143.1			Normalized		Picral	100
143.2(a)			Austenitized, quenched in water, tempered at 175°C		Nital (1%)	75
143.3(a)					Picral	75

(a) Arrow indicates total depth of decarburization judged from the data on page 123.

FIGURE 107. Micrographs illustrating decarburization in a 0.41% C steel. The amount of pearlite as a function of depth is superimposed on the microstructure (reprinted with permission from Samuels, L. E. 1980. *Optical Microscopy of Carbon Steels*. Metals Park, Ohio: American Society for Metals).

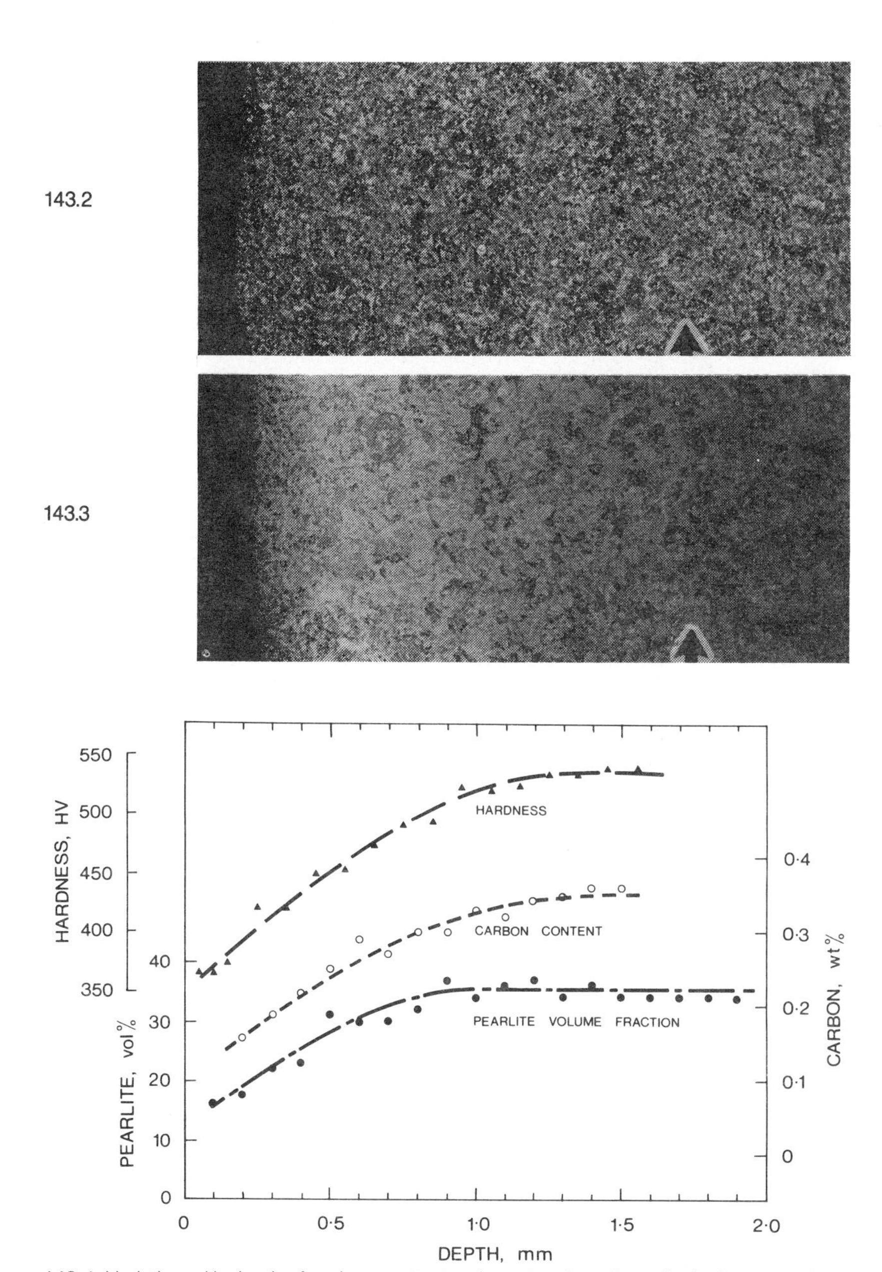

143.4 Variation with depth of carbon content, volume fraction of pearlite in the normalized condition, and hardness in the quenched-and-tempered condition for the decarburized 0.4% C steel illustrated in Figure 143.1 to 143.3.

FIGURE 107 (continued). Micrographs illustrating decarburization in a 0.41% C steel. The amount of pearlite as a function of depth is superimposed on the microstructure (reprinted with permission from Samuels, L. E. 1980. *Optical Microscopy of Carbon Steels*. Metals Park, Ohio: American Society for Metals).

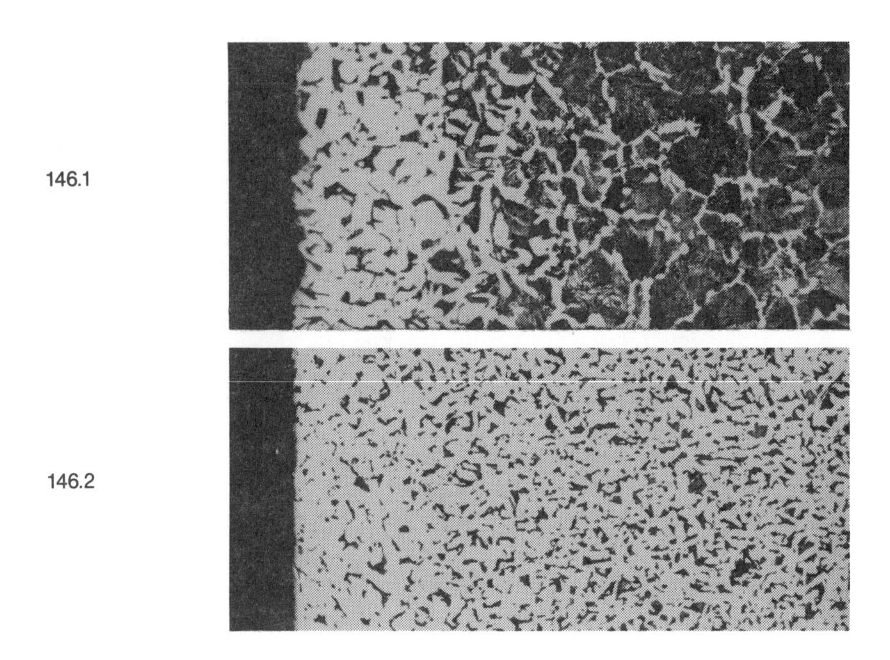

Figure No.	Material	Analysis, wt %	Condition	Hard-ness, HV	Etchant	Magnifi-cation
146.1	0.6%C hot rolled bar	0.55 C, 0.08 Si, 0.60 Mn	Normalized		Picral	250
146.2	0.2%C hot rolled plate	0.24 C, 0.02 Si, 0.80 Mn	Normalized		Picral	250

FIGURE 108. Micrographs comparing the effect of decarburization on the microstructure of a 0.55% C steel and a 0.2% C steel (reprinted with permission from Samuels, L. E. 1980. *Optical Microscopy of Carbon Steels*. Metals Park, Ohio: American Society for Metals).

148.1

148.2

Figure No.	Material	Analysis, wt %	Condition	Hard-ness, HV	Etchant	Magnifi-cation
	1.3%C	1.29 C, 0.17 Si, 0.40 Mn	Austenitized at 850°C, quenched in water, tempered at 175°C			
148.1(a)					Picral	100
148.2(a)					Picral	250
148.3(a)					Bisulfite	250
148.4(a)					Nital (1%)	250

NOTE: This material is shown in the spheroidized condition before hardening in Fig. 147.3 to 147.7.

(a) Arrow indicates total depth of decarburization estimated from the hardness traverse on the following page.

FIGURE 109. Micrographs illustrating the effect of decarburization on the microstructure of a quenched high-carbon (1.29% C) steel (reprinted with permission from Samuels, L. E. 1980. *Optical Microscopy of Carbon Steels*. Metals Park, Ohio: American Society for Metals).

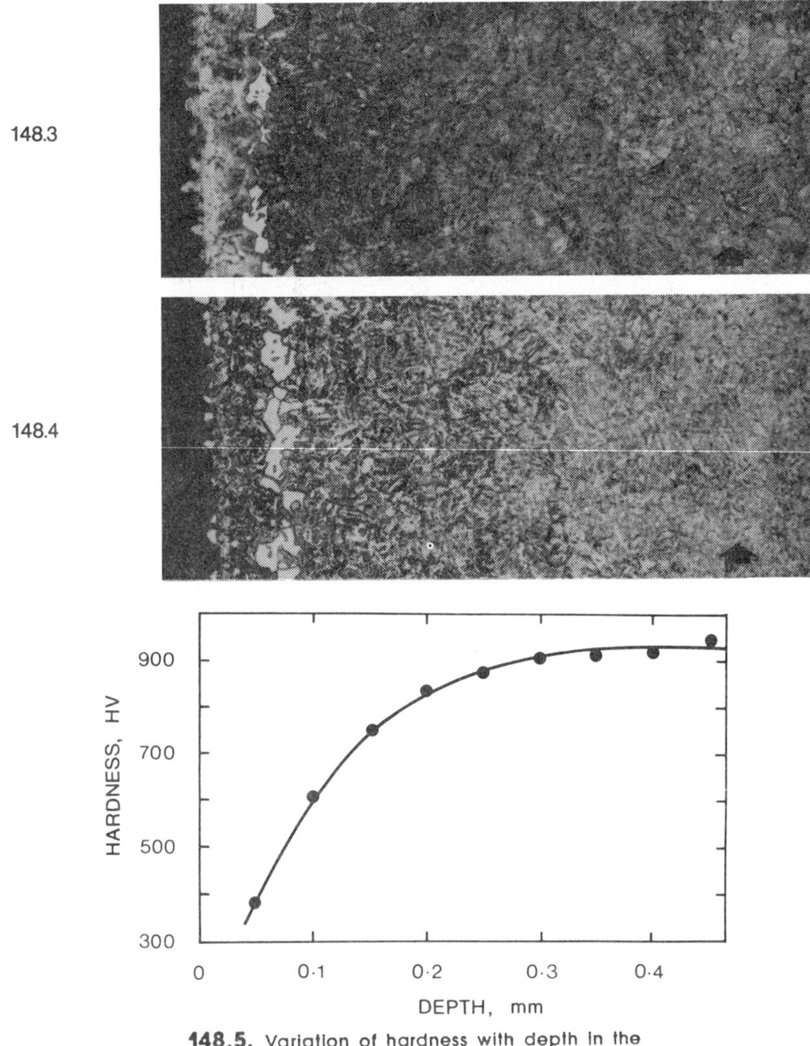

148.5. Variation of hardness with depth in the quenched-and-tempered hypereutectoid steel illustrated in Fig. 148.1 to 148.4.

FIGURE 109 (continued). Micrographs illustrating the effect of decarburization on the microstructure of a quenched high-carbon (1.29% C) steel (reprinted with permission from Samuels, L. E. 1980. *Optical Microscopy of Carbon Steels*. Metals Park, Ohio: American Society for Metals).

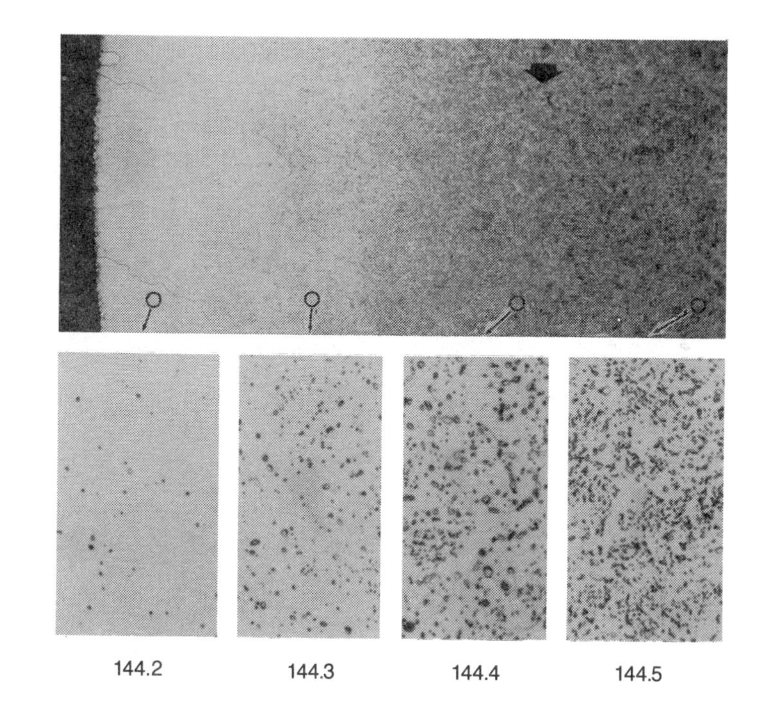

Figure No.	Material	Analysis, wt %	Condition	Hardness, HV	Etchant	Magnification
	0.4%C	0.41 C, 0.24 Si, 0.70 Mn	Austenitized; quenched in water, as in Fig. 143.2; tempered at 700°C for 4 h			
144.1					Nital (1%)	75
144.2					Nital (1%)	1000
144.3					Nital (1%)	1000
144.4					Nital (1%)	1000
144.5(a)					Nital (1%)	1000

(a) This is the structure of the unaffected core.

FIGURE 110. Microstructures illustrating the effect of decarburization on the spheroidized microstructure of a 0.4% C steel (reprinted with permission from Samuels, L. E. 1980. *Optical Microscopy of Carbon Steels*. Metals Park, Ohio: American Society for Metals).

147.1

147.2

Figure No.	Material	Analysis, wt %	Condition	Hard-ness, HV	Etchant	Magnifi-cation
	1.3%C	1.29 C, 0.17 Si, 0.40 Mn	Heated at 975°C for 5 h, cooled in air			
147.1					Picral	50
147.2					Picral	250
			Austenitized at 850°C, quenched in water, tempered at 700°C			
147.3(a)					Picral	250
147.4					Picral	1000
147.5					Picral	1000
147.6					Picral	1000
147.7					Picral	1000

NOTE: Although these two specimens are from the same batch of steel, they differ in depth of decarburization. The spheroidized material is illustrated after quench hardening in Fig. 109.

(a) Circled areas indicate locations of the fields illustrated in Fig. 147.4 to 147.7.

FIGURE 111. Microstructures illustrating the effect of decarburization on the spheroidized microstructure of a 1.3% C steel (reprinted with permission from Samuels, L. E. 1980. *Optical Microscopy of Carbon Steels*. Metals Park, Ohio: American Society for Metals).

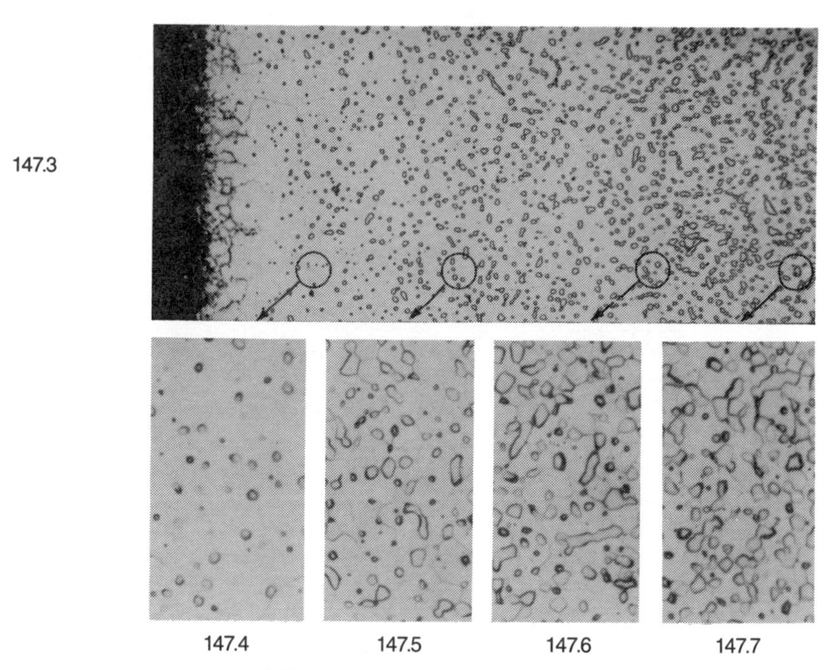

FIGURE 111 (continued). Microstructures illustrating the effect of decarburization on the spheroidized microstructure of a 1.3% C steel (reprinted with permission from Samuels, L. E. 1980. *Optical Microscopy of Carbon Steels*. Metals Park, Ohio: American Society for Metals).

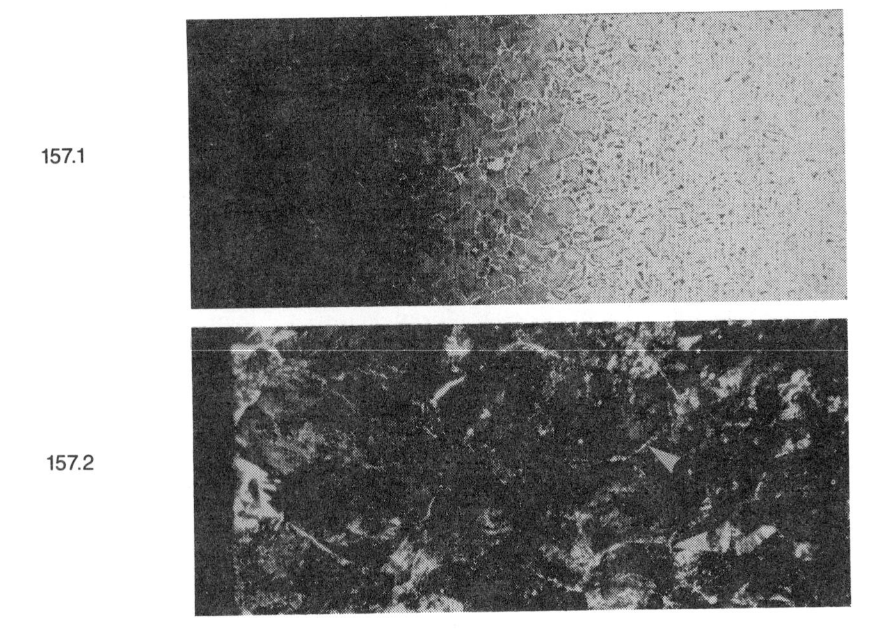

Figure No.	Material	Analysis, wt %	Condition	Hard-ness, HV	Etchant	Magnifi-cation
	0.15%C	0.17 C, 0.05 Si, 0.64 Mn	Pack carburized at 940°C for 2 h, decarburized, heat treated as indicated			
157.1			Austenitized at 850°C, cooled in air		Picral	75
157.2					Picral	500

FIGURE 112. Microstructures illustrating decarburization in a carburized steel due to heating with some air present at 850°C. The arrow in 157.2 denotes some Fe_3C in the austenite grain boundary. The white areas at and near the surface are ferrite (reprinted with permission from Samuels, L. E. 1980. *Optical Microscopy of Carbon Steels*. Metals Park, Ohio: American Society for Metals).

Decarburization following gas-carburization can occur. This can be caused by changes in the gas composition, or by temperature changes even if the gas composition remains the same. This latter effect is due to the shift in the equilibrium constant with temperature, which means that the thermodynamic equilibrium carbon content of the austenite shifts. Decarburization can also occur in heat treatments following carburizing if the atmosphere in which the heat treatment is carried out is not suitable; for example, air will certainly decarburize the steel.

Microstructures illustrating decarburization caused by heat treatments following the carburization are shown in Figure 112. The microstructure after carburizing and cooling slowly to 25°C is shown in Figure 72 (micrograph 149.3) and Figure 73 (micrograph 149.7); there is primary Fe_3C at the surface. The carbon gradient is shown in graph 149.9 of Figure 72, where it is seen that the carbon content is 1.1% at the surface. If this steel with this carbon gradient is austenitized at 850°C, then cooled in air, micrograph 157.2 in Figure 112 shows that a small amount of primary ferrite will be present in the surface layer, and that the carbon content will have been lowered to less than 0.8% C.

When the carburized steel is quenched from the carburizing temperature, considerable retained austenite will be present for the carbon gradient shown in Figure 72. This will be quite apparent because of the contrast between the white retained austenite and the martensite in the microstructure (see Figure 74). This contrast can be enhanced by a low temperature tempering treatment, not high enough nor long enough to affect the retained austenite, so that the martensite becomes tempered martensite (but with a very fine structure) which etches darker. Microstructures of the carburized steel after austenitizing following carburizing, then quenching to 25°C, are shown in Figure 113. Near the surface (micrograph 157.3) a region showing some faint white areas is seen (see arrow). At higher magnification (micrograph 157.4) it is seen that retained austenite is present. However, note that at the surface, there is *less* retained austenite, due to the fact that carbon content at the surface is lower than it is under the surface. The hardness profile is shown in graph 157.7 of Figure 113. Note that the hardness is still high for this "mild decarburization," showing no appreciable decrease at the surface because the carbon content has not been lowered sufficiently to lower the hardness of the martensite. However, greater decarburization ("moderate decarburization") is sufficient to lower the surface hardness.

Figure 114 shows micrographs of the same pack-carburized steel after more severe decarburization brought about by longer austenitizing times. Micrographs 158.1 and 158.2 show primary ferrite at the surface, and these can be compared to micrographs 157.1 and 157.2. Note that micrograph 158.2 shows that on the surface there is a layer of pure ferrite. When the

Figure No.	Material	Analysis, wt %	Condition	Hard-ness, HV	Etchant	Magnifi-cation
157.3 157.4			Austenitized at 850°C, quenched in water, tempered at 150°C		Bisulfite Bisulfite	75 500

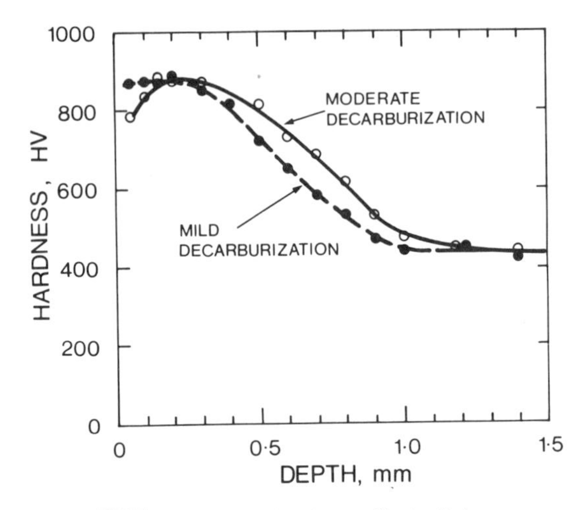

157.7. **Variation of hardness with depth in carburized bars subsequently decarburized and quench hardened. The structures of the mildly decarburized bar are illustrated in Fig. 157.1 to 157.6, and those of the moderately decarburized bar in Fig. 158.**

FIGURE 113. Microstructures of a sample that was water-quenched from the austenitization temperature. The steel was the same as for Figure 112 (reprinted with permission from Samuels, L. E. 1980. *Optical Microscopy of Carbon Steels*. Metals Park, Ohio: American Society for Metals).

158.1

158.2

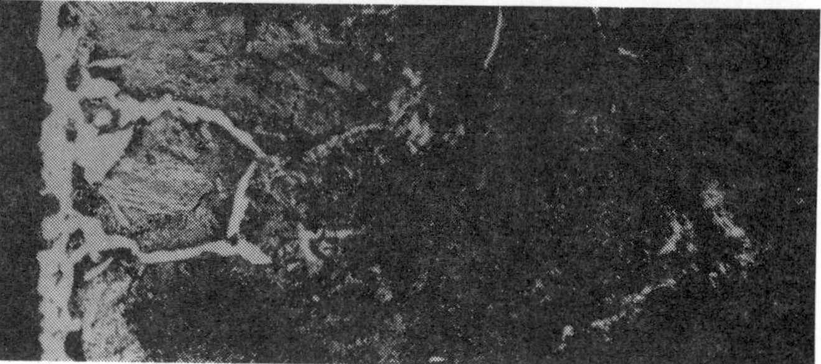

Figure No.	Material	Analysis, wt %	Condition	Hardness, HV	Etchant	Magnification
	0.15%C	0.17 C, 0.05 Si, 0.64 Mn	Carburized at 940°C for 2 h, decarburized, heat treated as indicated			
158.1			Austenitized at 850°C, cooled in air		Picral	75
158.2					Picral	500
158.3			Austenitized at 850°C, quenched in water, tempered at 150°C		Bisulfite	75
158.4					Bisulfite	500
158.5			Austenitized at 780°C, quenched in water, tempered at 150°C		Bisulfite	75
158.6					Bisulfite	500

FIGURE 114. Another example of microstructures showing decarburization of a carburized steel (reprinted with permission from Samuels, L. E. 1980. *Optical Microscopy of Carbon Steels*. Metals Park, Ohio: American Society for Metals).

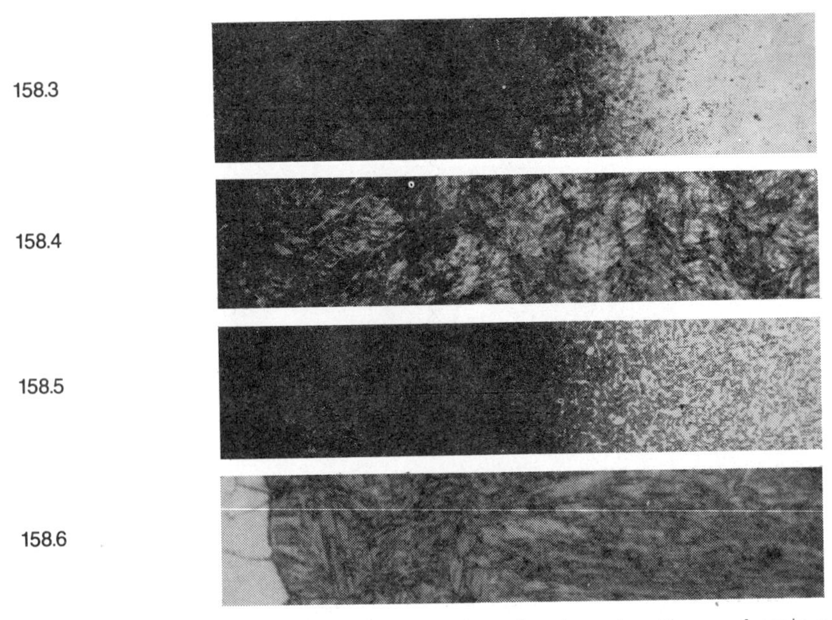

FIGURE 114 (continued). Another example of microstructures showing decarburization of a carburized steel (reprinted with permission from Samuels, L. E. 1980. *Optical Microscopy of Carbon Steels*. Metals Park, Ohio: American Society for Metals).

steel is quenched from an austenitizing temperature of 850°C, there is less evidence of retained austenite in the microstructure, as seen by comparing micrographs 158.3 and 158.4 to micrographs 157.3 and 157.4. When the steel is austenitized at a lower temperature of 780°C, the ferrite does not dissolve, so that upon quenching a thin layer of ferrite exists on the surface; this is shown in micrographs 158.5 and 158.6. At this low temperature and for the austenitizing time used, no significant carbon diffusion occurred, so that the ferrite layer on the surface was retained upon austenitizing, and hence retained upon quenching. The hardness profile in graph 157.7 of Figure 113 shows that the "moderate decarburization" in the microstructures in Figure 114 is sufficient to produce a lower hardness at the surface.

3g. HEAT TREATMENTS OF CARBURIZED STEELS

In carburizing, the desired structure in the surface layer is martensite (or tempered martensite), so that the carburized steel must be quenched from the austenite region. This can be accomplished by quenching directly from the carburizing temperature at the end of the carburizing time. However, this is not always done. One reason is that it may be difficult physically to remove the steel parts from the carburizing environment to quench them. Another reason is that the high carburizing temperature and long time may cause the austenite grain size to be too large. Thus, the carburized steel will be cooled to 25°C (e.g., air cooled), then austenitized at a lower temperature to attain a small austenite grain size, then quenched. An important consideration here is that the carbon gradient will be changed during this austenitization treatment. However, in some cases such changes in the gradient may be desirable. For example, if the carbon content is too high at the surface, and the case depth is not sufficient, proper austenitizing will allow diffusion of carbon and lower the carbon content at the surface, and also increase the depth of carbon penetration. The following microstructures illustrate some of these factors in austenitizing carburized steels. They are taken from *Optical Microscopy of Carbon Steels* by Samuels, and more details about them are given in his book.

Figure 115 shows micrographs illustrating the effect of quenching from different austenitizing temperatures. The microstructure of the slowly cooled steel after carburizing is shown in micrograph 149.3 of Figure 72. The carbon gradient is shown in graph 149.9 of Figure 72. Micrograph 150.1 of Figure 115 shows the microstructure after quenching from 940°C and micrograph 150.2 after quenching from 850°C. These are illustrative of the microstructure if the steel is quenched directly from the carburizing temperature, since for these temperatures the entire structure will be austenite (see the Fe–C phase diagram). The case depth is shown by the arrow in micrograph 150.1. This is better revealed by using a bisulfite etch, as shown in micrographs 150.5 and 150.6.

If the carburized steel is austenitized at 780°C, part of the steel is in the two-phase region. Fe_3C and austenite will be present at the surface, and austenite and ferrite at the core. Thus upon quenching, these regions will be only partly martensite. This effect is illustrated by micrographs 150.3 and 150.7. In 150.3, the faintly dark lines in the surface region are Fe_3C; the rest is martensite (and retained austenite). In the core, the martensite is similar in appearance to the ferrite with the nital etch, but the bisulfite etch distinguishes them clearly (micrograph 150.7).

If the carburized steel is reaustenitized and quenched, this is referred to as *single quenching*. If the carburized steel is reaustenitized twice, it is called *double quenching*. (Recommended heat treatments are given in Appendix 14 for some commercial carburizing steels.) The effect of this latter treatment on the microstructure is shown in micrographs 150.4 and 150.8. In this case, the carburized steel was austenitized at 900°C, quenched in oil, then austenitized at 780°C and water-quenched. This final austenitizing temperature is the same as that for micrograph 150.3 and 150.7, but

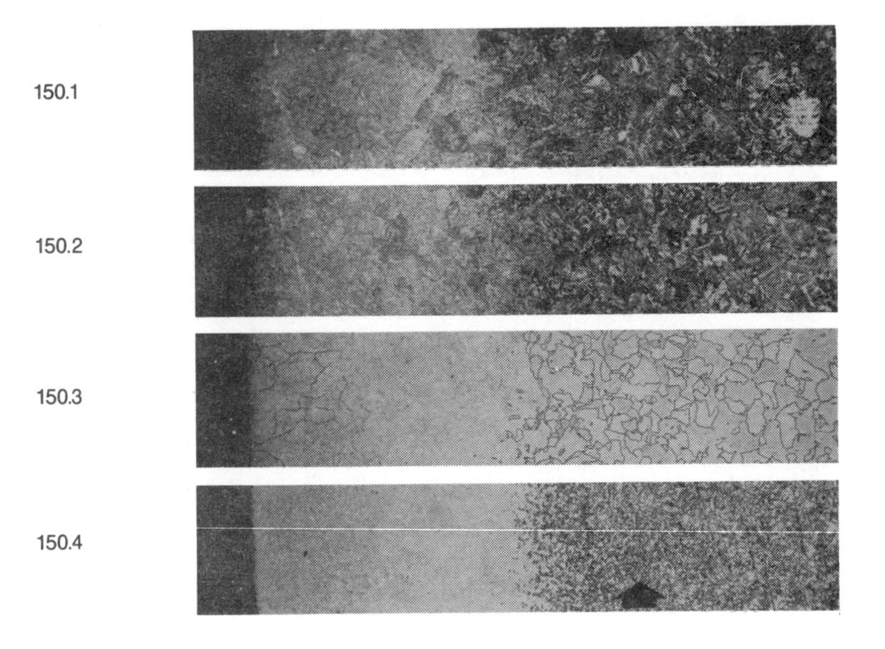

Figure No.	Material	Analysis, wt %	Condition	Hardness, HV	Etchant	Magnification
	0.15%C	0.17 C, 0.05 Si, 0.64 Mn	Pack carburized at 940°C for 2 h, cooled slowly to room temperature, then heat treated as indicated			
150.1			Austenitized at 940°C, quenched in water, tempered at 150°C		Nital (1%)	75
150.2			Austenitized at 850°C, quenched in water, tempered at 150°C		Nital (1%)	75
150.3			Austenitized at 780°C, quenched in water, tempered at 150°C		Nital (1%)	75
150.4			Austenitized at 900°C, quenched in oil, austenitized at 780°C, quenched in water, tempered at 150°C		Nital (1%)	75
150.5			As for Fig. 150.1		Bisulfite	75
150.6			As for Fig. 150.2		Bisulfite	75
150.7			As for Fig. 150.3		Bisulfite	75
150.8			As for Fig. 150.4		Bisulfite	75

NOTE: These illustrations are part of a series shown in Fig. 149 to 153. The arrows indicate the total depth of case estimated from Fig. 149.9

FIGURE 115. Microstructures of a carburized steel showing the effects of reaustenitizing then quenching (single quench) (reprinted with permission from Samuels, L. E. 1980. *Optical Microscopy of Carbon Steels.* Metals Park, Ohio: American Society for Metals).

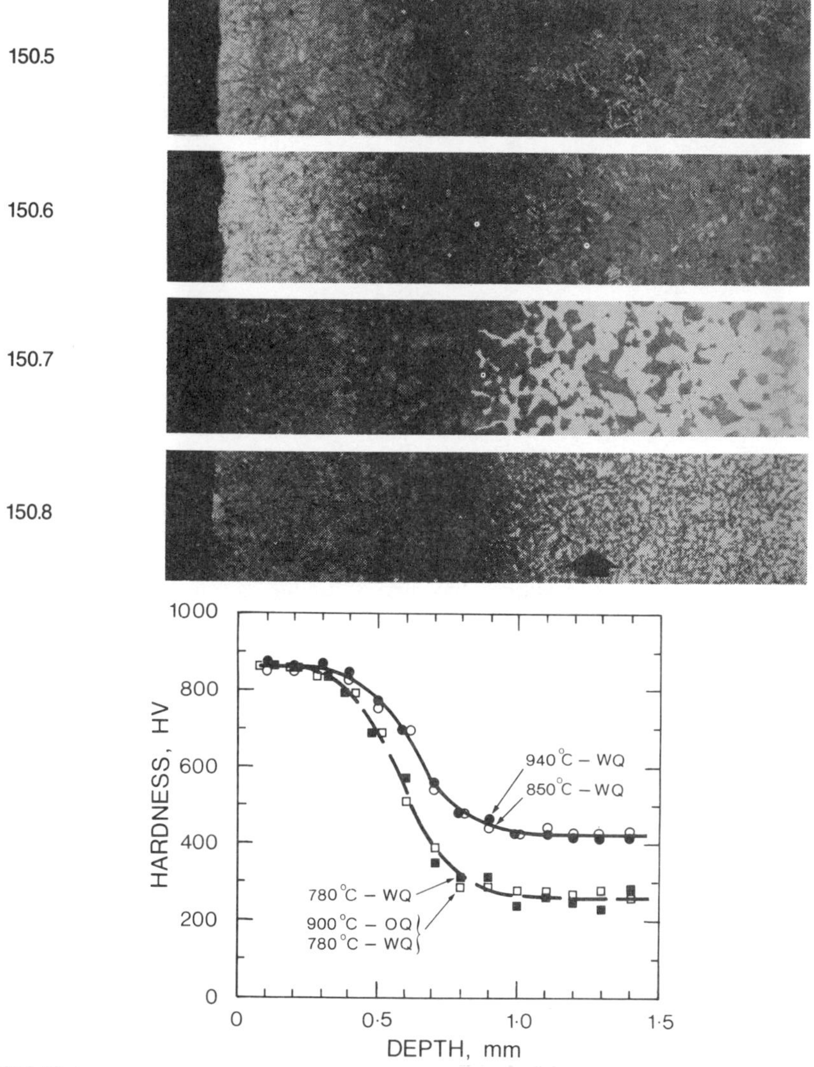

150.10. Variation of hardness with depth in a bar that has been pack carburized at 940°C for 2 h and then subjected to the quench-hardening treatments indicated. The associated microstructures are illustrated in Fig. 150, 151 and 153, as follows:

940°C – WQ: Fig. 150.1, 150.5, 151.1, 151.2, 153.1
850°C – WQ: Fig. 150.2, 150.6, 151.3, 151.4, 153.2
780°C – WQ: Fig. 150.3, 150.7, 151.5, 151.6, 153.3
900°C – OQ, 780°C – WQ: Fig. 150.4, 150.8, 151.7, 151.8, 153.4

FIGURE 115 (continued). Microstructures of a carburized steel showing the effects of reaustenitizing then quenching (single quench) (reprinted with permission from Samuels, L. E. 1980. *Optical Microscopy of Carbon Steels*. Metals Park, Ohio: American Society for Metals).

note that although the same microconstituents are present, the structure is finer. This is an important effect of multiple austenitizing of carburized steels.

The microstructural features are illustrated in more detail and at higher magnification in Figure 116.

The hardness profiles for these heat treatments are shown in graph 150.10 of Figure 115. Note that the surface hardness is the same for all treatments, but that austenitizing at 780°C produces a lower core hardness, since the structure there contained soft ferrite.

The problem of retained austenite after quenching must always be kept in mind. It can be converted to martensite by cooling below the M_f temperature, as illustrated in Figures 117 and 118. Note the improvement in hardness shown in graph 160.7.

The formation of high-carbon martensite is frequently accompanied by the formation of microcracks. This is illustrated by the micrographs in Figure 119. These are unetched microstructures, and the cracks are revealed as the dark lines. Note that there is less cracking when the carburized steel is reaustenitized (micrographs 153.3 and 153.4). This is because some of the carbon at the surface has diffused in towards the core, so that lower-carbon martensite is formed.

In carburizing, it is often beneficial to reaustenitize a carburized steel in order to alter the carbon gradient. This is referred to as a *diffusion* treatment. Figure 120 illustrates the principle. The microstructure of the carburized steel (after cooling slowly to 25°C) before the diffusion treatments is shown in Figure 72 (micrograph 149.3). The carbon content at the surface was 1.1% C (see graph 149.9 in Figure 72). If the carburized steel, with this gradient, is heated to 900°C for various times, carbon diffuses into the center of the steel. This is illustrated by the micrographs in Figure 120. (The samples were copper plated to prevent decarburization during this austenitization.) The reduction of the carbon gradient with austenitizing time is apparent in the microstructure (even at low magnification) by the increasing amounts of primary ferrite.

For these diffusion treatments, if the steel is quenched to 25°C, the microstructures attained are shown in Figure 121. This is a single quench treatment. The changes in the hardness profile with austenitizing time are shown in graph 156.9. With increasing austenitizing time, the carbon diffuses in, lowering the carbon content at and just below the surface. This raises the M_f temperature; hence less retained austenite is present after quenching, and the hardness increases. However, for longer times the hardness begins to decrease since the continued diminution of the carbon content lowers the hardness of the martensite. For the longest time used (8 hours), the maximum hardness is below the surface, indicating that some decarburization took place. Note also in this figure that the case depth increases with the austenitizing time.

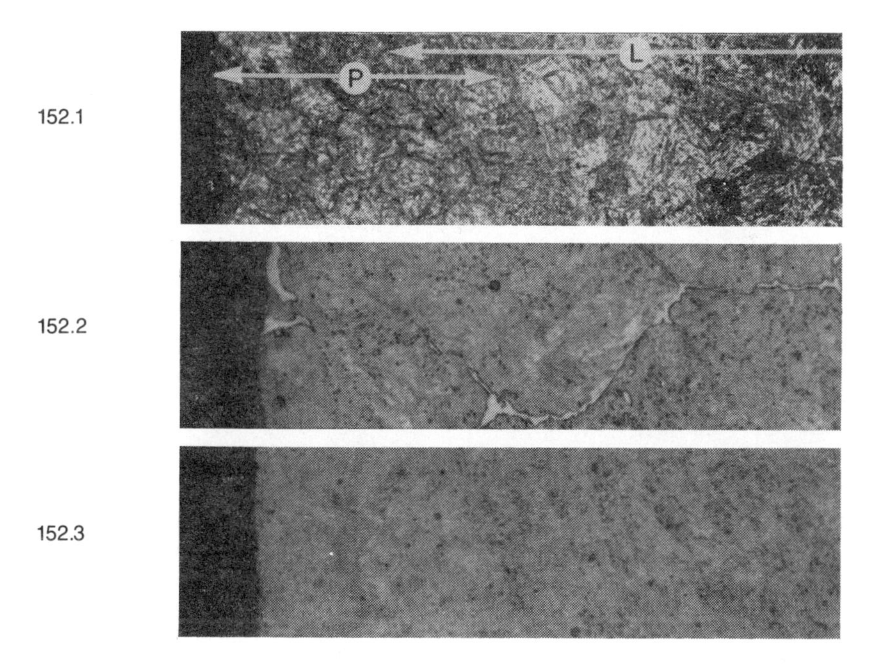

Figure No.	Material	Analysis, wt %	Condition	Hard-ness, HV	Etchant	Magnifi-cation
	0.15%C	0.17 C, 0.05 Si, 0.64 Mn	Pack carburized at 940°C for 2 h, cooled slowly to room temperature. heat treated as indicated			
152.1(a)(b)			Single quenched from 940°C (as for Fig. 150.1)		Nital (1%)	100
152.2(a)			Single quenched from 780°C (as for Fig. 150.3)		Nital (1%)	500
152.3(a)			Double quenched from 780°C (as for Fig. 150.4)		Nital (1%)	500

FIGURE 116. Microstructures of a carburized steel showing the effects of reaustenitizing then quenching (reprinted with permission from Samuels, L. E. 1980. *Optical Microscopy of Carbon Steels*. Metals Park, Ohio: American Society for Metals).

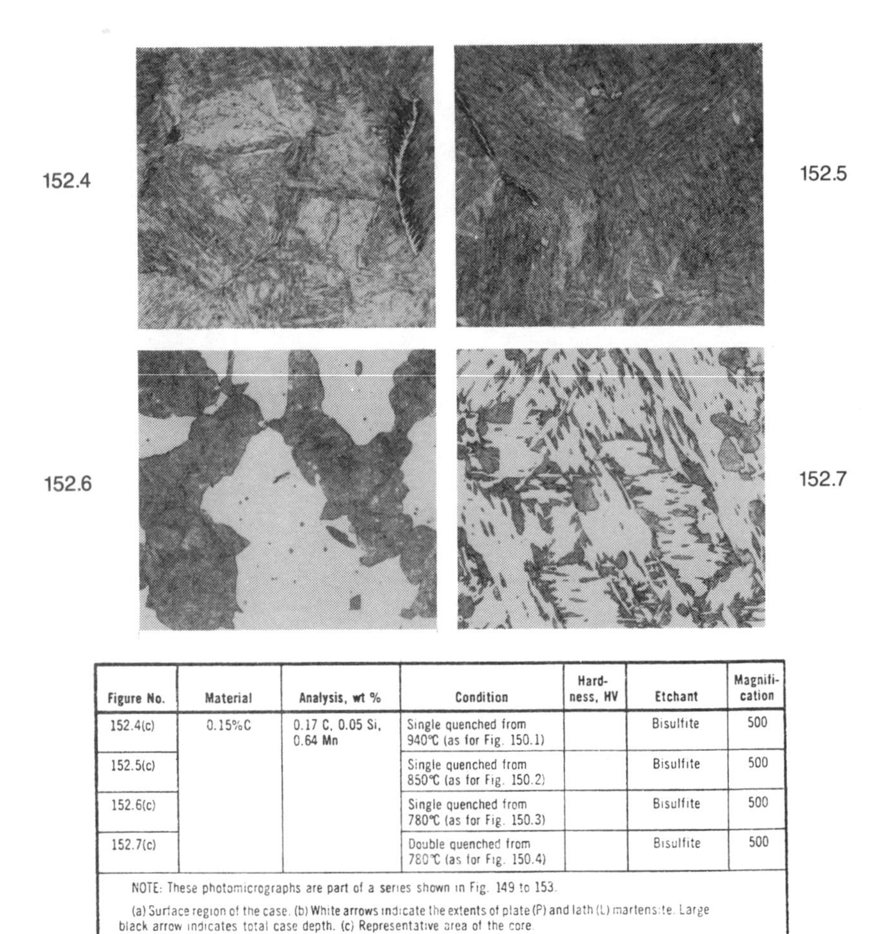

Figure No.	Material	Analysis, wt %	Condition	Hard-ness, HV	Etchant	Magnifi-cation
152.4(c)	0.15%C	0.17 C, 0.05 Si, 0.64 Mn	Single quenched from 940°C (as for Fig. 150.1)		Bisulfite	500
152.5(c)			Single quenched from 850°C (as for Fig. 150.2)		Bisulfite	500
152.6(c)			Single quenched from 780°C (as for Fig. 150.3)		Bisulfite	500
152.7(c)			Double quenched from 780°C (as for Fig. 150.4)		Bisulfite	500

NOTE: These photomicrographs are part of a series shown in Fig. 149 to 153.

(a) Surface region of the case. (b) White arrows indicate the extents of plate (P) and lath (L) martensite. Large black arrow indicates total case depth. (c) Representative area of the core.

FIGURE 116 (continued). Microstructures of a carburized steel showing the effects of reaustenitizing then quenching (reprinted with permission from Samuels, L. E. 1980. *Optical Microscopy of Carbon Steels*. Metals Park, Ohio: American Society for Metals).

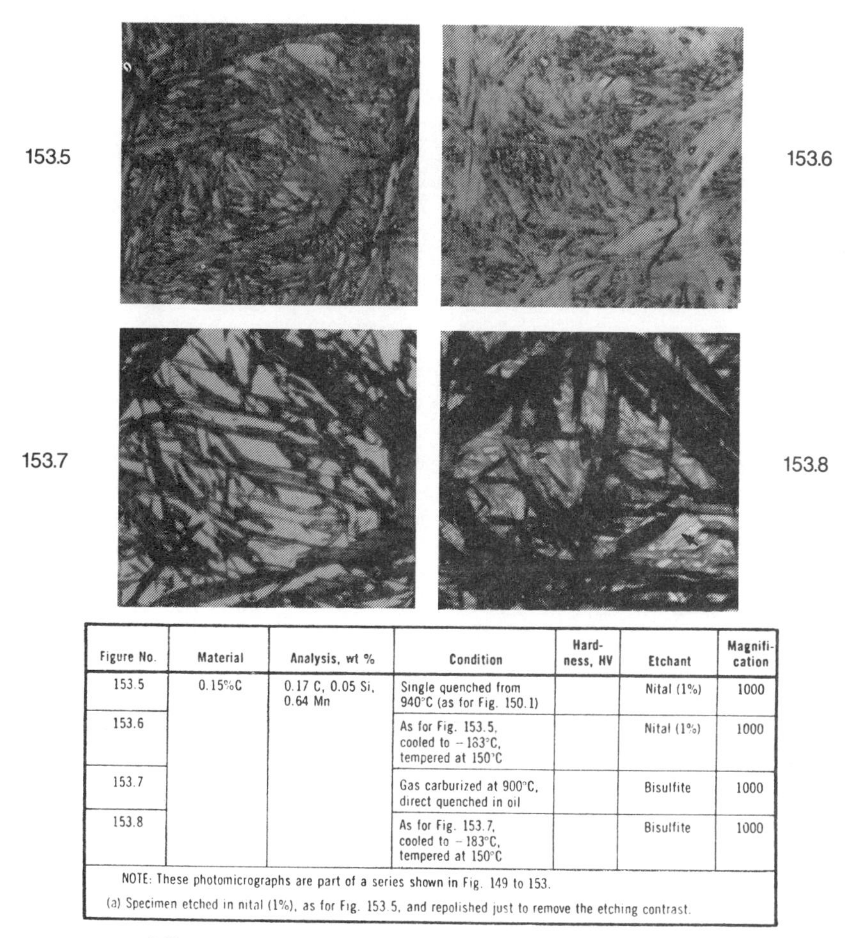

Figure No.	Material	Analysis, wt %	Condition	Hard-ness, HV	Etchant	Magnifi-cation
153.5	0.15%C	0.17 C, 0.05 Si, 0.64 Mn	Single quenched from 940°C (as for Fig. 150.1)		Nital (1%)	1000
153.6			As for Fig. 153.5, cooled to −183°C, tempered at 150°C		Nital (1%)	1000
153.7			Gas carburized at 900°C, direct quenched in oil		Bisulfite	1000
153.8			As for Fig. 153.7, cooled to −183°C, tempered at 150°C		Bisulfite	1000

NOTE: These photomicrographs are part of a series shown in Fig. 149 to 153.

(a) Specimen etched in nital (1%), as for Fig. 153.5, and repolished just to remove the etching contrast.

FIGURE 117. Microstructures showing the effect of sub-zero cooling on the conversion of retained austenite to martensite (reprinted with permission from Samuels, L. E. 1980. *Optical Microscopy of Carbon Steels*. Metals Park, Ohio: American Society for Metals).

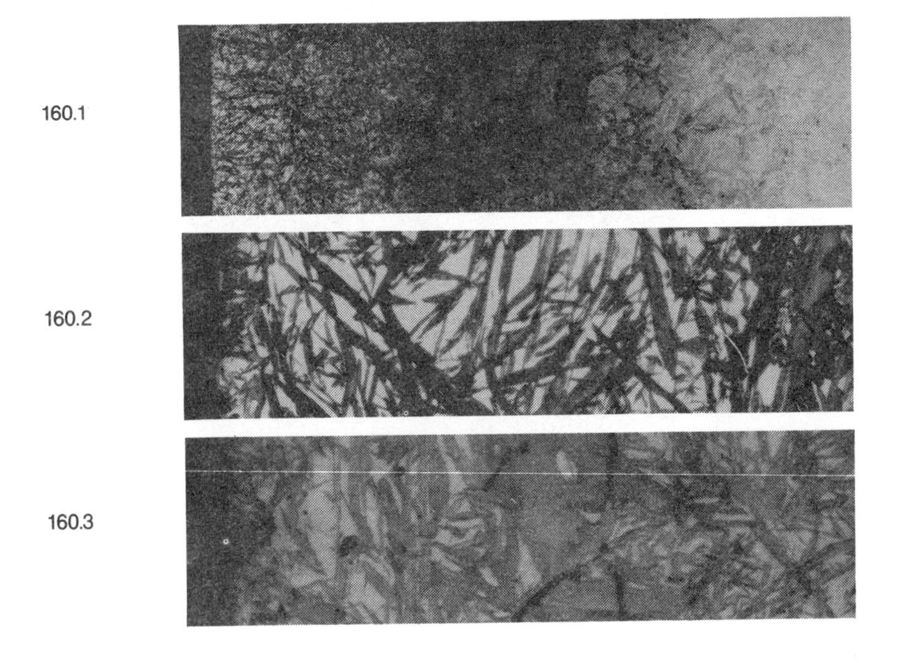

Figure No.	Material	Analysis, wt %	Condition	Hardness, HV	Etchant	Magnification
	0.15%C	0.17 C, 0.05 Si, 0.64 Mn	Gas carburized at 900°C for 4.5 h in sealed-quench furnace with excessive gas enrichment, quenched in oil, tempered at 150°C			
160.1			Direct from carburizing furnace: tempered at 150°C		Bisulfite	100
160.2					Bilsulfite	1000
160.3					Bisulfite(a)	1000
160.4			Cooled to −183°C after carburizing, tempered at 150°C		Bisulfite	100
160.5					Bisulfite	1000
160.6					Bisulfite(a)	1000

(a) Comparatively light etch.

FIGURE 118. Microstructures showing the effect of sub-zero cooling on the conversion of retained austenite to martensite (reprinted with permission from Samuels, L. E. 1980. *Optical Microscopy of Carbon Steels*. Metals Park, Ohio: American Society for Metals).

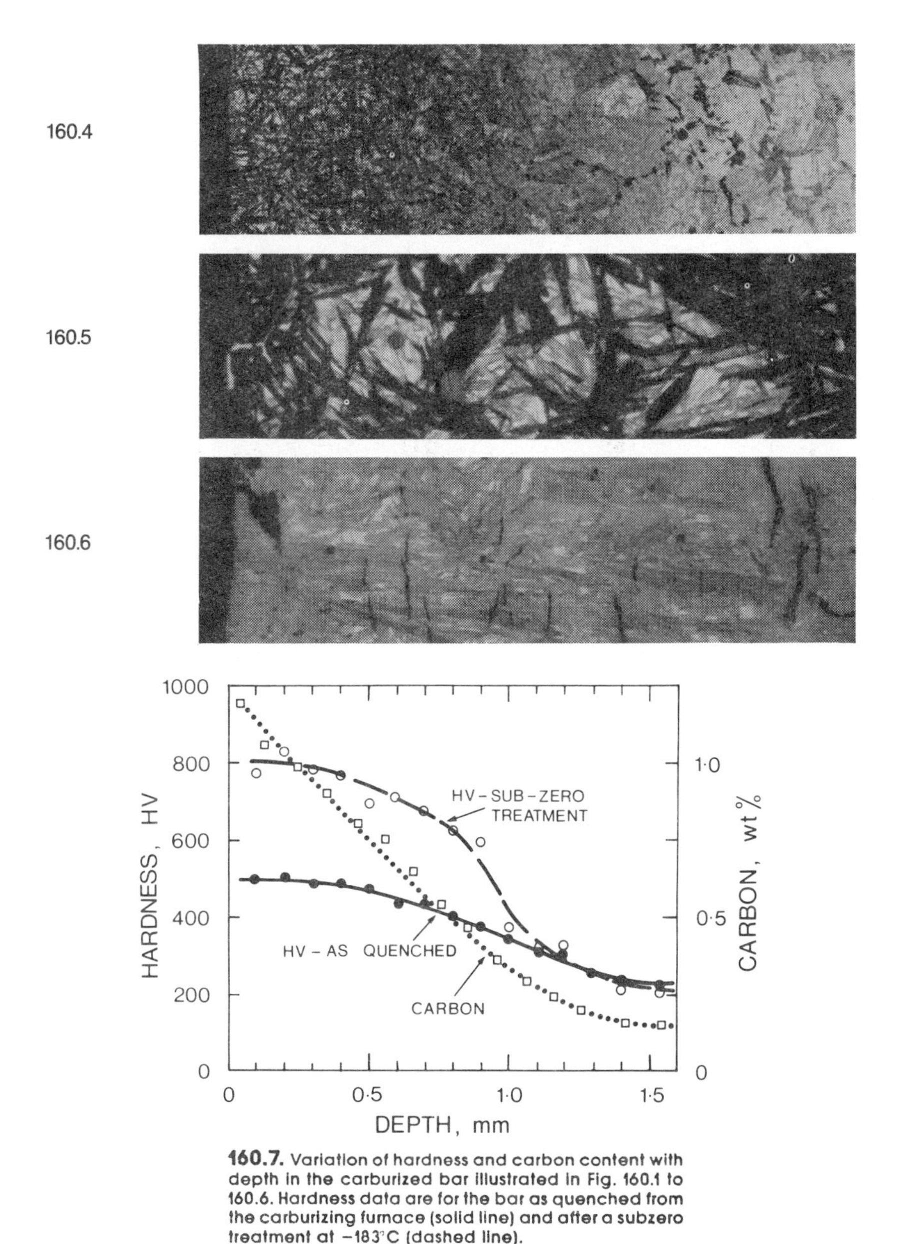

160.7. Variation of hardness and carbon content with depth in the carburized bar illustrated in Fig. 160.1 to 160.6. Hardness data are for the bar as quenched from the carburizing furnace (solid line) and after a subzero treatment at −183°C (dashed line).

FIGURE 118 **(continued).** Microstructures showing the effect of sub-zero cooling on the conversion of retained austenite to martensite (reprinted with permission from Samuels, L. E. 1980. *Optical Microscopy of Carbon Steels*. Metals Park, Ohio: American Society for Metals).

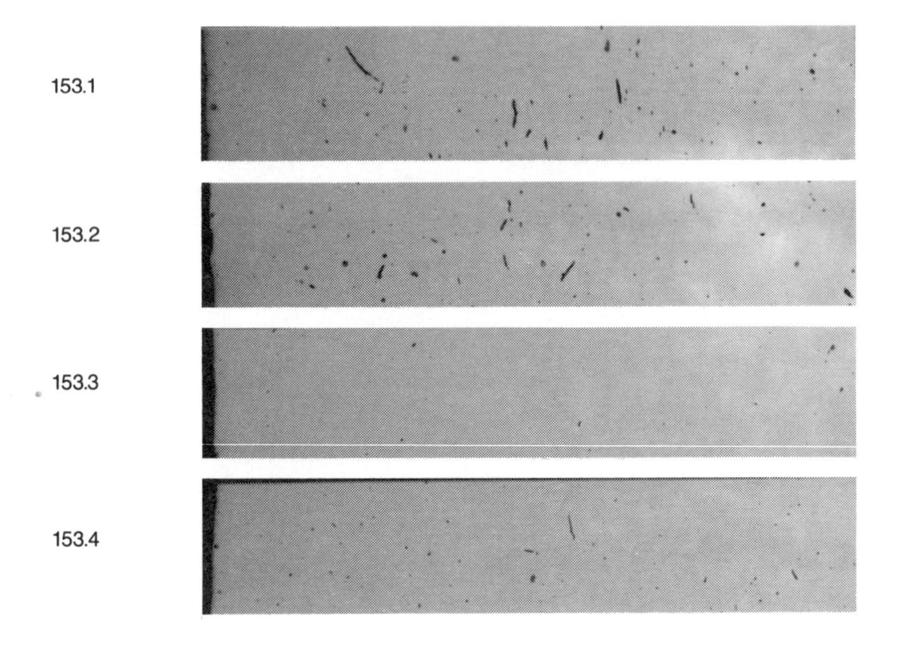

Figure No.	Material	Analysis, wt %	Condition	Hard-ness, HV	Etchant	Magnifi-cation
	0.15%C	0.17 C, 0.05 Si, 0.64 Mn	Pack carburized at 940°C for 2 h, cooled slowly to room temperature, heat treated as indicated			
153.1			Single quenched from 940°C (as for Fig. 150.1)		Unetched(a)	200
153.2			Single quenched from 850°C (as for Fig. 150.2)		Unetched(a)	200
153.3			Single quenched from 780°C (as for Fig. 150.3)		Unetched(a)	200
153.4			Double quenched from 780°C (as for Fig. 150.4)		Unetched(a)	200

FIGURE 119. Microstructures of a carburized steel which has been reaustenitized at different temperatures showing the effect of the degree of microcracking. Note that these microstructures are unetched (reprinted with permission from Samuels, L. E. 1980. *Optical Microscopy of Carbon Steels*. Metals Park, Ohio: American Society for Metals).

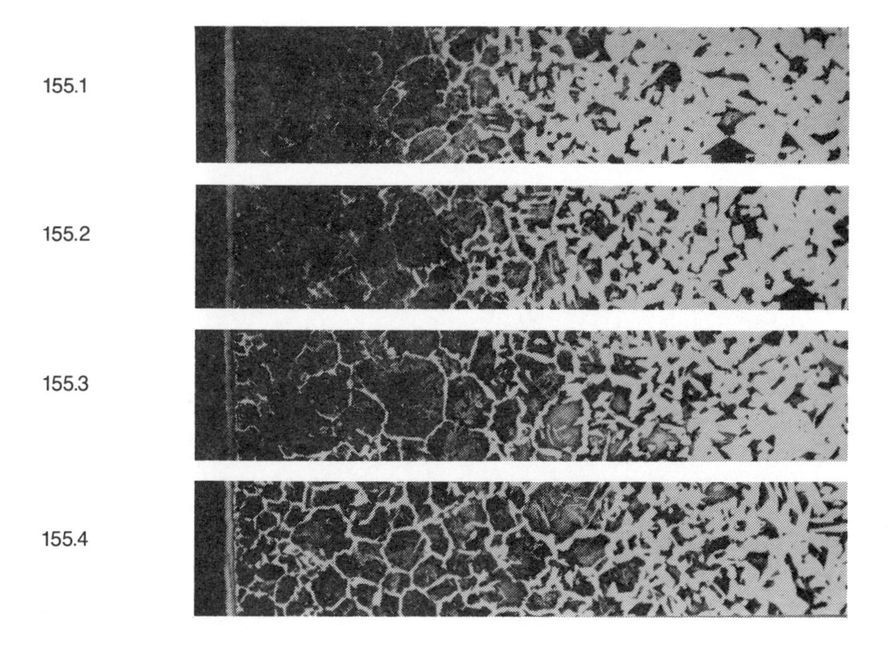

Figure No.	Material	Analysis, wt %	Condition	Hard-ness, HV	Etchant	Magnifi-cation
	0.15%C	0.17 C, 0.05 Si, 0.64 Mn	Pack carburized at 940°C for 2 h, heated at 900°C for time indicated, cooled slowly			
155.1(a)			1 h		Picral	75
155.2(a)			2 h		Picral	75
155.3(a)			4 h		Picral	75
155.4(a)			8 h		Picral	75
155.5			1 h		Picral	500
155.6			2 h		Picral	500
155.7			4 h		Picral	500
155.8			8 h		Picral	500

NOTE: The gray layer at the surfaces of these specimens is electrodeposited copper used to prevent decarburization during the diffusion treatment.

(a) The structure of the carburized case before diffusion treatment is shown in Fig. 149.3 and 149.7. The arrows in Fig. 155.1 and 155.2 indicate total case depth estimated from Fig. 156.9.

FIGURE 120. Microstructures showing the effects of diffusion treatments on the carbon gradient of a carburized steel (reprinted with permission from Samuels, L. E. 1980. *Optical Microscopy of Carbon Steels*. Metals Park, Ohio: American Society for Metals).

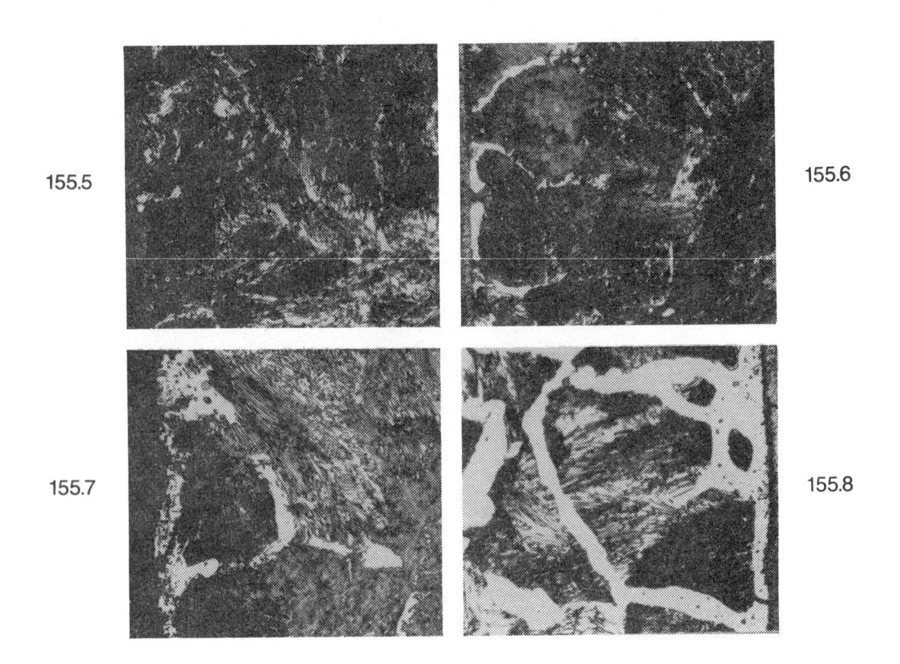

FIGURE 120 (continued). Microstructures showing the effects of diffusion treatments on the carbon gradient of a carburized steel (reprinted with permission from Samuels, L. E. 1980. *Optical Microscopy of Carbon Steels*. Metals Park, Ohio: American Society for Metals).

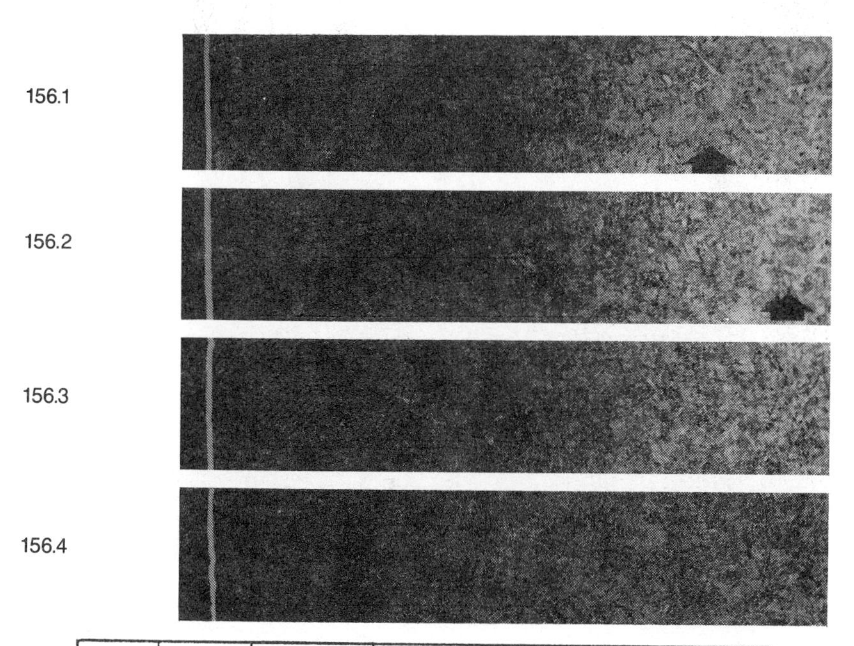

Figure No.	Material	Analysis, wt %	Condition	Hard-ness, HV	Etchant	Magnifi-cation
	0.15%C	0.17 C, 0.05 Si, 0.64 Mn	Pack carburized at 940°C for 2 h, heated at 900°C for time indicated, austenitized at 850°C, quenched in water			
156.1(a)			1 h		Bisulfite	75
156.2(a)			2 h		Bisulfite	75
156.3			4 h		Bisulfite	75
156.4			8 h		Bisulfite	75
156.5(b)			1 h		Bisulfite	500
156.6(b)			2 h		Bisulfite	500
156.7(b)			4 h		Bisulfite	500
156.8(b)			8 h		Bisulfite	500

NOTE: These are the same carburized and diffused bars as those illustrated in Fig. 155.

(a) Arrow indicates total case depth as determined from the hardness variation shown in Fig. 156.9. (b) The area shown is 15 μm below the surface.

FIGURE 121. Microstructures showing the effects of diffusion treatments on a carburized steel. The samples were water-quenched after reaustenitizing (the small arrows in 156.6–156.8 point out plate martensite) (reprinted with permission from Samuels, L. E. 1980. *Optical Microscopy of Carbon Steels*. Metals Park, Ohio: American Society for Metals).

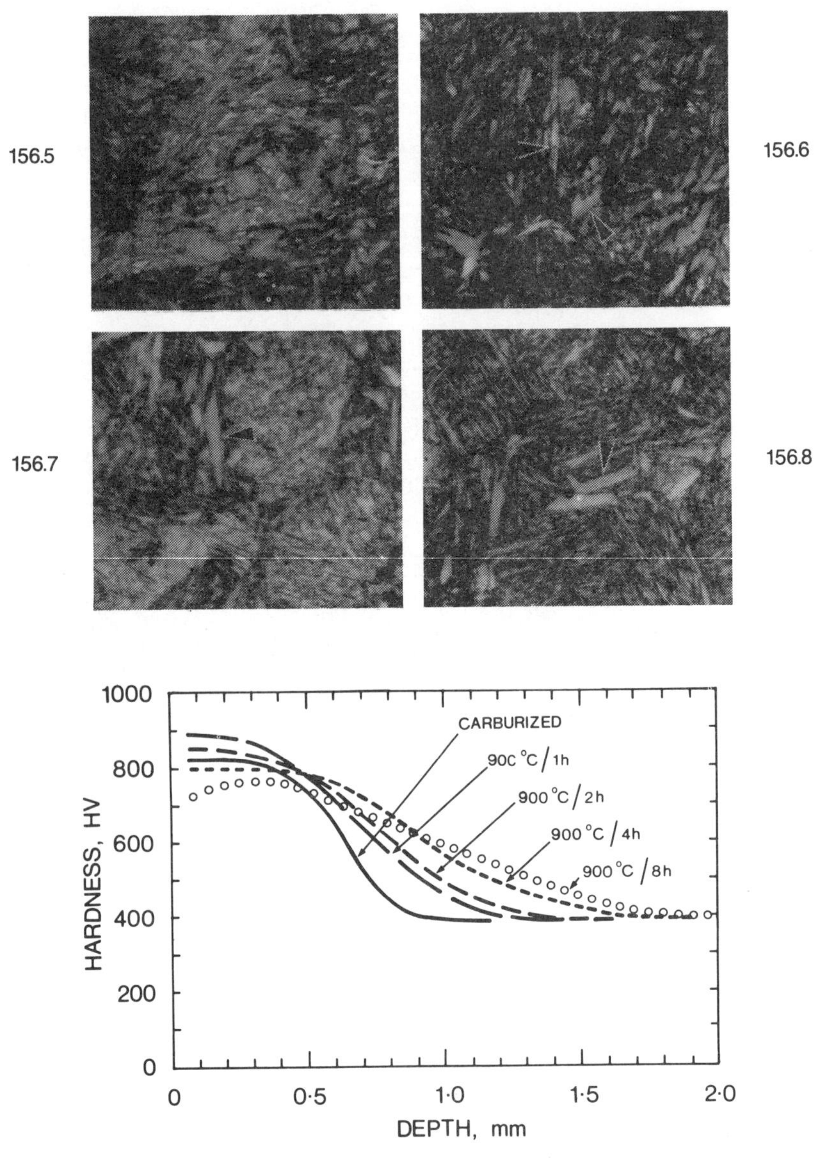

156.9. Variation of hardness with depth in bars pack carburized at 940°C for 2 h, given the diffusion treatments indicated, and then quench hardened.

FIGURE 121 (continued). Microstructures showing the effects of diffusion treatments on a carburized steel. The samples were water-quenched after reaustenitizing (the small arrows in 156.6–156.8 point out plate martensite) (reprinted with permission from Samuels, L. E. 1980. *Optical Microscopy of Carbon Steels*. Metals Park, Ohio: American Society for Metals).

Figure 122 shows hardness profiles illustrating the effect of diffusion treatments. Figure 122(a) shows the effect of carburizing time on the hardness profiles developed after carburizing followed by quenching to 25°C (direct quench). The low hardness at the surface is due to retained austenite. The curves in Figure 122(b) show the effect of single quenching on the hardness profile. Note that the hardness at the surface is increased considerably, although there is some retained austenite remaining. Also note that for these treatments the case depth does not change appreciably.

The effects of the austenitizing temperature for single quenching on the hardness profiles are shown in Figure 123. Also shown are the effects of sub-zero cooling following quenching, and the effects of tempering following quenching. The effect of sub-zero cooling on increasing the surface hardness is apparent. Figure 124 illustrates the effect of austenitizing temperature on the hardness profile. A lower austenitizing temperature will allow less diffusion of carbon, and hence will retain more nearly the original carbon gradient. This would give a higher carbon content at the surface, more retained austenite and a lower hardness. However, this assumes that only austenite is present. At 830 and 770°C, the surface layers will be in the two-phase austenite + carbide region if the carbon content is sufficiently high [see Figure 124(b)]. At 830°C the carbon content of the austenite will be about 0.95%, which will give less retained austenite upon quenching than for austenitizing at 940°C. Austenitizing at 770°C will give austenite containing about 0.75% C, allowing very little retained austenite upon quenching, and even higher hardness [Figure 124(a)]; however, some iron carbide will be present.

In Figure 125 are the carburizing and heat treatments recommended by Thelning. Figure 126 shows similar information from another source.

Current developments of computer programs to calculate the diffusion of carbon in austenite have allowed prediction of temperature–time sequences that will produce a given carbon gradient. For example, if the temperature or surface carbon content (based on the gas analysis) is altered from the initial specification, the computer programs can be used to determine a temperature, time, and gas analysis (e.g., moisture content of the gas) to develop the required carbon gradient.

3h. RESIDUAL STRESSES IN CARBURIZED STEELS

One of the main purposes of carburizing is to develop high compressive residual stresses on the surface. That this can occur in a carburized steel that is quenched and tempered, whereas the same steel when uncarburized then quenched and tempered may have tensile residual stresses at the surface, is illustrated by the data in Figure 127.

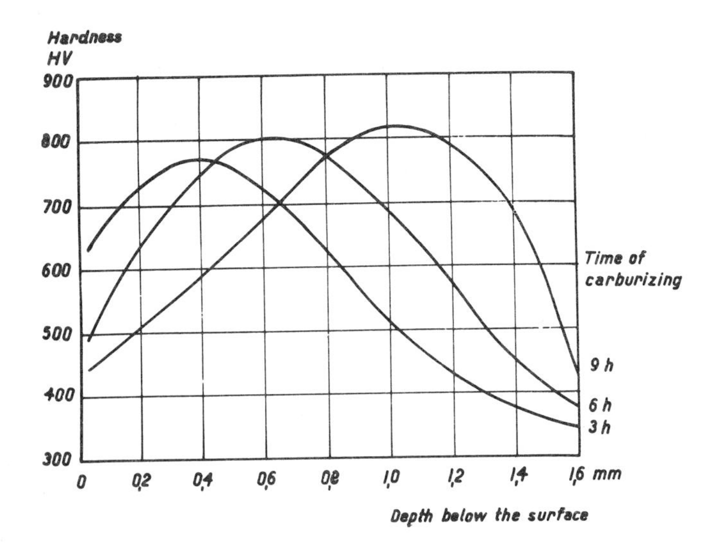

(a) *Direct quench from 925 °C*

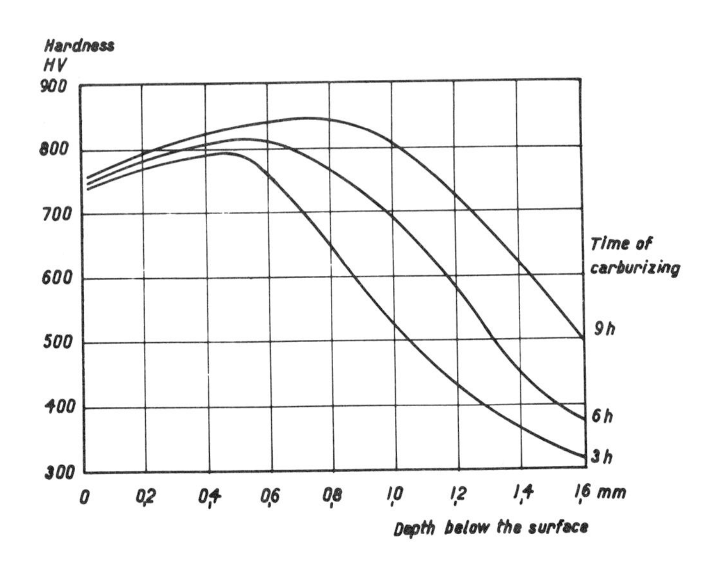

(b) *Single quench from 830 °C*

FIGURE 122. Hardness profiles of a carburized steel after (a) direct quenching from the carburizing temperature and (b) after reaustenitizing at 830°C then quenching following the treatment in (a). The steel was B S 637M17, En 352 (Bofors D R 34). The samples were pack-carburized at 925°C (reprinted with permission from Thelning, K.-E. 1975. *Steel and Its Heat Treatment.* Boston: Butterworths).

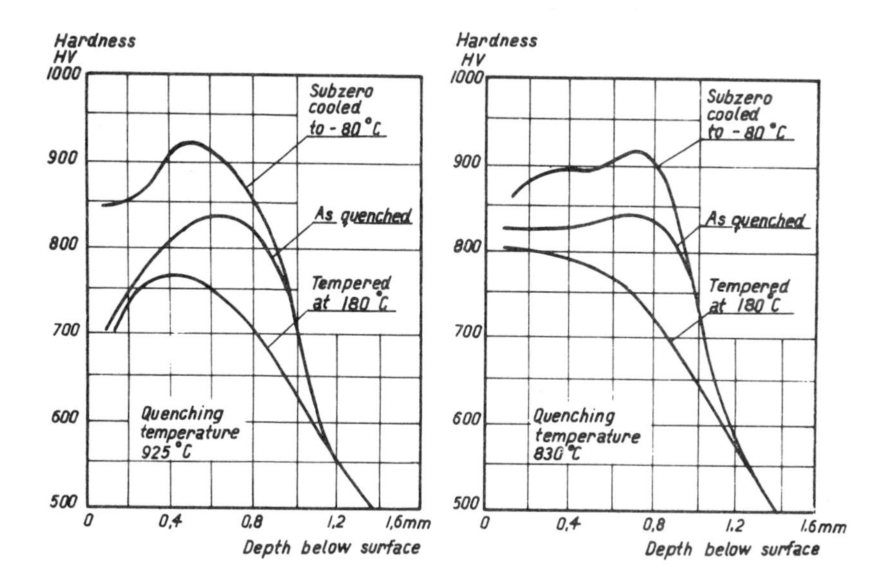

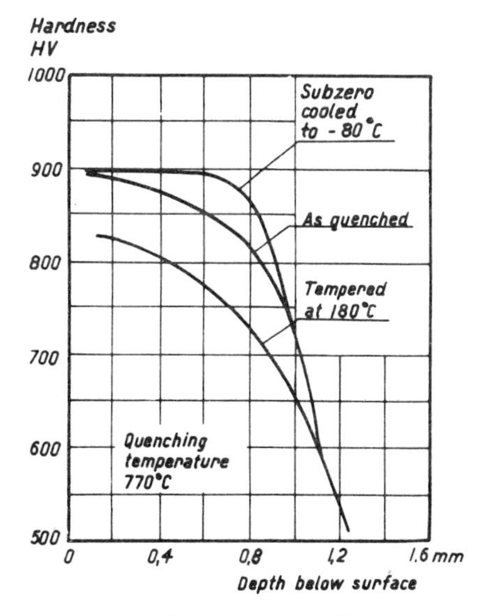

FIGURE 123. Hardness profiles of a carburized steel for different diffusion heat treatments. The samples were pack-carburized and single water-quenched from the temperatures indicated, followed by sub-zero cooling for 1 hr at −80°C, alternatively followed by tempering at 180°C (reprinted with permission from Thelning, K.-E. 1975. *Steel and Its Heat Treatment*. Boston: Butterworths).

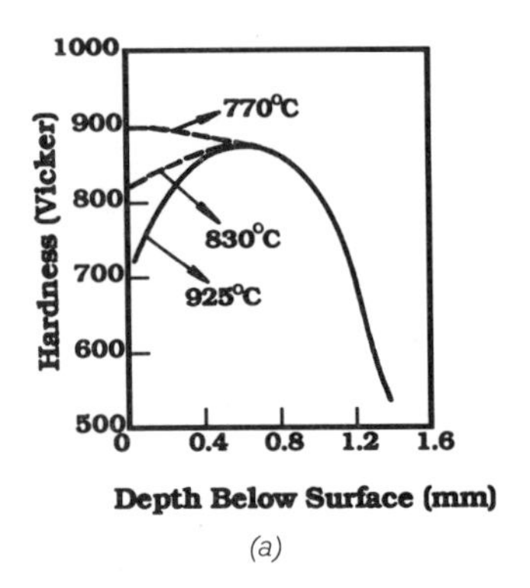

(a)

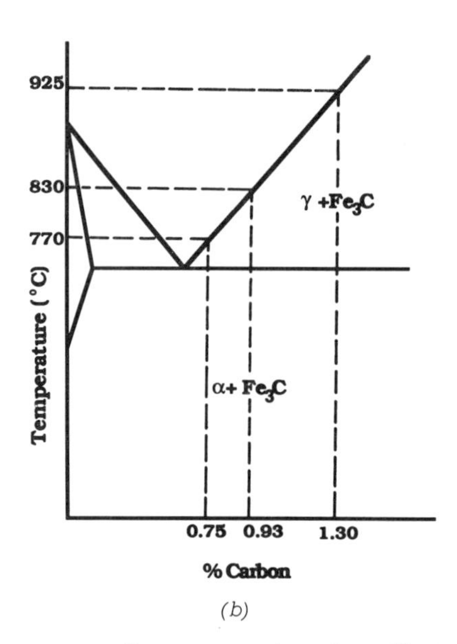

(b)

FIGURE 124. Hardness profiles comparing the effect of reaustenitizing temperature, then quenching (single quenching) (from data in Figure 123).

From practical experience gained in this field the following general heat-treatment recommendations are suggested for case-hardening steels. The treatment is designed to give a surface hardness of at least 60 HRC or 710 HV after quenching.

CARBURIZING IN SOLID COMPOUND (PACK CARBURIZING)

Temperature: 900–925 °C

Depth of case hardening mm	*Method of hardening*
< 0·40	Direct quench
0·40–1·25	Single quench from 800 to 820 °C
> 1·25	Double quench

When direct quenching is employed a carburizing temperature of 900 °C is recommended. If a case-hardening depth of more than 1·25 mm is aimed at it might be good practice to perform the traditional practice of a double quench, i.e. a first quench from about 880 °C and then a final quench. During the first heating for quenching, the carbon concentration is reduced by diffusion. In addition it breaks up and dissolves the carbide network that usually forms when pack carburizing is used to give a deep carbon penetration.

CARBURIZING IN A SALT BATH (LIQUID BATH CARBURIZING)

Temperature: 850–900 °C

Depth of case hardening mm	*Method of hardening*
< 0·50	Direct quench
0·50–0·75	Requench from 820 °C (Double quench)
0·75–1·00	Requench from 800 °C (Double quench)

For direct quenching, a carburizing temperature of 850–870 °C is recommended; for double quenching, 900 °C.

CARBURIZING IN GAS (GAS CARBURIZING)

Temperature: 900–940 °C

The carburization is carried out to the required depth of case hardening and to a surface carbon concentration of 0·70–0·80%. When the process is completed the temperature in the furnace is lowered to 830 °C and the part is then quenched in a suitable medium.

TEMPERING

Case-hardened steels are tempered at temperatures generally around 160–220 °C. Temperatures below 160 °C should not be used, particularly if a grinding operation is to follow, since grinding cracks develop very easily. Tempering is not necessary after a cyaniding treatment that gives a case depth of only a few tenths of a millimetre.

The hardness falls quite rapidly when the steel is tempered between 160 °C and 200 °C. If a hardness of 60 HRC is required the tempering temperature should not be higher than 180 °C.

FIGURE 125. Recommended heat treatments for case hardening of carburized steels (reprinted with permission from Thelning, K.-E. 1975. *Steel and Its Heat Treatment.* Boston: Butterworths).

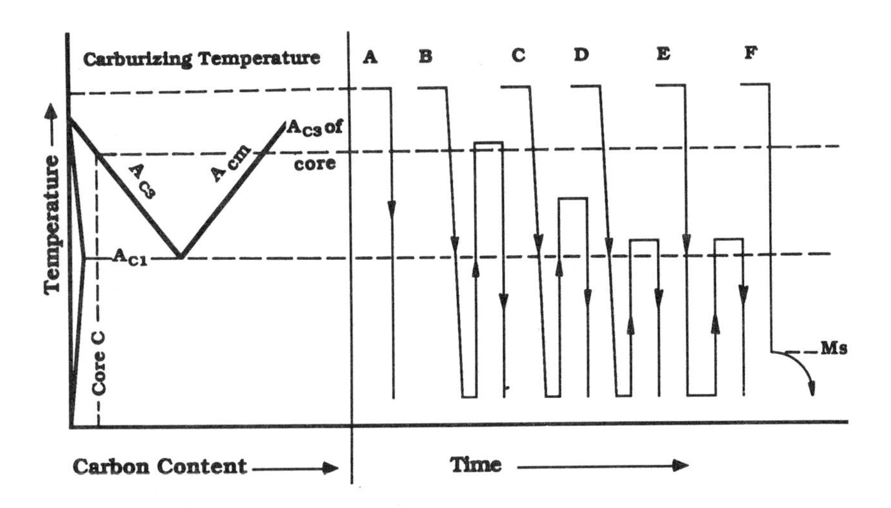

Treatment	Case	Core
A	Unrefined. Solution of excess carbide favored. Austenite retained. Distortion low to medium.	Unrefined but hardened.
B	Refined. Solution of excess carbide favored. Austenite retention promoted in higher alloyed steels.	Refined. Maximum core strength and hardness.
C	Refined. Some solution of excess carbide.	Partially refined. Stronger and tougher than Treatment D.
D	Refined. Excess carbide not dissolved.	Unrefined. Soft and machinable.
E	Refined. Some solution of excess carbide. Austenite retention minimized.	Low hardness. High toughness. Machinable.
F (Interrupted quench; marquenching)	Unrefined. Solution of excess carbide favored. Austenite retained. Distortion minimized.	Fully hardened.

FIGURE 126. Schematic representation of different heat treatments of carburized steels and their effects on case and core properties (adapted from *The Case Hardening of Nickel Alloy Steels*. 1946. New York: International Nickel Company).

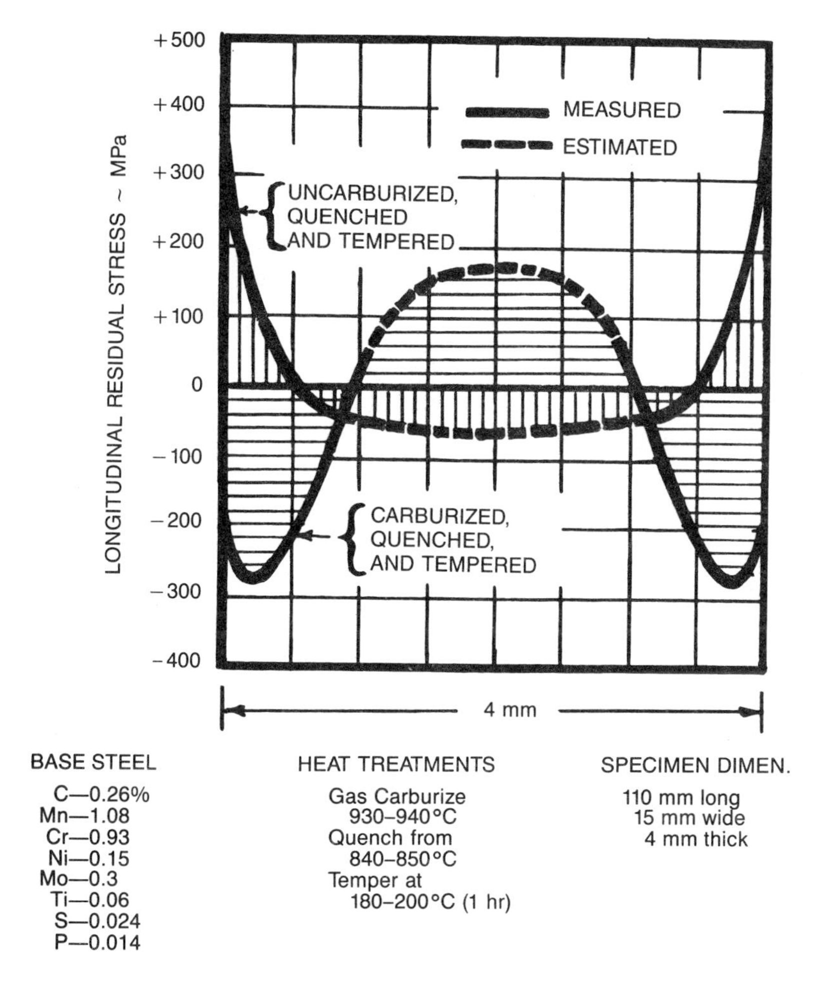

FIGURE 127. Longitudinal residual stress distribution in a quenched and tempered steel and in the same steel when carburized, quenched and tempered (reprinted with permission from Ebert, L. J. 1978. *Met. Trans.*, 9A:1537, a publication of The Minerals, Metals and Materials Society, Warrendale, PA).

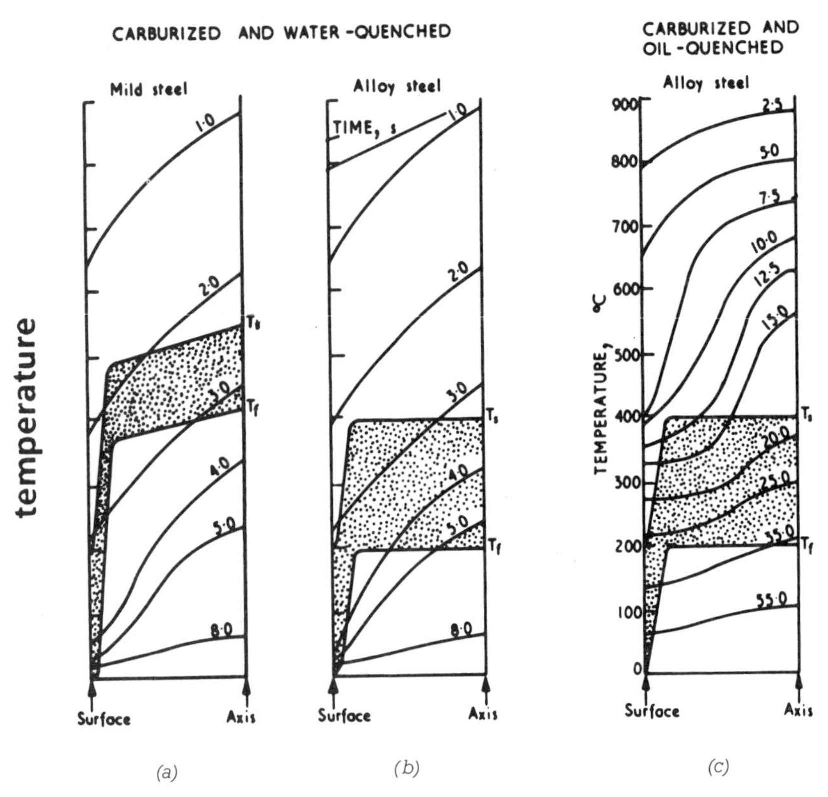

FIGURE 128. Progress of transformation during quenching carburized, 0.5 inch diameter steel bars (from Dawes, C. and R. J. Cooksey. 1966. *Heat Treatment of Metals, Special Report 95.* London: The Iron and Steel Institute, p. 77; by permission of The Institute of Metals).

In the following treatment, the formation of residual stresses during quenching of carburized steels is illustrated. For simplicity, the discussion is restricted to the development of longitudinal stresses in a cylinder. The final residual stresses depend upon several factors: whether the core goes to martensite or the softer products, which depends on the local hardenability and the cooling rate; the time sequence at which various locations transform, which depends upon the cooling rate and the continuous cooling TTT diagram; and the M_s and M_f temperatures, which vary with carbon content and hence location in the case.

We first consider the case of quenching cylinders small enough so that the entire structure becomes martensite. Figure 128(a) shows temperature curves across the radius of a cylinder at different times typical of those encountered during water quenching. The M_s and M_f temperatures are also shown; these are lowered as the surface is approached due to the increasing carbon content. Note that at 2 sec, the center and the surface are both still austenite, but at the edge of the case (just under the surface), the temperature is below M_s, and hence martensite begins to form here first. This causes an expansion, which is resisted by the surface and the center. Thus at this instant the residual stresses will be tension at the surface and the center, and compression in this underlying layer. Then stress relaxation occurs, and the stresses decrease. As cooling progresses, the formation of martensite increases from the case–core towards the center. At about 3 sec, almost all of the core is martensite, and the surface is almost at the M_f temperature. During this time (from 2 to 3 sec), more plastic deformation of the soft, austenitic surface occurs, causing a reduction in the residual stresses. Then as the surface cools, it transforms to martensite, with an expansion. However, the underlying layer and core are all martensite (see 3 sec curve) and strong, resisting this expansion. This puts the surface in compression and the underlying layers in tension, as shown in Figure 128(a).

The M_s and M_f temperatures are lower for alloy steels than for plain carbon steels. This is shown in Figure 128(b), where cooling an alloy steel is considered. In the plain carbon steel, at 4 sec the transformation to martensite is almost complete in the case and the surface, whereas in the alloy steel only the surface has formed almost all martensite. Note that when the region just below the surface is all martensite (at about 3.5 sec), the surface and the core are about 50% martensite. If about this time plastic deformation occurs, reducing the stresses, then when the surface and the center transform, expansion occurs, placing these regions in compression, and the underlying layer in tension, which may cause cracking in this region.

The effect of cooling rate on the sequence of martensite formation is seen by comparing Figures 128(b) and 128(c). In oil-quenching, the transformation is completed first at the core–case interface, then it progresses outwards to the surface. When the surface begins to transform (at about 25 sec), the underlying layers and the core contain a lot of martensite, and are almost all martensite when the surface is only 50% transformed. If the stresses are relaxed, then when the surface completes the formation of martensite, it will be in compression. Note that sub-surface cracking should not occur here.

We now consider the case where the core transforms to non-martensitic products before the case transforms. As shown in Figure 129(a), the reduction in the M_s and M_f temperatures in the case due to the higher carbon contents allows the core to begin to transform (to primary ferrite, pearlite, bainite) before transformation occurs in the case. Thus the core expands while the outside is still austenite, placing it in tension and the core in compression. However, plastic flow of the softer austenite will reduce these stresses. Then, when the surface expands upon forming martensite, it is placed in compression and the core in tension. The development of the residual stresses in this carburized steel when the surface transforms *before* the core is illustrated in Figure 129(b). The case forms martensite, placing it in compression and the center in tension. Relaxation occurs, then the surface transforms to non-martensite products and perhaps also to martensite, which places the surface in tension and the core in compression. Thus such a steel would be susceptible to cracking in the case.

The above descriptions should be modified if retained austenite is present after cooling to 25°C, which it will be, in most carburized steels. This reduces the amount of expansion, and will reduce the magnitude of the compressive residual stresses (see Figure 130).

Models have been developed that allow calculation of the residual stresses as a function of cooling time. (This was mentioned when induction-hardening was discussed). These take into account the amount of phases present and their mechanical properties as a function of temperature. Figure 130 shows the results of such a calculation for carburized steels. The amount of austenite present as a function of quenching time is shown in Figure 130(a). At the end of quenching (estimated by 4423 sec), there is 40% retained austenite at the surface. For this steel, the distribution of the microconstituents in the final quenched steel is shown in Figure 130(b). Note that in this case the cooling rate at the surface was sufficiently low to form a small amount of bainite and pearlite. The residual stresses as a function of time are shown in Figure 130(c). Note that at short times (e.g., 4 and 65 sec) the surface has a tensile residual stress. However, this reverses as cooling continues, so that finally the surface is in compression.

The success of this type of calculation is shown by Figure 131, which compares calculated to measured residual stresses in a 3.0 cm diameter carburized steel after quenching in oil from 830°C. (This was a single quench treatment.)

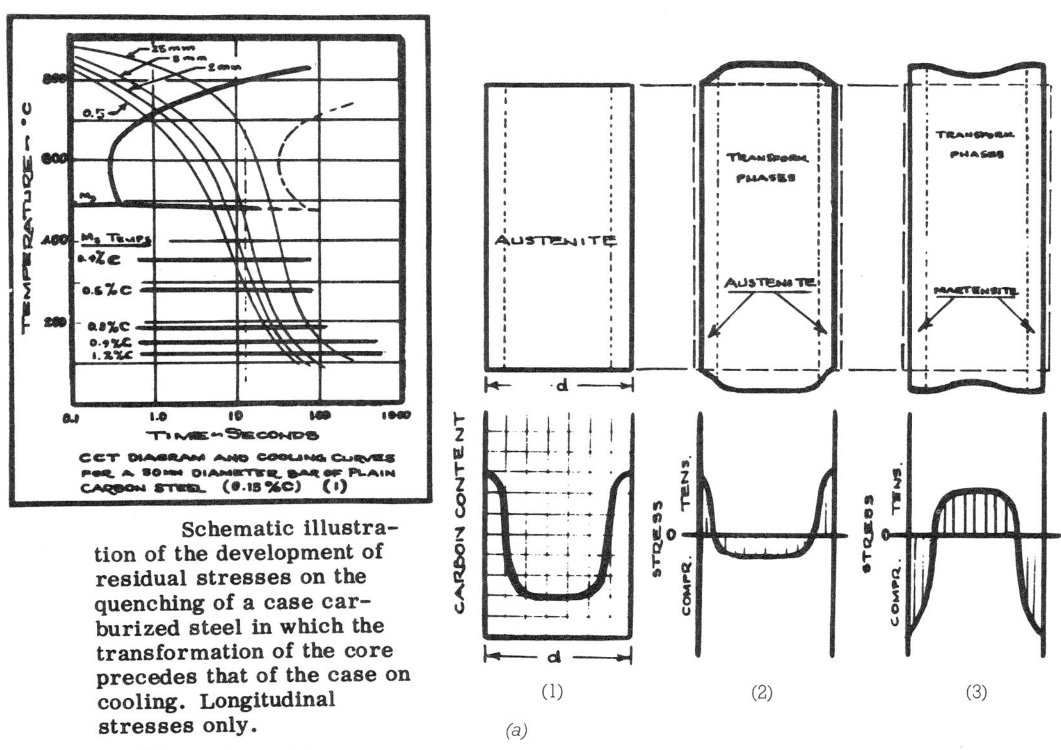

FIGURE 129. Illustration of the development in carburized steels of (a) compressive residual stresses at the surface (reprinted with permission from Ebert, L. J. 1978. *Met. Trans.*, 9A:1537, a publication of The Minerals, Metals and Materials Society, Warrendale, PA).

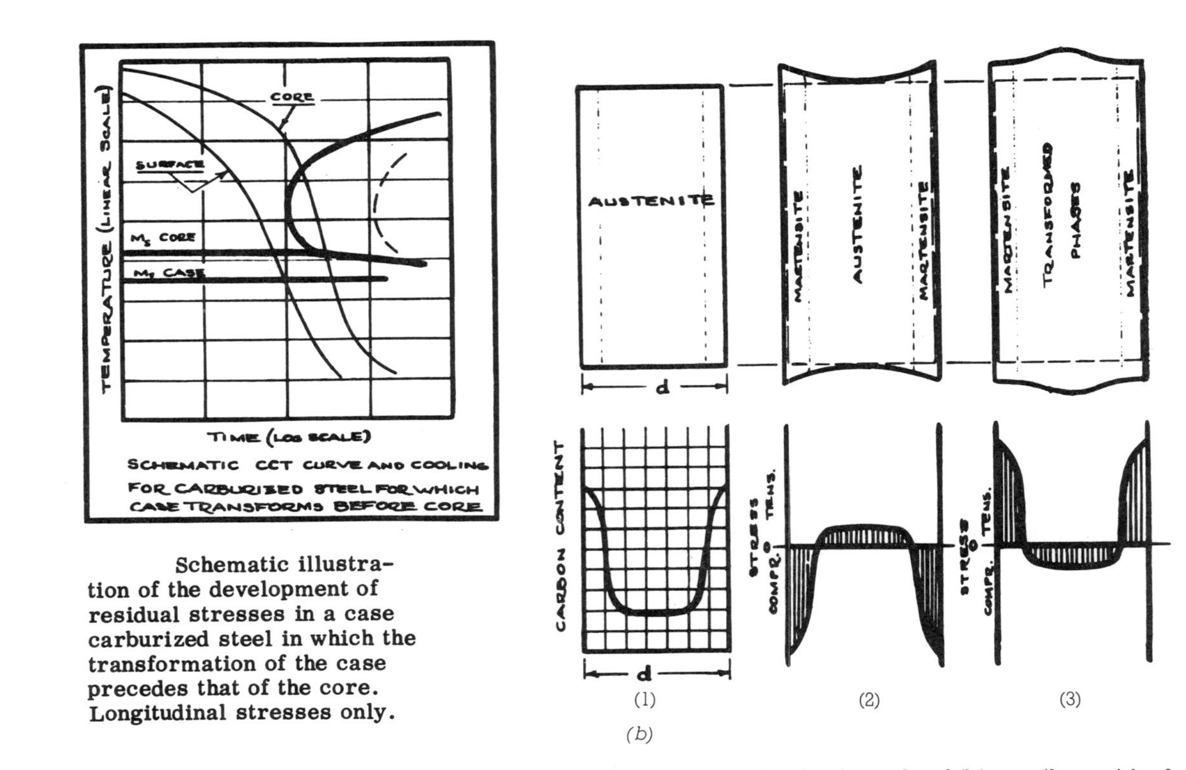

FIGURE 129 (continued). Illustration of the development in carburized steels of (b) tensile residual stresses at the surface (reprinted with permission from Ebert, L. J. 1978. *Met. Trans.*, 9A:1537, a publication of The Minerals, Metals and Materials Society, Warrendale, PA).

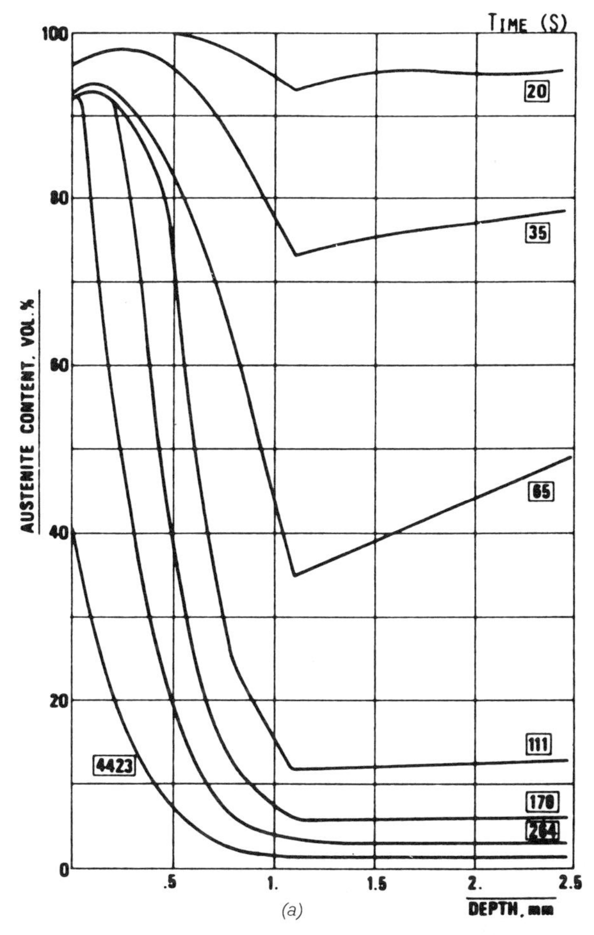

Retained austenite content vs. depth after various times from the beginning of the quenching of carburized SAE 1321 (C_s=0.8%C, total case depth = 1 mm).

FIGURE 130. The calculated amount of phases and microconstituents and the calculated residual stresses as a function of time during cooling of a carburized steel (reprinted with permission from Hildenwall, B. 1978. In *Hardenability Concepts with Applications to Steel*, D. V. Doane and J. S. Kirkaldy, editors, The Minerals, Metals and Materials Society, Warrendale, PA, p. 579).

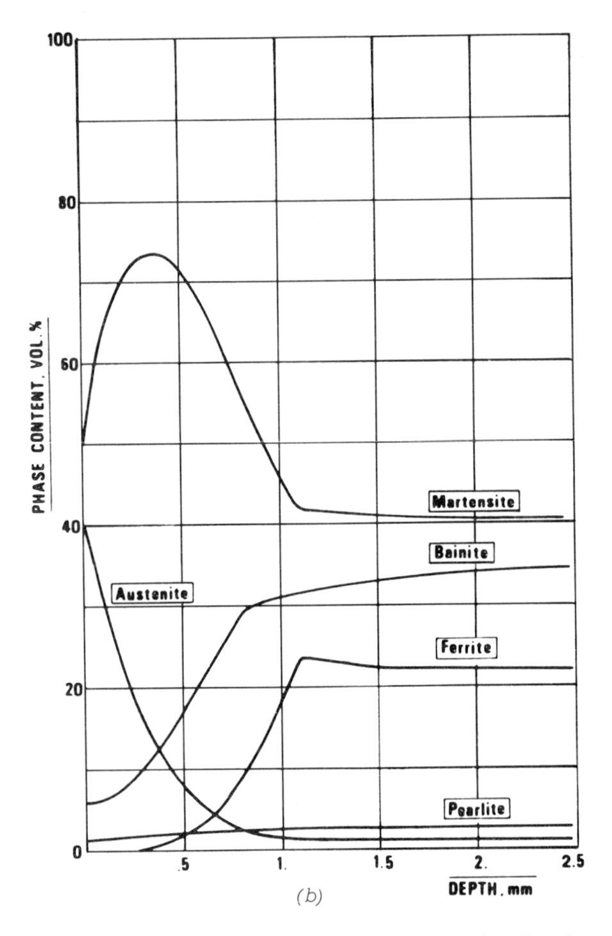

(b)

The calculated phase content after completed hardening of carburized SAE 1321. Surface carbon content, C_s=0.8%C, and total case depth=1 mm.

FIGURE 130 (continued). The calculated amount of phases and microconstituents and the calculated residual stresses as a function of time during cooling of a carburized steel (reprinted with permission from Hildenwall, B. 1978. In *Hardenability Concepts with Applications to Steel*, D. V. Doane and J. S. Kirkaldy, editors, The Minerals, Metals and Materials Society, Warrendale, PA, p. 579).

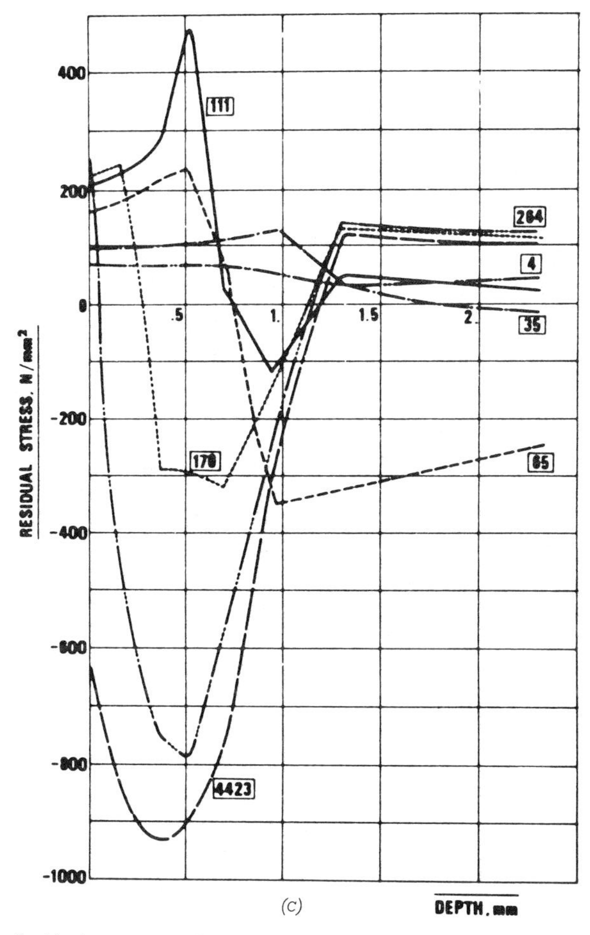

(c)

Residual stress vs. depth after various times from the beginning of the quenching of carburized SAE 1321 (C_s=0.8%C, total case depth = 1 mm).

FIGURE 130 (continued). The calculated amount of phases and microconstituents and the calculated residual stresses as a function of time during cooling of a carburized steel (reprinted with permission from Hildenwall, B. 1978. In *Hardenability Concepts with Applications to Steel*, D. V. Doane and J. S. Kirkaldy, editors, The Minerals, Metals and Materials Society, Warrendale, PA, p. 579).

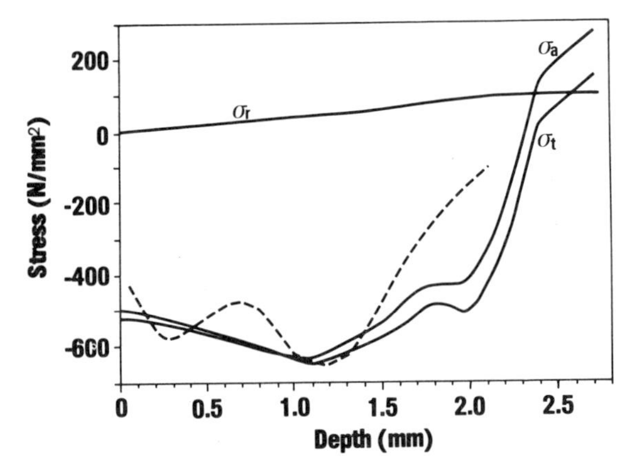

Calculated (full lines) and measured (dashed lines) residual stresses in a carburized cylinder (∅ = 30 mm) in steel quenched in oil (60 °C) from 820 °C.

FIGURE 131. Comparison of the calculated residual stress to that measured for a carburized 30 mm diameter steel cylinder which was quenched into oil from 820°C (single quench) (reprinted with permission from Ericsson, T. and B. Hildenwall. 1981. *Residual Stress and Stress Relaxation*, E. Kula and V. Weiss, editors, New York: Plenum, p. 19).

The lower residual stress in the surface layers in Figure 131(c) is due to retained austenite. This effect, predicted by the calculation, is substantiated by the experimental data shown in Figure 132. (The effect at the surface is due to decarburization, as will be discussed.)

The carbon content at the surface is an important parameter in carburizing because of its effect on the case depth. This is shown in Figure 133. These are calculated longitudinal stresses. Note that for the uncarburized steel (0.2% C) the stresses are tensile at the surface. Here the steels were quenched to 15 °C, with no sub-zero cooling, so that the effect of higher-carbon content on retained austenite is reflected in the reduction in the compressive stress at the surface being greatest for the highest-carbon content.

The case depth also is an important parameter in the residual stress distribution. Some data are shown in Figure 134. The greater the case depth, the deeper the region of compressive residual stress. However, the above effect of retained austenite must be kept in mind, because if the increasing case depth has associated with it a higher-carbon content at the surface, then more retained austenite will be present. This can be alleviated by a diffusion heat treatment. It can also be regulated by setting the carbon potential of the carburizing medium (e.g., gas composition). As shown in Figure 134, there is less of an effect of retained austenite for the case of carburizing with a carbon potential of 0.7% (left-hand figure).

If the carbon gradient has been altered by subsequent decarburization, so that the carbon content of the austenite at the surface is less than that in the underlying case, then the residual stresses may be adversely affected. The reason is shown in Figure 135. In this description, it is assumed that all martensite forms throughout the cylinder. The surface forms martensite before the underlying layer, as shown, for example, by the curve at 20 sec. Thus, at this time the surface has expanded, placing it in compression and the underlying austenite in tension. Plastic deformation of this austenite occurs, and hence so does stress relaxation. Then, when the underlying layer of austenite transforms to martensite, it places the surface in tension (or at least in less compression than would occur if the surface were not decarburized).

3i. EFFECT OF CARBURIZING ON FATIGUE PROPERTIES

In general, it is expected that carburizing would improve the fatigue strength of steel components for many loading conditions, such as reversed bending. An example is shown in Figure 136. Since for steels the fatigue strength increases with tensile strength, the stronger surface should have a higher fatigue strength than if it were not carburized. The harder martens-

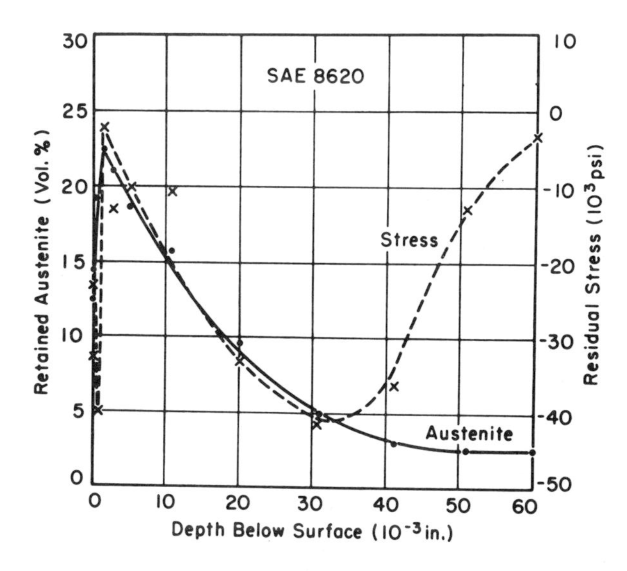

FIGURE 132. The measured residual stress and amount of retained austenite for a carburized 8620 steel as a function of depth. The sample was gas carburized (0.85–0.95% C potential) to a total case depth of about 0.050 inches, quenched into oil, then tempered at 300°F for 30 min (from Koistinen, D. P. 1958. *Trans. ASM*, 50:227).

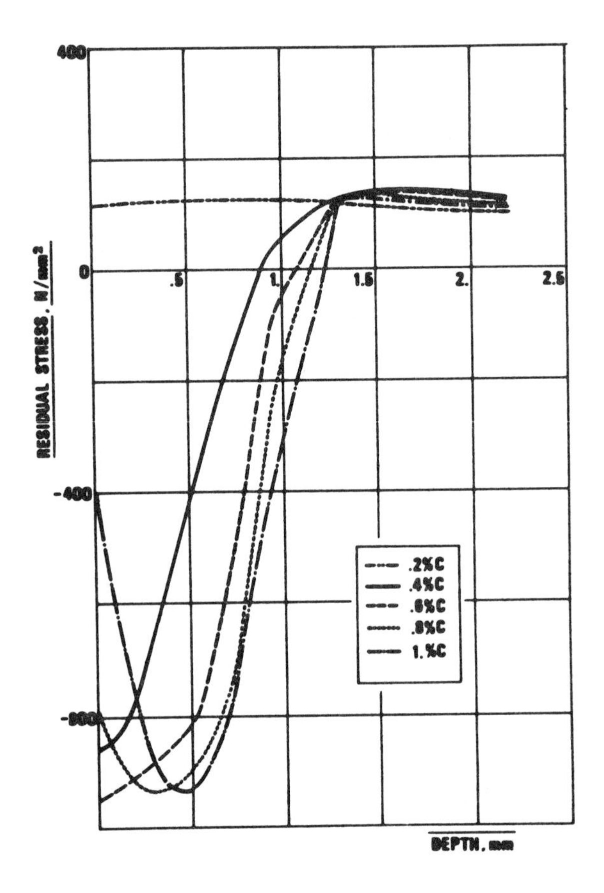

The calculated residual stresses in a carburized plate of SAE 1321 with different surface carbon contents.

FIGURE 133. The effect of surface carbon content on the longitudinal residual stresses in a carburized steel (reprinted with permission from Hildenwall, B. 1978. In *Hardenability Concepts with Applications to Steel*, D. V. Doane and J. S. Kirkaldy, editors, The Minerals, Metals and Materials Society, PA, p. 579).

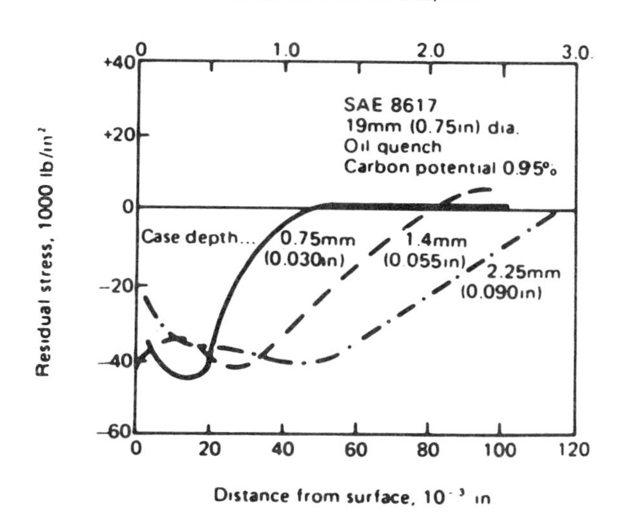

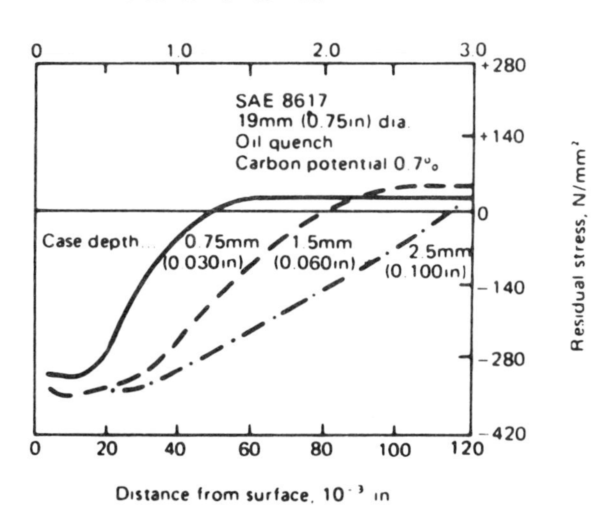

FIGURE 134. The effect of case depth on the longitudinal residual stresses in carburized steels (adapted with permission from Coleman, W. S. and M. Simpson. 1957. In *Fatigue Durability of Carburized Steel*. Metals Park, Ohio: American Society for Metals, p. 47).

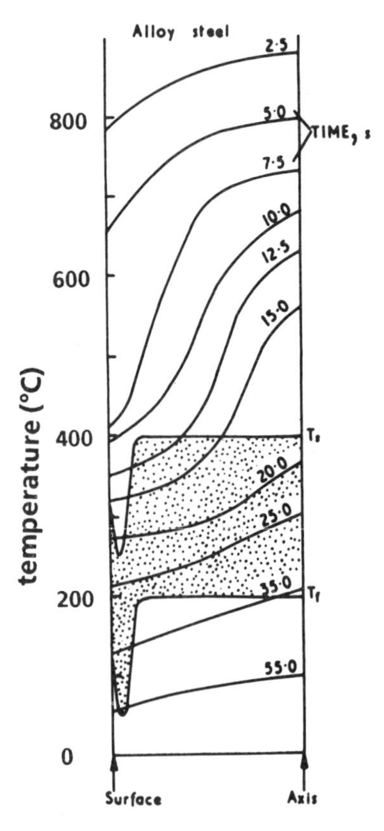

FIGURE 135. Progress of transformation during quenching a carburized, 0.5 inch diameter steel bar that has some surface decarburization (from Dawes, C. and R. J. Cooksey. 1966. *Heat Treatment of Metals, Special Report 95*. London: The Iron and Steel Institute, p. 77; by permission of The Institute of Metals).

itic surface should resist surface abrasion, reducing the number of fatigue crack initiating sites. And probably most important, the compressive residual stresses at the surface should increase the fatigue strength since fatigue cracks are initiated and grow when, during the loading cycle, the stress is tensile. However, there are synergistic effects here, so that it is difficult to do experiments that isolate specific variables.

Figure 137(a) shows that a carburized 8620 steel has better fatigue properties than a carburized 8640 and 1018 steel. All were carburized in the same atmosphere. The hardness profiles are shown in Figure 137(b). The surface hardness is the same for the three steels, but the core hardness is highest for the higher-carbon 8640 steel. The oil-quenched 8620 steel and the brine-quenched 1018 steel have the same profile, but the lower harden-

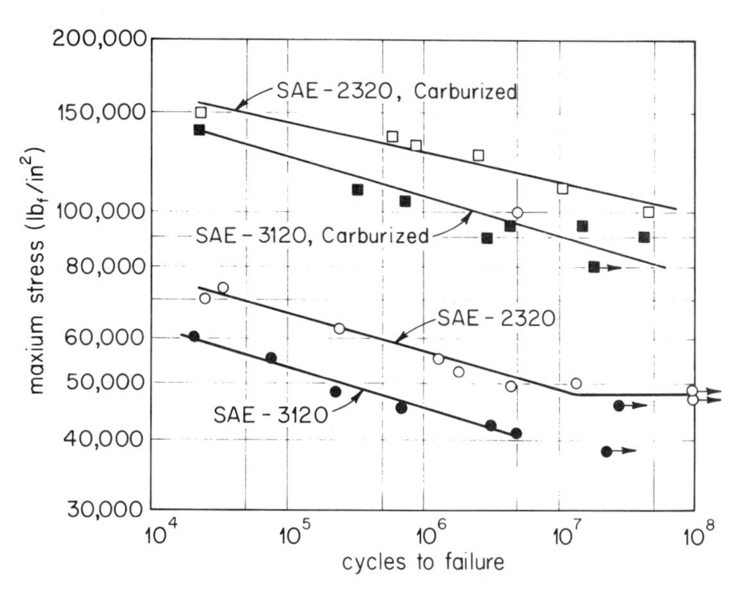

FIGURE 136. The effect of carburizing on the fatigue properties of two steels (from Moore, M. F. and N. J. Alleman. 1928. *Trans. ASST*, 13:405).

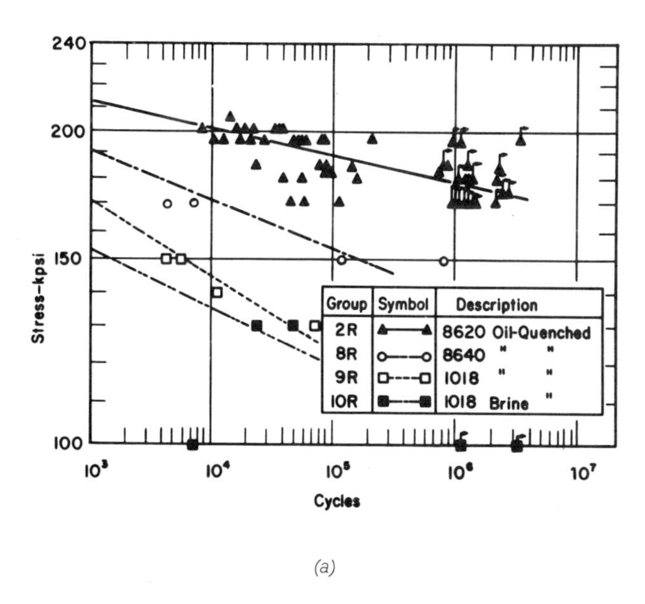

Group	Symbol	Description
2R	▲—▲	8620 Oil-Quenched
8R	○—○	8640 " "
9R	□---□	1018 " "
10R	■—■	1018 Brine "

(a)

FIGURE 137. Effect of carburizing on the fatigue properties and residual stresses of three steels. (a) Fatigue data for 1/4 inch round specimens (reprinted with permission from Roberts, J. G. and R. L. Mattson. 1957. *Fatigue Durability of Carburized Steel*. Metals Park, Ohio: American Society for Metals, p. 68).

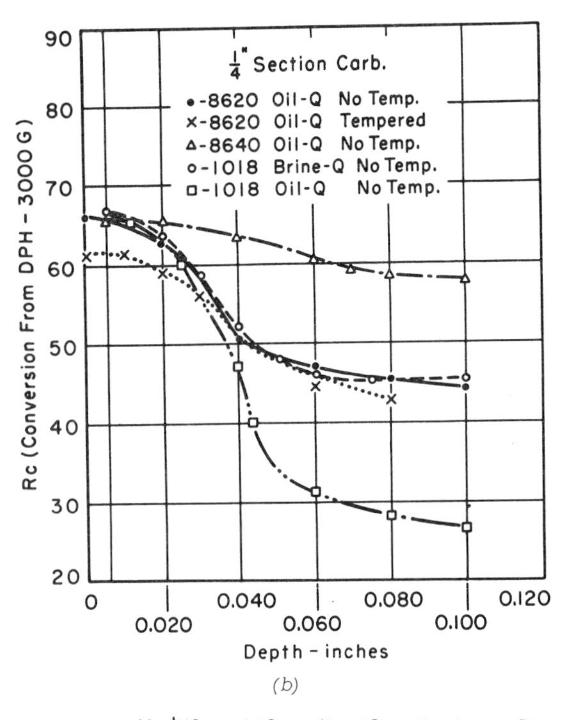

(b)

Hardness Patterns of Carburized ¼-Inch Round Specimens.

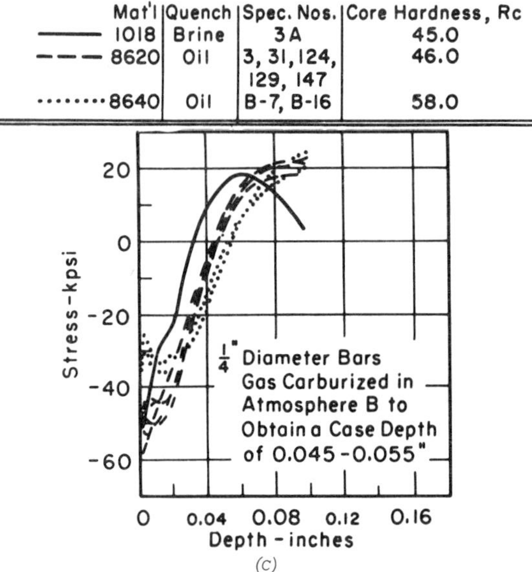

(c)

Effect of Material and Quench on Residual Stress Distributions in Carburized Bars. Carburized 1675–1700°F (920°C), to obtain case depths indicated and quenched as shown. Not tempered.

FIGURE 137 (continued). Effect of carburizing on the fatigue properties and residual stresses of three steels. (b) Hardness profiles for the samples of the fatigue data in (a). (c) Longitudinal residual stresses with depth for samples of fatigue data shown in (a) (reprinted with permission from Roberts, J. G. and R. L. Mattson. 1957. *Fatigue Durability of Carburized Steel*. Metals Park, Ohio: American Society for Metals, p. 68).

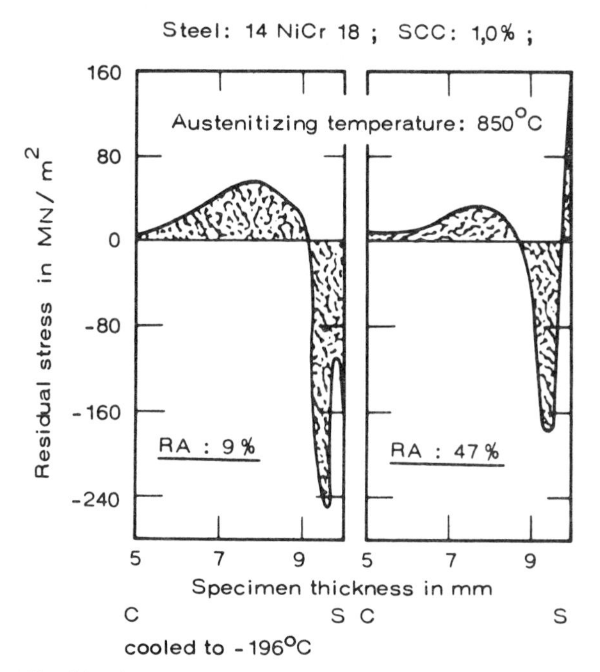

FIGURE 138. Residual stress distribution for a carburized steel in the as-quenched condition and in the sub-zero cooled condition (from Brandis, H. and W. Schmidt. 1984. In *Case-Hardened Steels: Microstructural and Residual Stress Effects*, D. E. Diesburg, editor, Warrendale, PA: The Metallurgical Society of AIME, p. 189).

ability of the 1018 steel produces a lower core hardness when oil-quenched. The carburized 8620 steel has a higher compressive residual stress at the surface than the carburized 8640 steel [Figure 137(c)]. Although the carburized 1018 steel has a compressive residual stress at the surface that is about the same as that of the 8620 steel, it is less compressive with depth.

The effect of retained austenite on the fatigue strength has long been a subject of examination. As discussed in the section on residual stresses in carburized steels, retained austenite does lower the compressive residual stress at the surface. An example illustrating this is shown in Figure 138. In this case, the surface had 47% retained austenite and a tensile residual stress; this was changed to compressive stress by a sub-zero treatment that reduced the amount of retained austenite at the surface to only 7%. The corresponding carbon gradient and the hardness profiles are shown in Figure 139.

Fatigue curves for this steel with various amounts of retained austenite

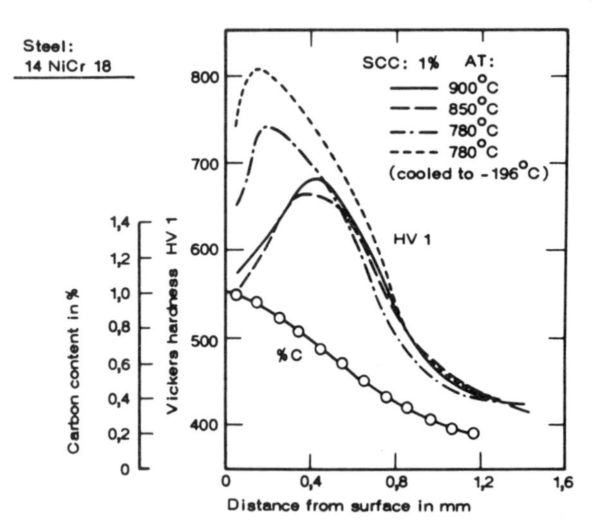

FIGURE 139. Carbon gradient and hardness profile for a carburized steel (from Brandis, H. and W. Schmidt. 1984. In *Case-Hardened Steels: Microstructural and Residual Stress Effects*, D. E. Diesburg, editor, Warrendale, PA: The Metallurgical Society of AIME, p. 189).

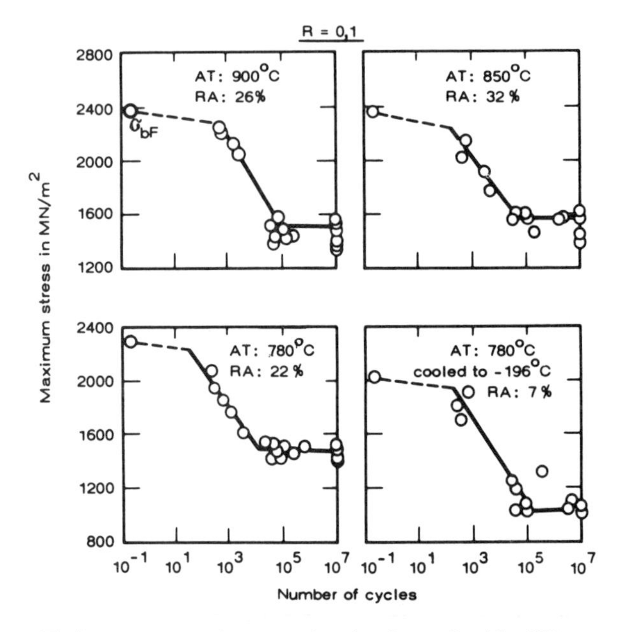

FIGURE 140. Fatigue curves for a carburized steel with different amounts of retained austenite at the surface (from Brandis, H. and W. Schmidt. 1984. In *Case-Hardened Steels: Microstructural and Residual Stress Effects*, D. E. Diesburg, editor, Warrendale, PA: The Metallurgical Society of AIME, p. 189).

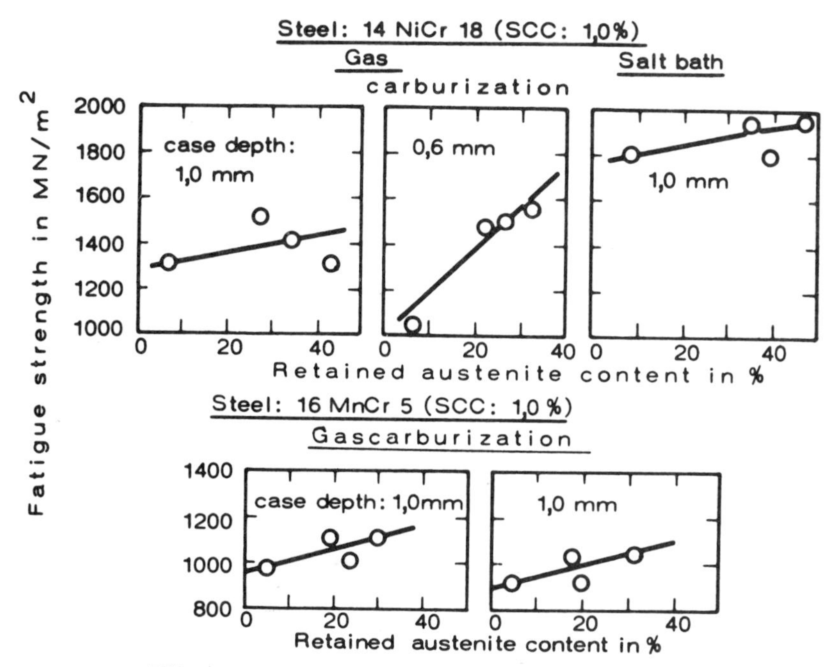

FIGURE 141. Effect on the fatigue strength of the amount of retained austenite at the surface of carburized steels (from Brandis, H. and W. Schmidt. 1984. In *Case-Hardened Steels: Microstructural and Residual Stress Effects*, D. E. Diesburg, editor, Warrendale, PA: The Metallurgical Society of AIME, p. 189).

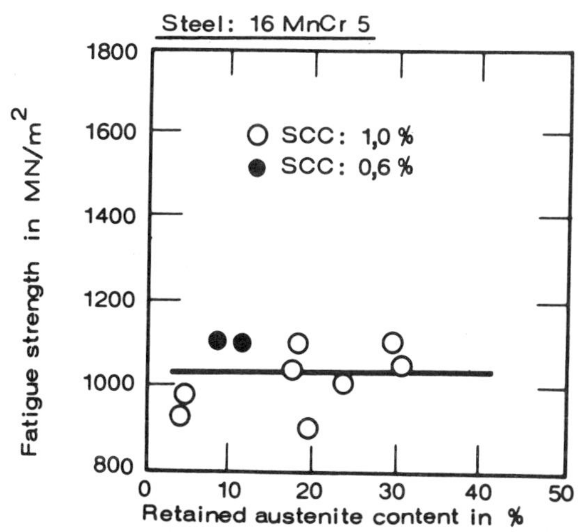

FIGURE 142. The effect on the fatigue strength of the amount of retained austenite at the surface of a carburized steel. The tests were in bending (R = 0.1) (from Brandis, H. and W. Schmidt. 1984. In *Case-Hardened Steels: Microstructural and Residual Stress Effects*, D. E. Diesburg, editor, Warrendale, PA: The Metallurgical Society of AIME, p. 189).

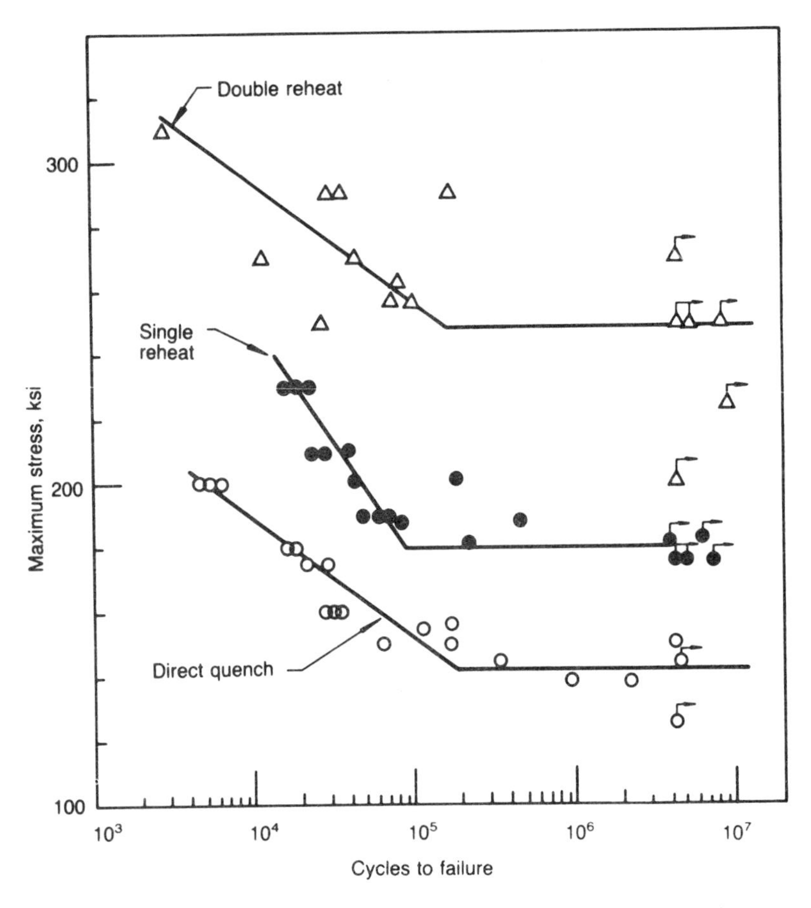

Maximum applied stress versus cycles to failure (S-N curve) for four-point bend fatigue specimens of 8620 steel given three different carburizing treatments.

FIGURE 143. Fatigue curves for a carburized 8620 steel for direct quenching and for single and double quenching (from Apple, C. A. and G. Krauss. 1973. *Met. Trans.*, 4:1195).

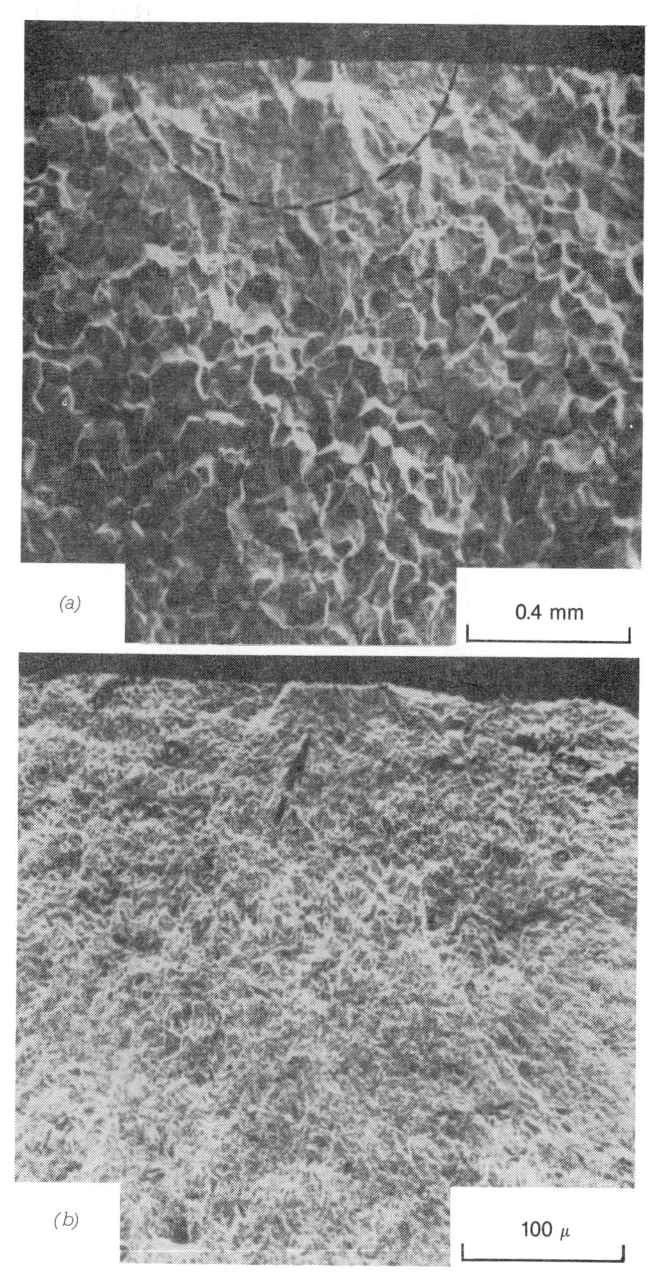

Fatigue crack initiation in carburized coarse-grained 8620 steel (a) quenched directly from carburizing at 927 °C (1700 °F) and (b) reheated after carburizing to 788 °C (1450 °F). Both specimens tempered at 145°C (300 °F). Scanning electron micrographs.

FIGURE 144. Fractographs of the fracture surface of fatigue samples of a carburized 8620 steel, for direct quenching and for single quenching (from Apple, C. A. and G. Krauss. 1973. *Met. Trans.*, 4:1195).

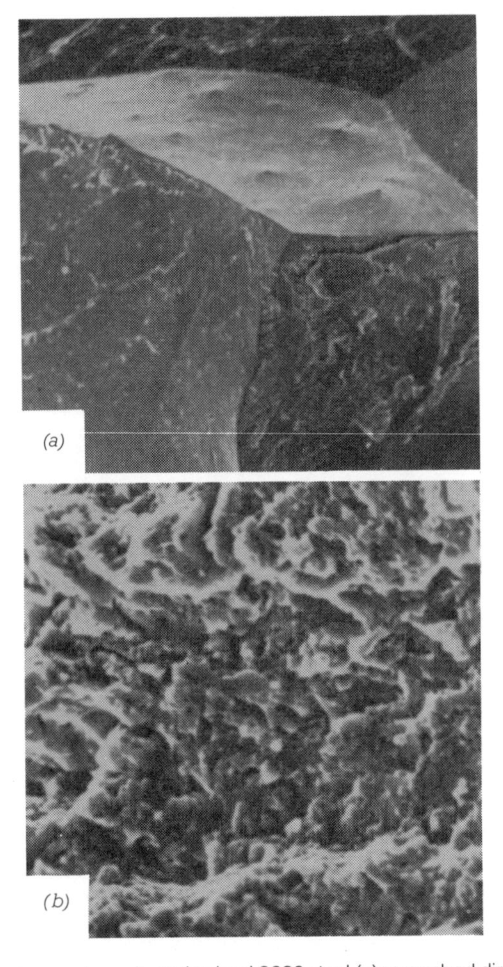

Overload case fracture surfaces in carburized 8620 steel (a) quenched directly after carburizing at 927 °C (1700 °F) and (b) reheated to 708 °C (1450 °F). Both specimens tempered at 145 °C (300 °F). Scanning electron micrographs.

FIGURE 145. Fractographs of the overload region of fatigue samples of an 8620 carburized steel, for direct quenching and for single quenching. These are higher magnifications than those in Figure 144 (from Krauss, G. 1978. *Met. Trans.*, 9A:1527).

are shown in Figure 140. Note that the fatigue strength is less for the lowest amount of retained austenite at the surface. The effect of retained austenite at the surface on the fatigue strength of several carburized steels is shown in Figure 141, where it is seen that the fatigue strength increases with increasing amounts of retained austenite. However, not all steels behave like this, as is shown by the data in Figure 142. The improvement of fatigue properties with the presence of retained austenite in the surface of carburized steels is reported to be due to stress-induced martensite formation from the austenite.

Single or double quenching can increase the fatigue strength, as is illustrated by the data in Figure 143. The multiple quench produced a finer structure, with fewer microcracks. Also, the prior austenite grain size was reduced. That this may be important in fracture that eventually follows fatigue crack growth is shown by the fractographs in Figures 144 and 145.

CHAPTER 4 **Nitriding**

4a. INTRODUCTION

Nitriding is an important method of altering the chemistry of the surfaces of steels. Nitrogen is a relatively small atom, and dissolves interstitially in iron. The solubility is higher in austenite than in ferrite, and the nitrogen forms nitrides with iron. The iron–nitrogen phase diagram is shown in Figure 146. As in heat-treating steels, quenching from the austenite range will produce a hard martensite. Nitrogen can be added to the surface of a steel in the austenite range by placing it in contact with an appropriate nitrogen-containing environment in the austenite range. Although this is a surface-hardening method, experience has shown that the properties developed cannot compete with those of carburized steels. (There is a treatment, called carbonitriding that adds both carbon and nitrogen to the surface of an austenitic steel, which in some cases is advantageous.)

The nitriding process that is the subject of this section involves the addition of nitrogen to the steel *below* the eutectoid temperature. Thus, the hardening does not involve the formation of hard martensite. Instead, it is associated with the formation of nitrides, which impart high hardness to the surface and high wear resistance and improved fatigue properties to the steel.

4b. METHODS OF NITRIDING

Pure N_2 will not dissociate, even if the steel is in the austenite range, which is why nitrogen can be used as an inert gas for heat-treating. To provide nascent nitrogen to the surface, one of the most common methods is to use a mixture of gaseous nitrogen, hydrogen and ammonia, and rely on the chemical reaction

$$2NH_3 = 2N_2 + 3H_2$$

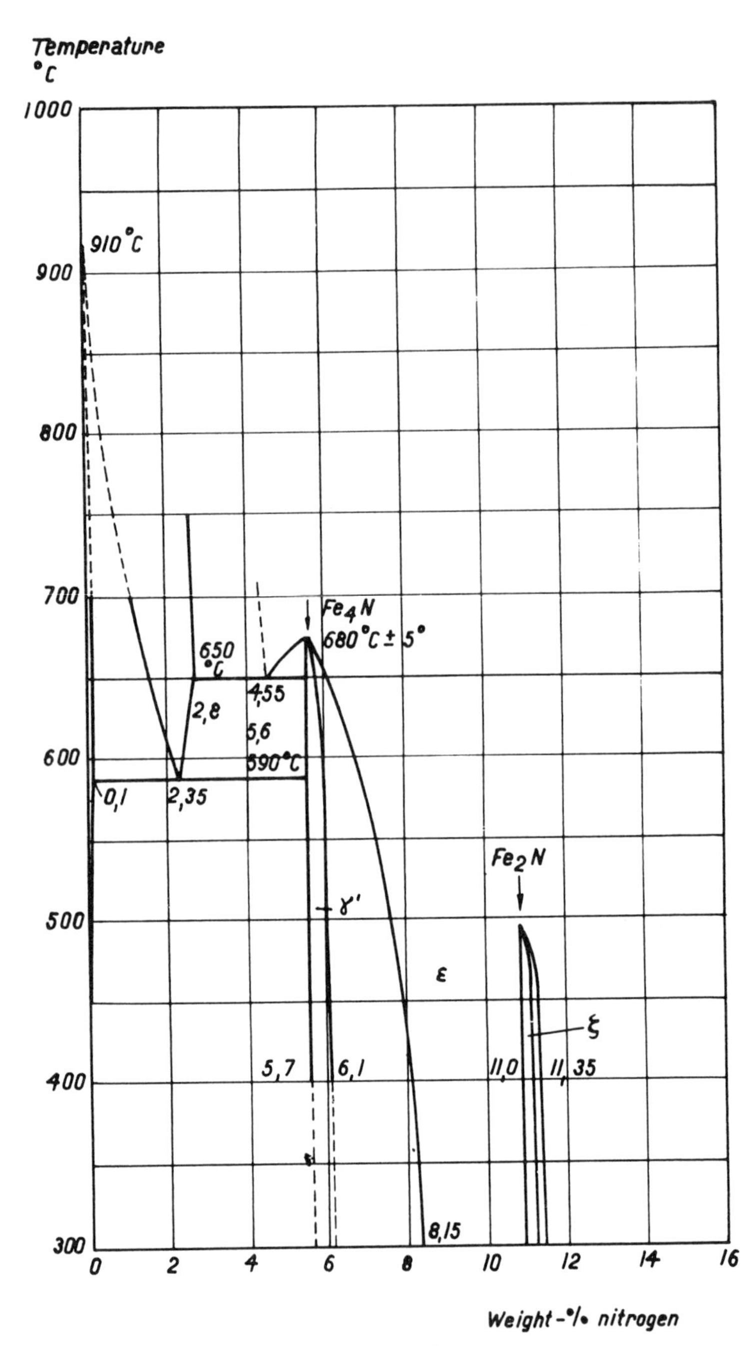

FIGURE 146. The iron–nitrogen phase diagram (adapted from Hansen, M. 1958. *Constitution of Binary Alloys, 2nd Edition*. New York: McGraw-Hill.)

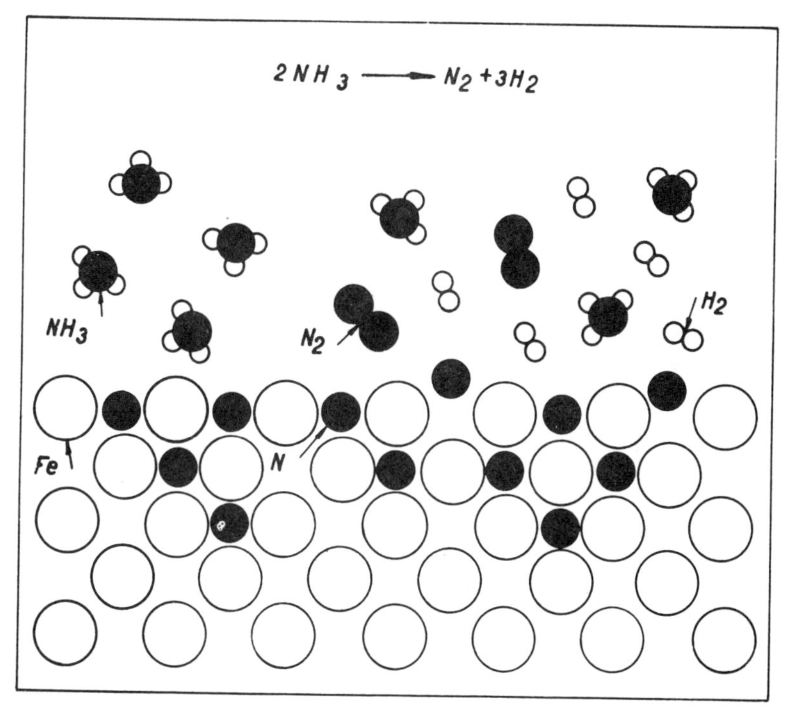

FIGURE 147. Schematic illustration of the nitriding of a steel by use of ammonia (reprinted with permission from Thelning, K.-E. 1975. *Steel and Its Heat Treatment*. Boston: Butterworths).

The nitriding mechanism is illustrated schematically in Figure 147. The thermodynamics of this reaction can be treated in a fashion similar to that developed for gas-carburizing.

A method that is becoming increasingly popular is ion-nitriding, also called plasma-nitriding and glow-discharge nitriding. In this method the steel part is made a few thousand volts negative. In this potential field, nitrogen gas at relatively low pressure (e.g., 1 to 10 torr) is ionized, forming a plasma gas. The positive nitrogen ions are attracted to the negatively charged part, where they are deposited on the surface. The bombardment is sufficiently intense that the steel surface heats; thus, an independent method of heating the steel may not be needed. The excitation of the gas produces visible radiation, so that the part is surrounded by a glow. An important advantage of this method over gas-nitriding is that deposition of nitrogen is rather uniform over surfaces of complex geometry, whereas in gas-nitriding the deposition can depend upon the gas flow pattern. The advantages and disadvantages of ion-nitriding are listed in Table 2.

TABLE 2.
Some Advantages of Ion-Nitriding.

Advantage
Absence of pollution
Relatively uniform nitriding of surfaces of complex shapes
Selective nitriding by simple masking
Reduced nitriding time
Cleaning of surfaces

Nitriding can also be done in liquid salts. There are many variations on this process, and descriptions can be found in H. E. Boyer, editor, 1987. *Case Hardening of Steel*. Metals Park, Ohio: ASM International.

4c. CASE DEPTH

The formation of the case in nitriding is a diffusion process, controlled by the same factors as are discussed in carburizing. The diffusion process involves the movement of nitrogen into the *ferrite* lattice but also involves the reaction of the nitrogen with iron and alloying elements present (e.g., aluminum) to form nitrides. Thus, diffusion modeling to predict the case depth is more complicated than for the case of carburizing.

Figure 148 shows examples of nitrogen profiles. Note that the nitriding temperatures used were between 500 and 600°C (*below* the eutectoid temperature) and that the times required to attain a case depth comparable to that used in carburizing are considerably longer than those used in carburizing. (Also note that after long nitriding times the surface hardness may begin to decrease; see the next section.)

4d. MICROSTRUCTURES

The solubility of nitrogen in ferrite is low (see Figure 146), so that as nitrogen diffuses into the steel, iron nitrides (or other nitrides) will form. These may be sufficiently large to be quite resolvable, as shown in Figure 149. The rate of diffusion of nitrogen into the steel is sufficiently low that a layer of nitrides builds up on the surface of the steel, forming a clearly visible "white layer," such as the one shown in Figure 149. (This layer is sometimes referred to as the "compound layer.") The layer here is about 0.02 mm thick, which is much less than the case depth (see Figure 148). Thus, the nitrogen is not confined to the white layer. The white layer will continue to thicken with nitriding time, as shown in Figure 150. The thickness also depends on the nitriding temperature, as shown in Figure 151.

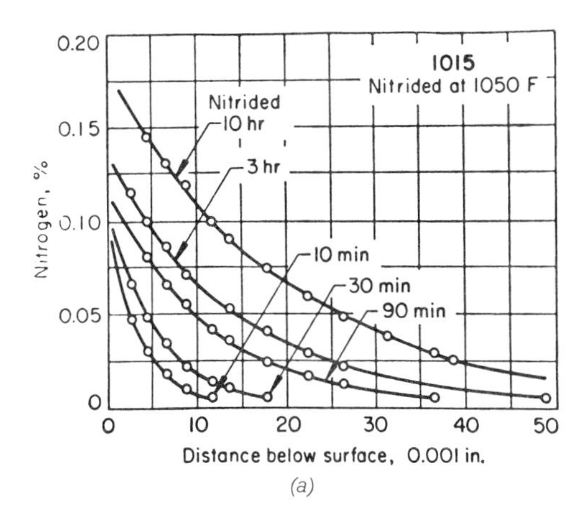

(a)

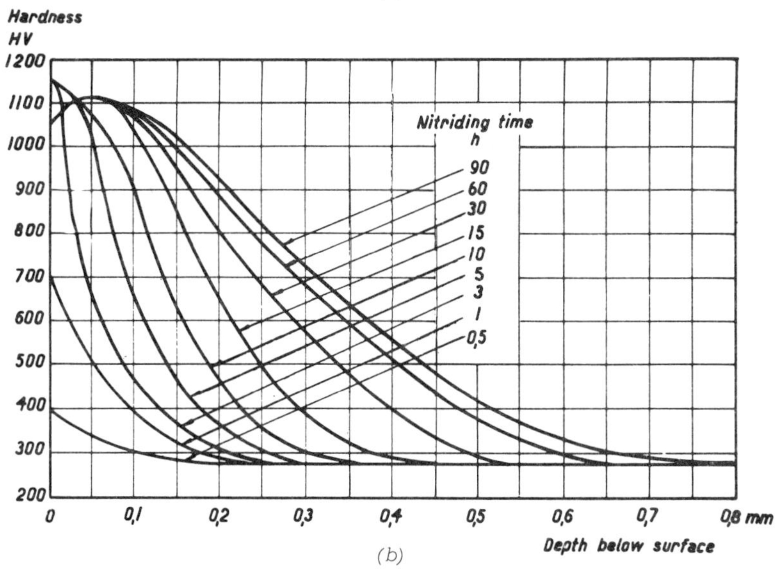

(b)

Hardness profile through surface layer of BS 905M39 (Bofors ARO 75) after gas nitriding for 0.5–90 h at 510°C.

FIGURE 148. (a) Nitrogen concentration profiles in a 1015 steel nitrided at 566°C, using the aerated bath process. (b) Hardness profiles in a B S 905M39 (Bofors A R O 75) steel gas-nitrided at 510°C [(a) adapted with permission from *Metals Handbook, Vol. 2*. 1964. Metals Park, Ohio: American Society for Metals; (b) adapted with permission from Thelning, K.-E. 1975. *Steel and Its Heat Treatment*. Boston: Butterworths].

FIGURE 149. The microstructure of a nitrided steel, showing the white layer (at top of micrograph) and diffusion layer with precipitated acicular iron nitrides (black thin lines). The steel contained 0.15% C and was gas-nitrided for 10 hours at 500°C. 500× (reprinted with permission from Thelning, K.-E. 1975. *Steel and Its Heat Treatment*. Boston: Butterworths).

Figure 152 shows another microstructure of a nitrided steel. The etching reveals generally the region in which nitrides are formed, although they are not resolved here. Beyond this zone there is a region that contains no nitride precipitates, but in which nitrogen is in solid solution in the ferrite. The compound layer is clearly visible on the surface.

The degree of formation of the white layer can be controlled by controlling gas composition. This is illustrated by the microstructures in Figure 153.

The type of nitrides that may be present and the structure of the white layer in nitrided steels is complicated. The phase diagram (Figure 146) shows that three nitrides may form if the nitrogen content is high enough. Since carbon is present in the steels, these nitrides may contain appreciable amounts of carbon, and thus may best be called carbonitrides (or nitrocarbides). The effect of the carbon/nitrogen ratio of the nitrides on the hardness is shown in Figure 154. Note that these nitrides are not especially hard. The "white layer" consists of mixtures of these two nitrides, and this implies that this layer will be soft. However, in alloy steels, this

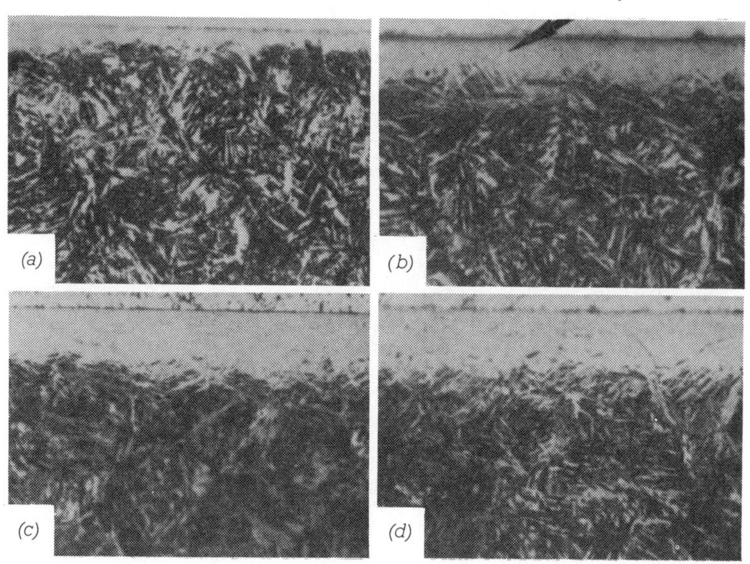

FIGURE 150. Microstructures of a nitrided steel showing the increase in thickness of the white layer (top of micrographs) with nitriding time. The steel was En 41 B (Bofors A R O 75) and was nitrided at 500°C in a 30% dissociated ammonia gas for (a) 10, (b) 30, (c) 60 and (d) 90 hours. 200× (reprinted with permission from Thelning, K.-E. 1975. *Steel and Its Heat Treatment*. Boston: Butterworths).

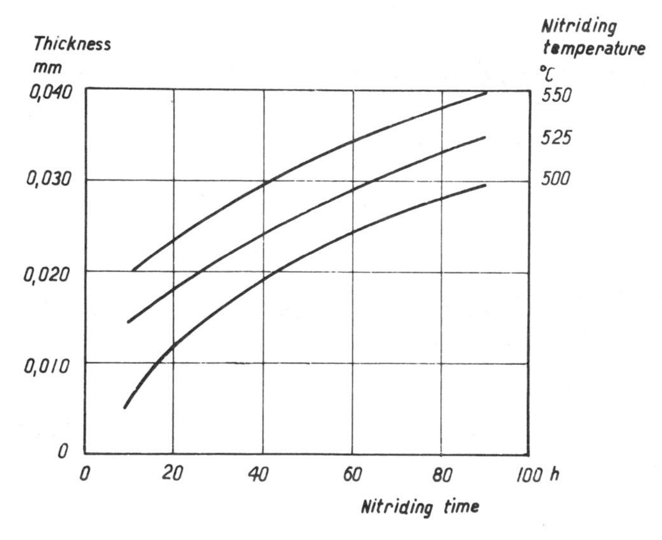

FIGURE 151. The effect of nitriding temperature and time on the thickness of the white layer. The steel was B S 905M39, En 41 B (Bofors A R O 75) (reprinted with permission from Thelning, K.-E. 1975. *Steel and Its Heat Treatment*. Boston: Butterworths).

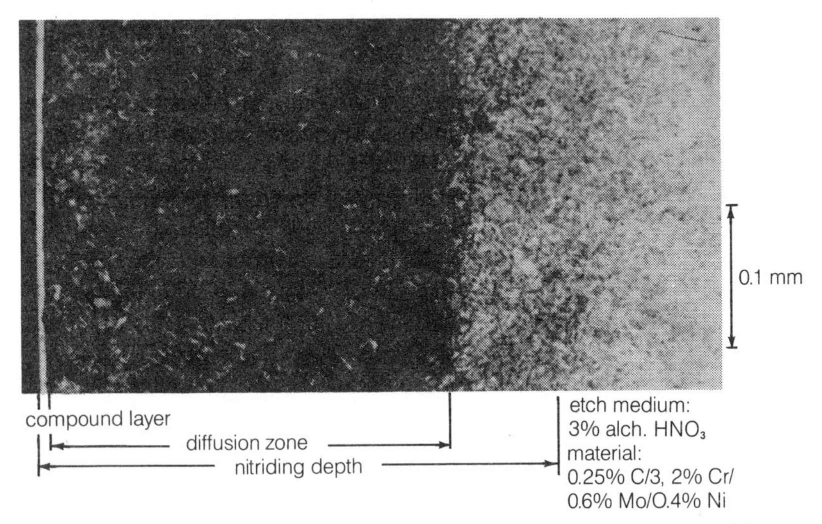

FIGURE 152. Microstructure of an ion-nitrided steel (reprinted with permission from *Metals Handbook, 9th Edition, Vol. 4, Heat Treating*. 1981. Metals Park, Ohio: American Society for Metals).

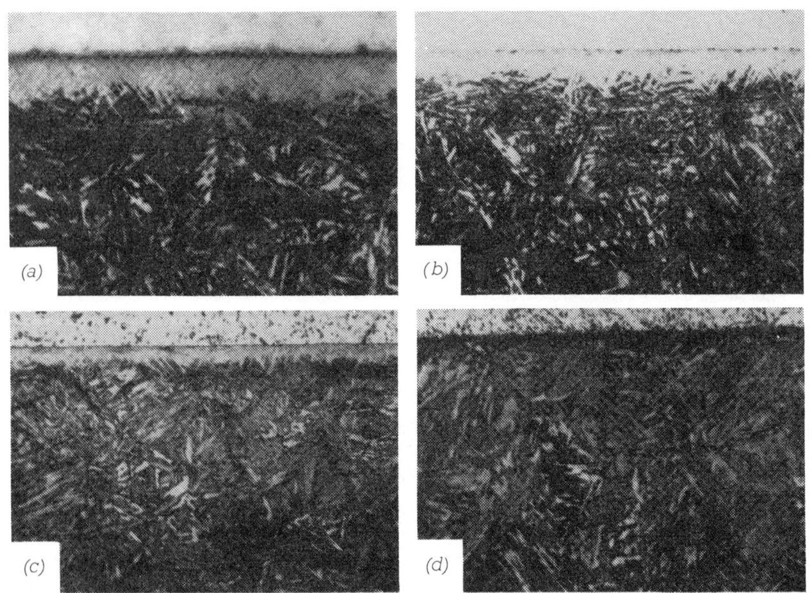

Thickness of white layer resulting from different methods of nitriding B S 905M39, En 41 B (Bofors A R O 75): (a) 30 h at 500°C and 20% degree of dissociation; (b) 5 h at 500°C and 20% degree of dissociation then 25 h at 525°C and 85% degree of dissociation; (c) 5 h at 500°C and 20% degree of dissociation then 25 h at 500°C and 55% degree of dissociation; (d) 5 h at 500°C and 20% degree of dissociation then 25 h at 550°C and 90% degree of dissociation.

FIGURE 153. The effect of gas-nitriding variables on the thickness of the white layer (reprinted with permission from Thelning, K.-E. 1975. *Steel and Its Heat Treatment*. Boston: Butterworths).

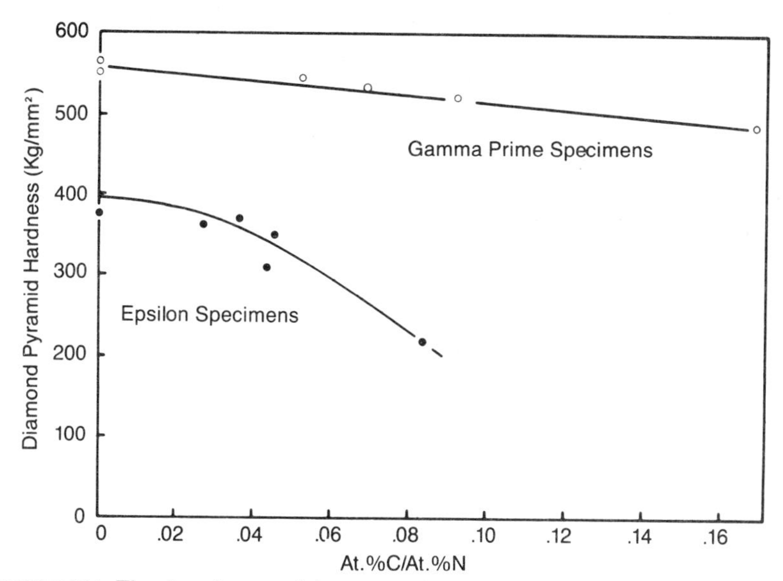

FIGURE 154. The hardness of the γ' and ϵ nitrides with different carbon contents (reprinted with permission from Levy, S. A., J. F. Libsch, and J. D. Wood. 1969. *Trans. AIME*, 245:753, The Minerals, Metals and Materials Society, Warrendale, PA).

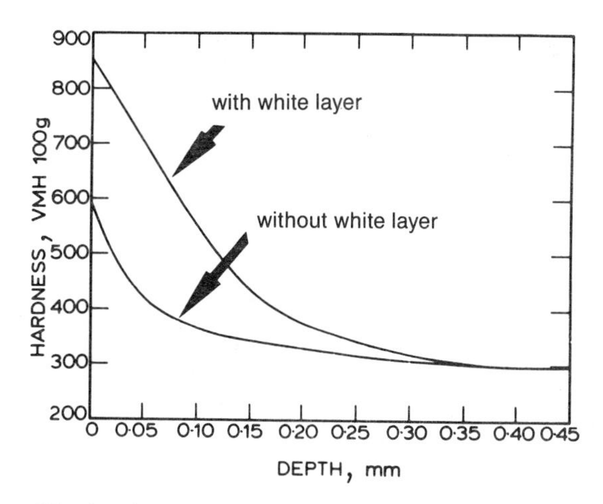

Microhardness profiles on the same sample of 722 M24 steel nitrided without "white layer" in H_2—13% NH_3 atmosphere for 24 h at 500 °C.

FIGURE 155. Hardness profiles for a nitrided steel with and without a white layer (from Clayton, D. B. and K. Sachs. 1977. In *Heat Treatment '76*. London: The Institute of Metals, p. 1; by permission of The Institute of Metals).

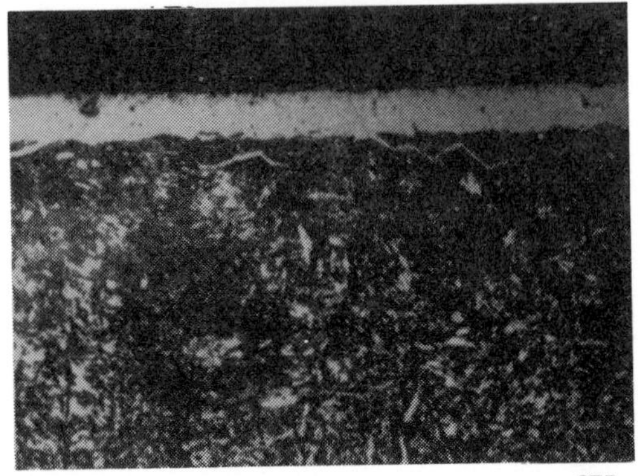

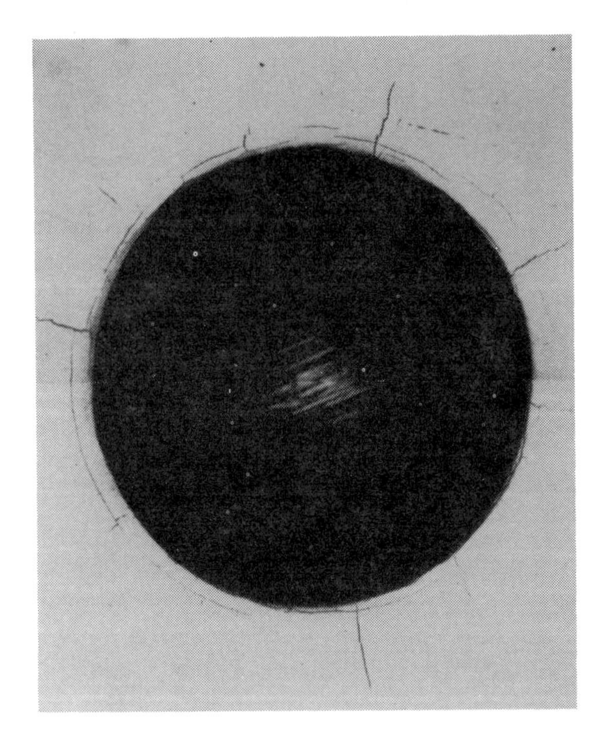

FIGURE 156. The formation of cracks around a hardness indentation made on the surface of a nitrided steel which had a white layer (reprinted with permission from Thelning, K.-E. 1975. *Steel and Its Heat Treatment*. Boston: Butterworths).

may not be the case, as is shown by the data in Figure 155. These hardness profiles were obtained under conditions where a white layer formed and where it did not form. Note that the presence of the white layer gave the higher hardness. However, this was an aluminum-containing steel, and Al-nitrides formed instead of the iron-nitrides. In Figure 148(b) the reduction in hardness at the surface for the longer nitriding times is due to the formation of the softer white layer.

The white layer is also brittle, which is illustrated by Figure 156. The depth of the white layer is shown by the cross-sectional microstructure. Upon making a low-load hardness impression on the nitrided surface, cracks are produced from around the indentation. In many applications, if the white layer forms, it must be removed before installation of the nitrided part.

If the nitrides that form are inherently soft, then the hardening associated with nitriding and known nitride formation must be caused by a fine dispersion of the nitrides in the ferrite. This dispersion could be formed during nitriding, or upon cooling from the nitriding temperature. The phase diagram in Figure 146 shows that nitrogen has a low solubility in iron, and thus to obtain significant hardening by the formation of nitrides in ferrite would require a very fine dispersion. However, it is important

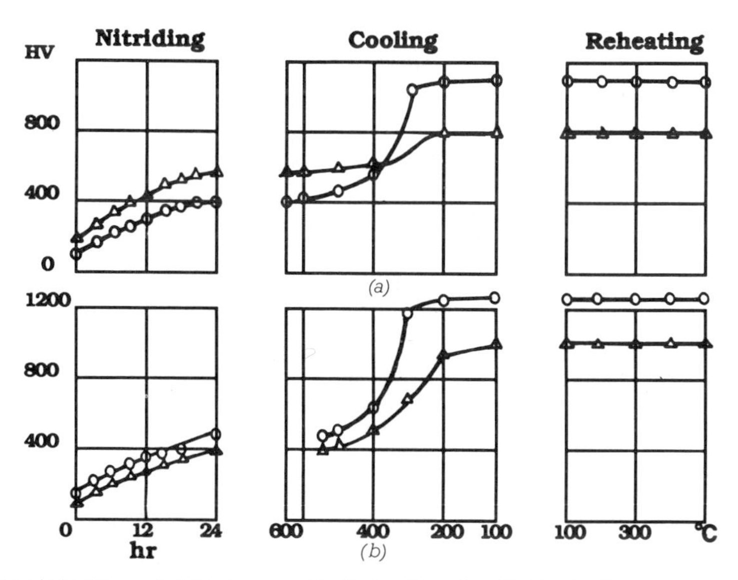

Change in hardness of nitrided case on Fe-Al alloy (○—○—○), Fe-Mo alloy (△—△—△).
a) Nitrided at 620°C for 24 hrs; b) 540°C for 24 hrs.

FIGURE 157. The surface hardness of two steels during nitriding, subsequent cooling, and the subsequent reheating (adapted from Yakhina, V. D. and V. V. Nikitin. 1975. *Metal Sci. and Heat Treatment,* 17:125).

to keep in mind that the commercial nitriding steels contains elements such as Al that are strong nitride formers. Thus it is the formation of these types of nitrides that is important.

A study by Yakhnia and Nikitin showed that the hardness of nitrided steels develops upon cooling from the nitriding temperature and not during nitriding. They used a hot hardness tester so that they could measure the hardness during nitriding, during cooling following nitriding, and during subsequent reheating of the nitrided steels. They used an Fe–8% Al and an Fe–8% Mo alloy. Their results are shown in Figure 157. There is an increase in hardness during nitriding due to the formation of some nitrides. However, it is clear that the steels developed the high hardness only during cooling. Also note that the hardness remained constant at the high values upon reheating, showing the resistance of the nitrides to coarsening. The applicability of these data to commercial nitriding steels of lower Al and Mo contents is not so direct. In these high Al- and Mo-containing alloys, some austenite that was present at the nitriding temperature transformed to ferrite and nitrides upon cooling, and this complicated the interpretation of the data. However, these authors do report that an effect similar to the one found for these two alloys was observed for a commercial nitriding steel.

4e. NITRIDING STEELS

Steels recommended for nitriding have elements present that are strong nitride-formers and that have been found to develop the required high hardness. Figure 158 shows the potent effect of Al and Ti on the hardness of a steel after nitriding. Some hardness profiles showing the effect of Al and other elements on the nitriding response are shown in Figure 159. The influence of Al is very apparent in these data. The chemical composition of some nitriding steels is listed in Figure 160. Also see Appendix 13.

4f. HEAT TREATMENTS OF NITRIDING STEELS

There are two considerations in the heat treatment of nitriding steels. One is the choice of heat treatment to be given to the steel *prior* to nitriding (see Appendix 13), and the other is the choice of nitriding temperature and time.

The required nitriding time is based on the desired case depth and data such as is shown in Figure 148. The nitriding temperature affects the nitriding time; the lower the temperature, the longer the time, as shown in Figure 161. The lower the temperature, the higher the hardness, but, of course, the longer the time required.

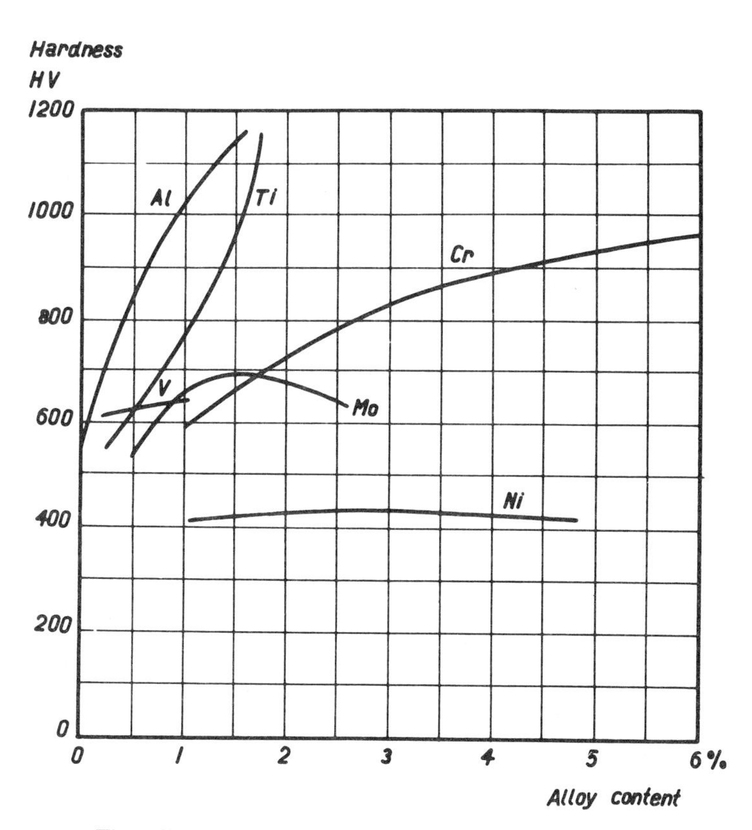

FIGURE 158. The effect of alloying element on the hardness after nitriding of a steel. The base alloy was 0.35% C, 0.30% Si and 0.70% Mo (reprinted with permission from Thelning, K.-E. 1975. *Steel and Its Heat Treatment*. Boston: Butterworths).

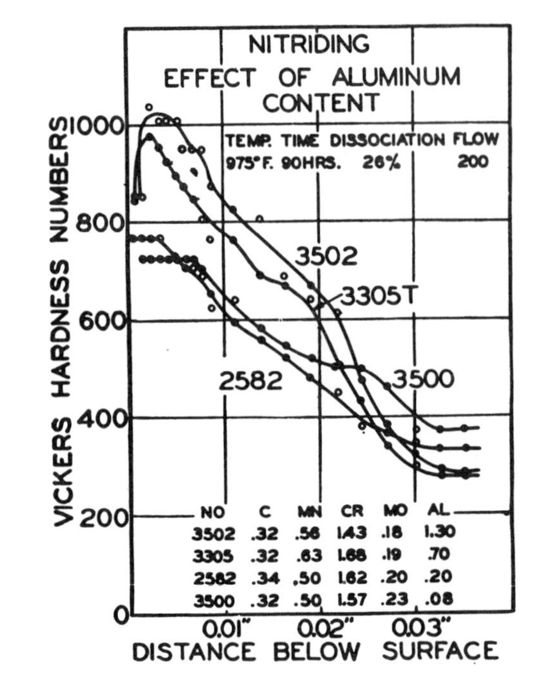

Effect of Aluminum Content on Hardness and Penetration.

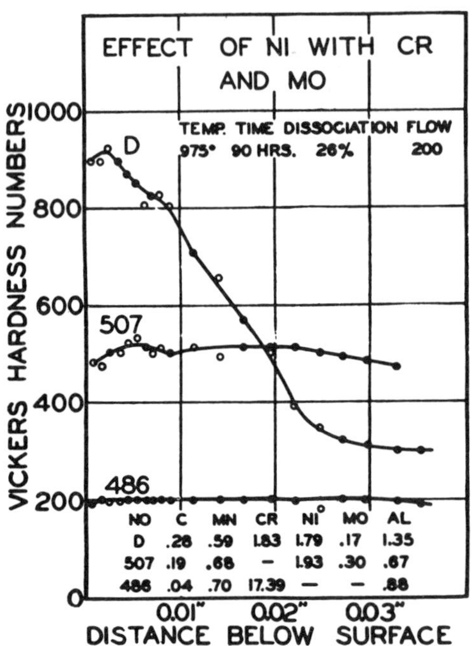

Effect of Nickel with Chromium and with Molybdenum.

FIGURE 159. Hardness profiles for nitrided steels, showing the effect of aluminum (reprinted with permission from Sergeson, R. 1929. *Nitriding Symposium of American Society for Steel Treating*. Reproduced in *Source Book on Nitriding*. 1977. Metals Park, Ohio: American Society for Metals).

Nitriding steels in accordance with ISO/R 683/X-1970

Type of steel	C %	Si %	Mn %	P % max	S % max	Al %	Cr %	Mo %	Ni % max	V %
1	0·28–0·35	0·15–0·40	0·40–0·70	0·030	0·035	—	2·80–3·30	0·30–0·50	0·30	—
2	0·35–0·42	0·15–0·40	0·40–0·70	0·030	0·035	—	3·00–3·50	0·80–1·10	—	0·15–0·25
3	0·30–0·37	0·20–0·50	0·50–0·80	0·030	0·035	0·80–1·20	1·00–1·30	0·15–0·25	—	—
4	0·38–0·45	0·20–0·50	0·50–0·80	0·030	0·035	0·80–1·20	1·50–1·80	0·25–0·40	—	—

(a)

Nominal composition and preliminary heat treating cycles for aluminum-containing low-alloy steels commonly gas nitrided.

Steel			Composition, %								Austenitizing temperature(a)		Tempering temperature(a)	
SAE	AMS	Nitralloy	C	Mn	Si	Cr	Ni	Mo	Al	Se	°C	°F	°C	°F
...	...	G	0.35	0.55	0.30	1.2	...	0.20	1.0	...	955	1750	565 to 705	1050 to 1300
7140	6470	135M	0.42	0.55	0.30	1.6	...	0.38	1.0	...	955	1750	565 to 705	1050 to 1300
...	6475	N	0.24	0.55	0.30	1.15	3.5	0.25	1.0	...	900	1650	650 to 675	1200 to 1250
...	...	EZ	0.35	0.80	0.30	1.25	...	0.20	1.0	0.20	955	1750	565 to 705	1050 to 1300

(a) Sections up to 50 mm (2 in.) in diameter, quenched in oil; larger sections may be water quenched.

(b)

FIGURE 160. The chemical composition of some commercial nitriding steels. (a) Reprinted with permission from Thelning, K.-E. 1975. *Steel and Its Heat Treatment*. Boston: Butterworths. (b) From *Metals Handbook, 9th Edition, Vol. 4, Heat Treating*. 1981. Metals Park, Ohio: American Society for Metals.

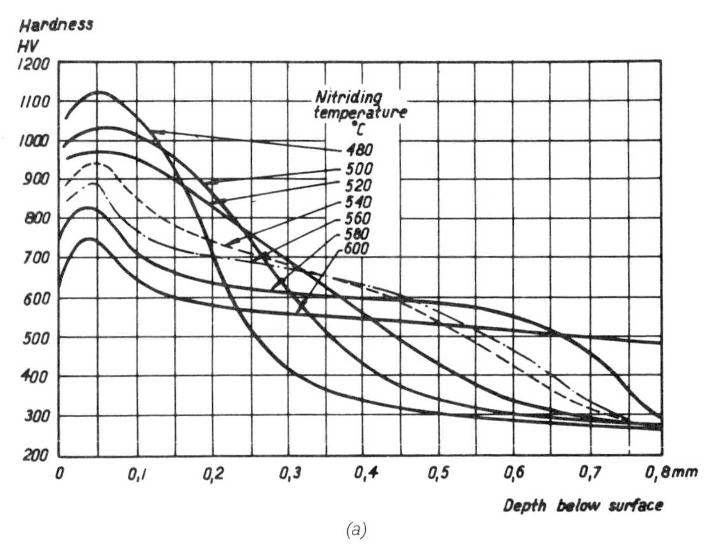

(a)

Influence of nitriding temperature on hardness and depth of nitriding of BS 905M39, En 41 B (Bofors ARO 75). Nitriding period 60 h.

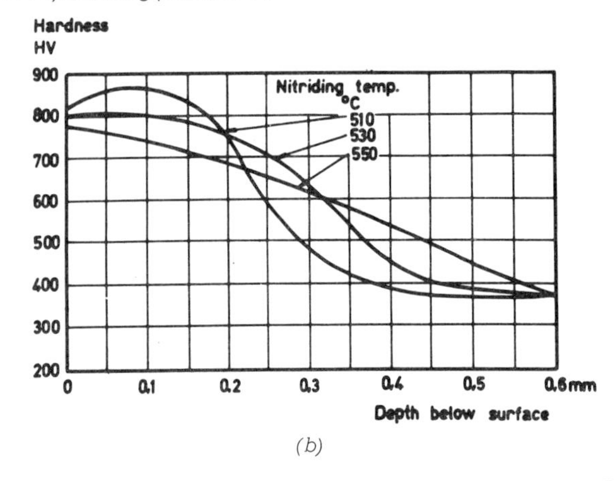

(b)

Effect of nitriding temperature on hardness and depth of nitriding of En 29 B (Bofors RO 7155). Nitriding period 60 h.

FIGURE 161. The effect of nitriding temperature on hardness profiles (reprinted with permission from Thelning, K.-E. 1975. *Steel and Its Heat Treatment*. Boston: Butterworths).

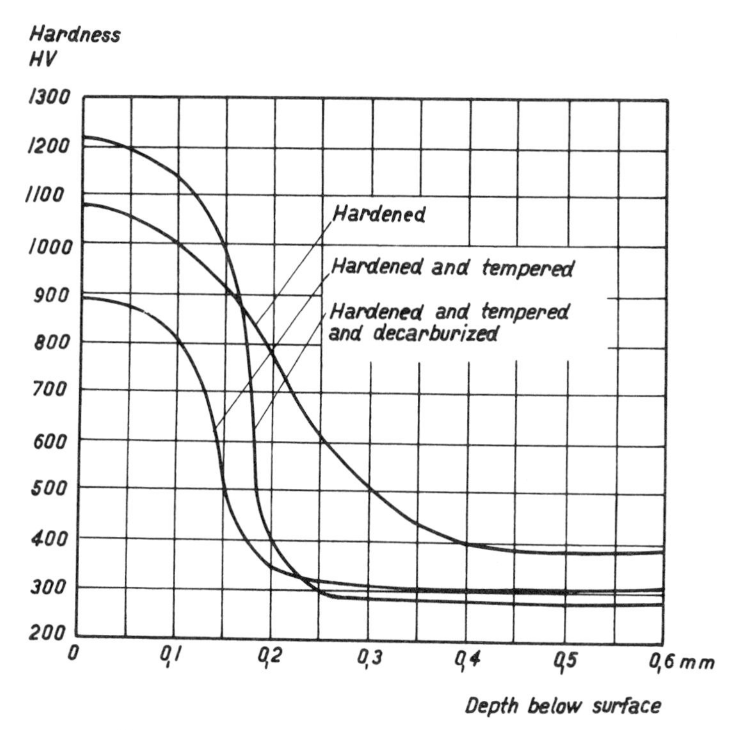

FIGURE 162. Effect of microstructure prior to nitriding on the hardness profile after nitriding. The steel was En 29 B (Bofors R O 7155) and was glow discharge nitrided for 8 hr at 520°C (adapted from Noren, T. M. and L. Kindbom. 1958. *Stahl und Eisen*, 78:1881; reprinted with permission from Thelning, K.-E. 1975. *Steel and Its Heat Treatment*. Boston: Butterworths).

The microstructure of the steel prior to nitriding does influence the nitriding process. Structures containing more ferrite will nitride more rapidly. This effect is illustrated by the data in Figure 162. Note that the steel when hardened (induction-hardened) prior to nitriding has a hardness greater than that produced by hardening then tempering prior to nitriding. If the steel is hardened, tempered, then decarburized prior to nitriding, a higher hardness is attained. This is probably due to a more rapid increase in nitrogen allowed by the faster diffusion of nitrogen in the all-ferrite layer at the surface.

It is seen that nitriding is carried out around 500°C for several hours. Thus the structure of the core, which is determined by the heat treatment prior to nitriding, must be such that no appreciable change occurs during nitriding. If the core is to be a tempered martensite structure, then the tempering must be carried out at a temperature sufficiently greater than

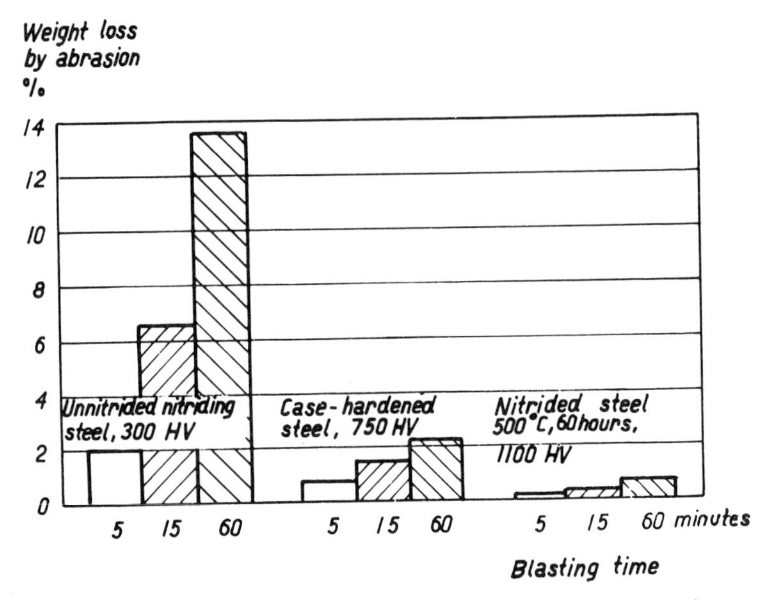

FIGURE 163. Illustration of the effect of nitriding on the wear resistance of a steel. The samples were blasted with steel grit (adapted from Lambert, R. 1955. *Revue de Metallurgie*, 52:553; reprinted with permission from Thelning, K.-E. 1975. *Steel and Its Heat Treatment*. Boston: Butterworths).

the nitriding temperature so that no change occurs during nitriding. The required temperature can be estimated by assuming that the tempering time required to achieve a given tempered hardness is inversely proportional to an Arrhenius expression,

$$1/t = A \exp(-Q/RT)$$

Here A is a constant, R is the ideal gas constant, T is the absolute temperature, and Q is the activation energy for the tempering process. The equation above can be applied to obtain the tempering time at one temperature that will give the same hardness obtained in a given time at another temperature. To illustrate the calculation, the value of Q is taken as that for diffusion of carbon in ferrite (84,000 J/K-Mole). Thus, if we obtain a certain hardness after tempering for $t_1 = 2$ hours at $T_1 = 625$°C (898 K), then the time t_2 to give the same hardness at $T_2 = 500$°C (773 K) will be given by use of the relation

$$[(1/t_1)/(1/t_2)] = [\exp(Q/RT_1)]/[\exp(Q/RT_2)]$$

(Note that the constant A cancels.) The calculated time is about 60 hours. Thus if the tempered hardness is set by tempering for two hours at 625°C, then no significant additional softening will occur upon nitriding at 500°C for 60 hours. Such calculations show that tempering prior to nitriding should be carried out about 100°C above the nitriding temperature in order to prevent further softening of the core during nitriding.

4g. PROPERTIES OF NITRIDED STEELS

One of the main uses of nitriding is to increase the wear resistance of steels. Figure 163 illustrates the advantage of using nitrided steels for this property.

Figure 164 shows the marked improvement in fatigue strength which can be developed by nitriding. The fatigue strength of this steel is about

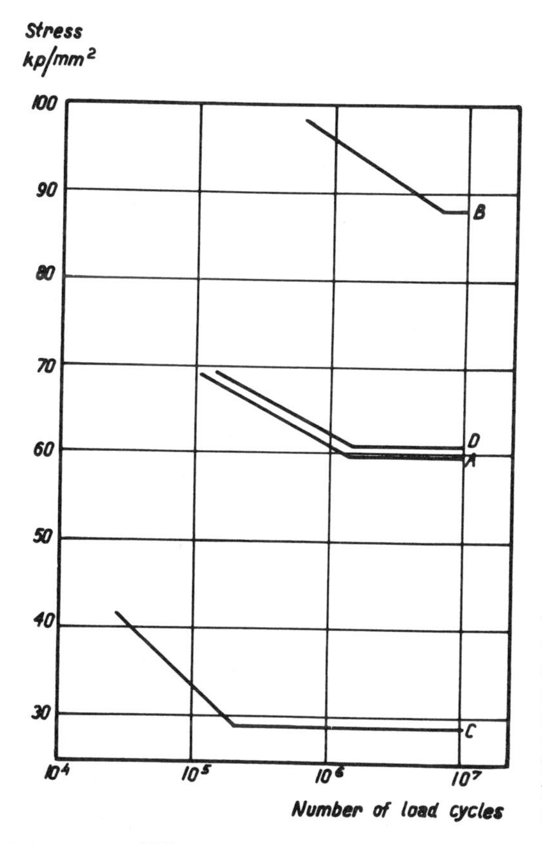

Effect of tufftriding on rotating-beam fatigue strength. Steel DIN 14 CrMoV 69 (0.14% C, 1.5% Cr, 0.90% Mo + V). *A*, smooth test specimen, hardened and tempered at 570°C; *B*, smooth test specimen, hardened and tempered at 600°C then tufftrided for 90 min at 570°C; *C*, notched test specimen, hardened and tempered at 570°C; *D*, notched test specimen, hardened and tempered at 600°C then tufftrided for 90 min at 570°C; notched factor $\alpha_k = 2$.

FIGURE 164. Effect of nitriding on the fatigue properties of a steel (adapted from Finnern, B., K. Vetter and H. Jesper. 1965. *Z. Wirtschaftliche Fertigung*, 60:444).

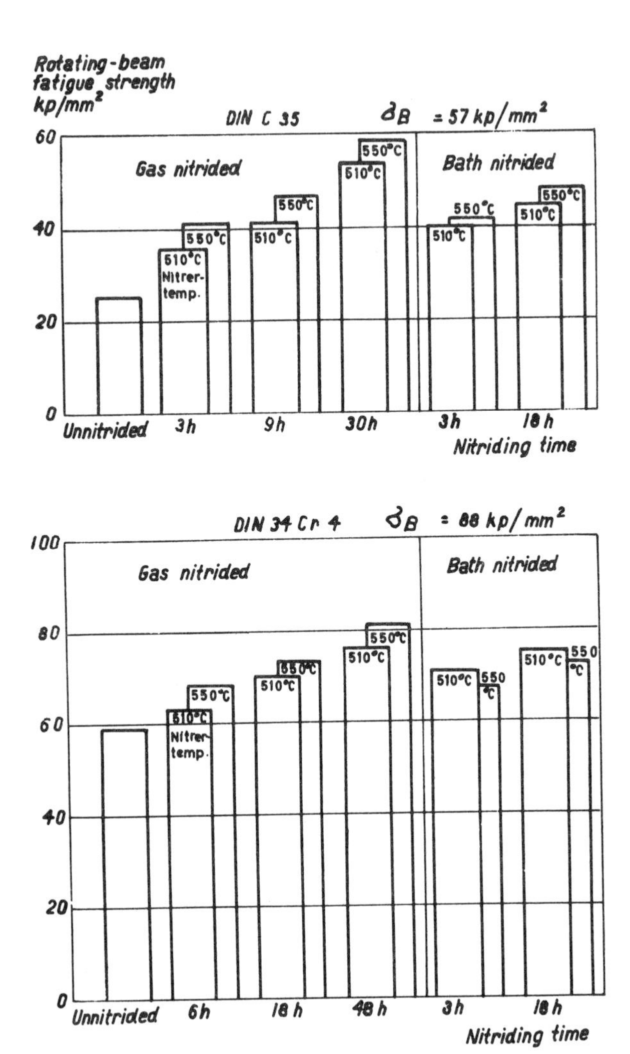

FIGURE 165. The effect of nitriding temperature and time on the fatigue strength of two steels. The test bars were 5.9 mm in diameter and made of hardened and tempered steel; they were gas and salt-bath nitrided at 510 and 530°C (adapted from Wiegard, H. 1966. *Harterei-Techn. Mitt.*, 21:263; reprinted with permission from Thelning, K.-E. 1975. *Steel and Its Heat Treatment*. Boston: Butterworths).

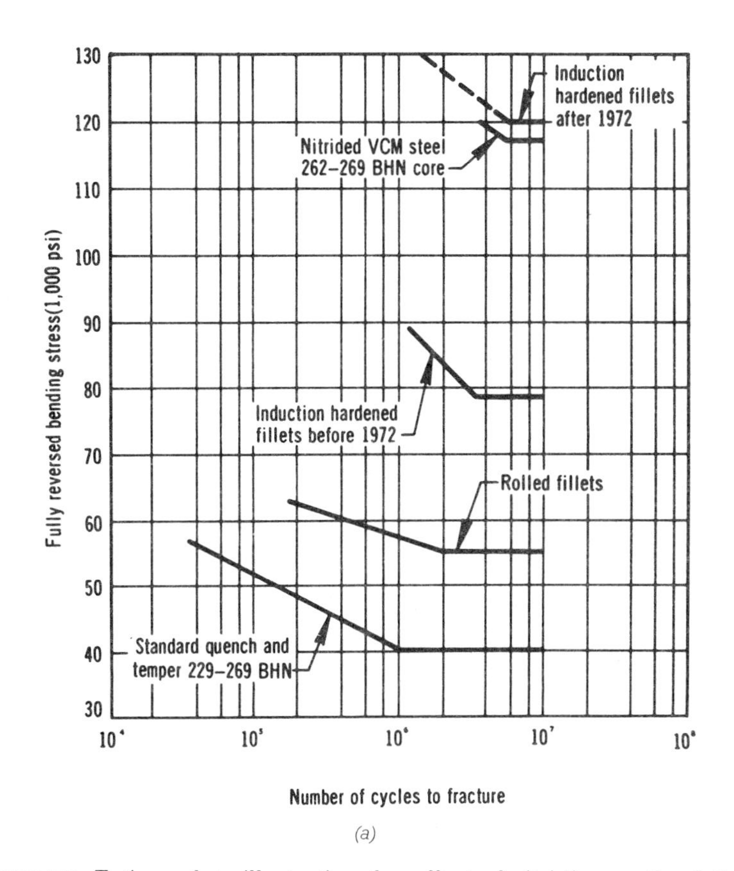

(a)

FIGURE 166. Fatigue data illustrating the effect of nitriding on the fatigue properties. (a) Effect of surface treatment on fatigue life of crankshafts (courtesy Allison Detroit Diesel Division, General Motors Corp.; reprinted with permission from Kern, R. F. and M. E. Suess. *Steel Selection.* © 1979, John Wiley & Sons, Inc.).

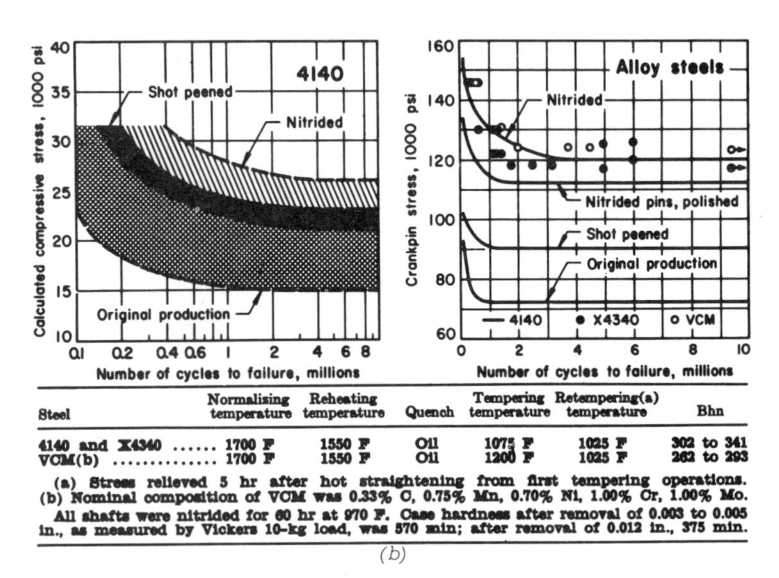

Steel	Normalising temperature	Reheating temperature	Quench	Tempering temperature	Retempering(a) temperature	Bhn
4140 and X4340	1700 F	1550 F	Oil	1075 F	1025 F	302 to 341
VCM(b)	1700 F	1550 F	Oil	1200 F	1025 F	262 to 293

(a) Stress relieved 5 hr after hot straightening from first tempering operations. (b) Nominal composition of VCM was 0.33% C, 0.75% Mn, 0.70% Ni, 1.00% Cr, 1.00% Mo.

All shafts were nitrided for 60 hr at 970 F. Case hardness after removal of 0.003 to 0.005 in., as measured by Vickers 10-kg load, was 570 min; after removal of 0.012 in., 375 min.

(b)

FIGURE 166 (continued). Fatigue data illustrating the effect of nitriding on the fatigue properties. (b) (left) Torsional fatigue test results on nitrided aircraft engine crankshafts. (right) Bending fatigue test results on sections from same nitrided crankshafts (reprinted with permission from *Metals Handbook, Vol. 1, Properties and Selection of Metals.* 1961. Metals Park, Ohio: American Society for Metals).

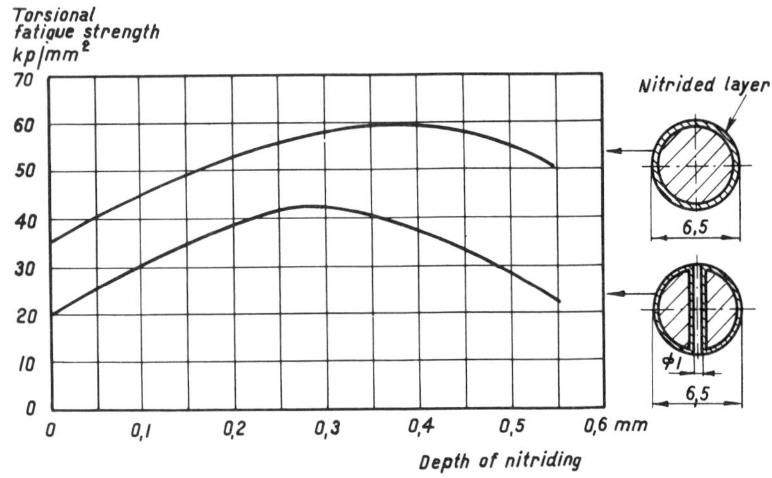

Torsional fatigue strength as function of depth of nitriding of a smooth test specimen and a drilled test specimen, respectively, made from a Cr—Mo—V steel having the following composition: 0.3% C, 2.5% Cr, 0.2% Mo, 0.25% V; σ_B = 110 kp/mm².

FIGURE 167. Data illustrating the effect of case depth in nitriding on the fatigue strength (adapted from Wiegard, H. 1966. *Harteri-Techn. Mitt.*, 21:263; reprinted with permission from Thelning, K.-E. 1975. *Steel and Its Heat Treatment.* Boston: Butterworths).

60 kp/mm^2 when in the quenched and tempered condition. (Note that the tempering temperature was 600°C). When the steel in this condition was nitrided, the fatigue strength increased to about 88. These data also show the improvement that nitriding provides for notched samples. When the quenched and tempered steel was fatigue tested with a notch present, the fatigue strength was about 28 kp/mm^2, but in the nitrided condition it was about 62. The effect of the type of nitriding and the nitriding temperature and time on the fatigue strength is shown in Figure 165.

The effect of nitriding on improving the fatigue properties of crankshafts is illustrated by the data in Figure 166.

Figure 167 shows the influence of the depth of nitriding on the fatigue strength. Such data are used to determine the required case depth in nitriding. (These data also show that the presence of a hole lowers the fatigue strength.)

CHAPTER 5 **Ion Implantation**

5a. INTRODUCTION

In preceding chapters, the carbon or nitrogen contents of steel surfaces were altered by appropriate chemical reactions to release carbon or nitrogen atoms on the surface, which then diffused from the surface towards the interior to build up a concentration profile (or case). Another newer technique of changing the chemistry of surfaces is by bombardment with elemental ions of sufficient energy that they penetrate the surface and become embedded. This is called *ion implantation*. It is an important method of surface chemistry control and also of the control of the structure (for example, the bombarding ions may be sufficiently energetic that an amorphous layer is formed). This method has wide applications in the manufacture of electronic devices.

There are definite specific advantages to ion implantation of metals and alloys, and some of these are listed in Table 3. However, some of the disadvantages listed must be taken into account in applying this method.

The development of this technique for the surface treatment of metals and alloys has now reached the stage of commercial application. Of specific interest here is the possibility of adding carbon or nitrogen to the surfaces of steels in order to develop a case with useful properties. Thus, in this chapter the principles of ion implantation are briefly reviewed, and then the application of this method to the surface modification of steels, and the accruing properties, are described.

5b. THE PROCESS OF IMPLANTATION

Implantation of atoms or ions requires that they have sufficient energy to penetrate the lattice to sufficient depths to be retained. Ions are used since they are charged and hence can be accelerated to sufficient energy levels. Thus a device is required that will ionize the atoms to be implanted and then accelerate them. The implantation process then requires that a suitable quantity of the ions be retained in the solid. In this section the gen-

TABLE 3.
Some Advantages and Disadvantages of Ion Implantation of the Surface of Steels.

Advantages
Only little dimensional change
No layer to delaminate
Sample usually at or near room temperature
Structures can be created that are not attainable by conventional means.
Bulk properties are not changed.
Little effect on surface finish
Disadvantages
Expensive and complicated equipment required
Implantation heating (but <200°C) may occur
May be difficult to uniformly implant desired surface
Complex microstructure required may be difficult to reproduce.
Sputtering may be a problem.

eral method of producing the ion beam is briefly described, and then general features of the interaction of the ion beam with the material are discussed.

METHOD OF PRODUCING ION BEAMS

An ion implantation system requires: (1) an ion source, (2) an extractor to remove the ions from the source, (3) a device to separate the ions to recover only those desired for implantation, (4) an accelerator to impart the necessary energy for implantation into the lattice, and (5) a target chamber in which the part to be implanted can be adequately scanned and maneuvered to achieve proper surface coverage. One such device is shown schematically in Figure 168.

The material to be ionized is introduced into a chamber as a gas. It may already be in the form of a gas (e.g., nitrogen), or the gas may be produced by vaporization from a source, or it can be produced by other means. The atoms must then be ionized. This can be accomplished by several means in principle, by supplying sufficient energy to remove an electron. Thus, the gas can be heated to a very high temperature, it can be radiated with sufficient energy, etc. However, the ionization must be accomplished in a fashion that also produces sufficient quantity of ions to make the subsequent implantation process feasible. It has been found that a plasma discharge is usually required to do this, and thus most systems rely on this method. To produce the plasma, the gas is subjected to a high potential gradient (a few kV) to produce an electrical discharge, forming ionized gas atoms. The ionization process is enhanced by the introduction of thermal electrons from a heated filament (see Figure 168). The plasma is formed into a suitable configuration by an applied magnetic field.

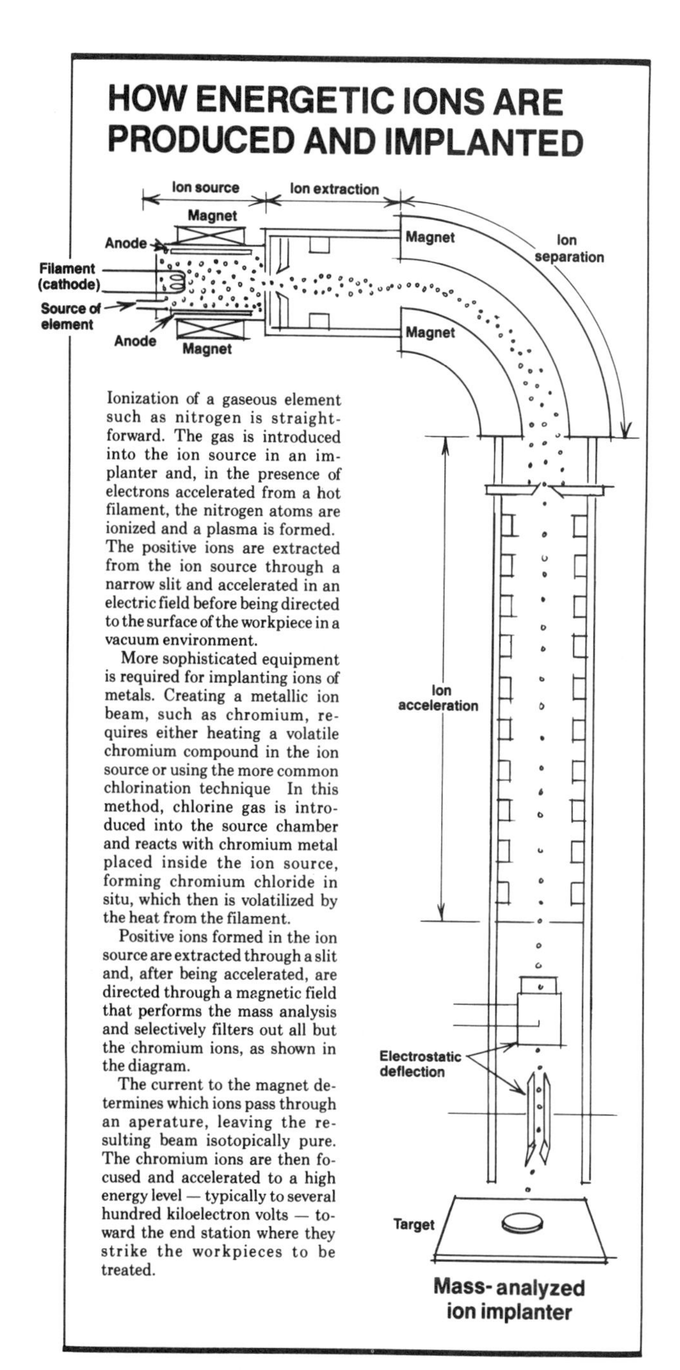

HOW ENERGETIC IONS ARE PRODUCED AND IMPLANTED

Ionization of a gaseous element such as nitrogen is straightforward. The gas is introduced into the ion source in an implanter and, in the presence of electrons accelerated from a hot filament, the nitrogen atoms are ionized and a plasma is formed. The positive ions are extracted from the ion source through a narrow slit and accelerated in an electric field before being directed to the surface of the workpiece in a vacuum environment.

More sophisticated equipment is required for implanting ions of metals. Creating a metallic ion beam, such as chromium, requires either heating a volatile chromium compound in the ion source or using the more common chlorination technique In this method, chlorine gas is introduced into the source chamber and reacts with chromium metal placed inside the ion source, forming chromium chloride in situ, which then is volatilized by the heat from the filament.

Positive ions formed in the ion source are extracted through a slit and, after being accelerated, are directed through a magnetic field that performs the mass analysis and selectively filters out all but the chromium ions, as shown in the diagram.

The current to the magnet determines which ions pass through an aperature, leaving the resulting beam isotopically pure. The chromium ions are then focused and accelerated to a high energy level — typically to several hundred kiloelectron volts — toward the end station where they strike the workpieces to be treated.

FIGURE 168. Schematic representation of a device to produce ions for implantation (reprinted from Sioshansi, P. *Machine Design*. © 1986, Cleveland, OH: Penton Publishing Inc.).

To extract the ions, a plate with a small hole is charged negatively a few keV. This attracts the ions towards the plate, and many pass through the hole into the magnetic separator. The ions can be separated according to charge and mass, by passing through a suitably controlled magnetic or electrostatic field. Commonly a magnetic field is used. As illustrated in Figure 168, the ions are forced into an arc path, and the field adjusted to only allow those of the desired mass or charge to pass through the opening at the entrance to the accelerator.

The selected ions then enter an accelerator. This device is basically a series of plates that are progressively charged to a high potential. Thus, the ions that enter encounter initially the high potential of the first plate and are accelerated towards it. When they reach the vicinity of this plate, the next plate is charged with the high voltage, and the ions are accelerated again. Thus the energy of the ions is raised to the desired value. For proper implantation, the final energy must be in the range of mega-electron volts. (Note that this is much higher than that required for ionization or separation.)

The accelerated ion beam then enters the target chamber. Here the beam can be manipulated by electrostatic or electromagnetic deflection systems, so that it can be focused and moved to give a proper ion intensity and coverage of the surface of the part. Also the part itself may have to be moved by use of a manipulative support system.

An important problem with the system described above is that the ion beam must be directed to the surface of the part (see Figure 168), and best implantation is obtained when the beam is normal to the surface. This is difficult to do with complex shapes. In a new method, the part is directly in the plasma source and is pulse-biased (relative to the chamber walls) to accelerate the ions normal to the surface of the part. This gives relatively uniform implantation. This process is called plasma source ion implantation (PSII). (See Conrad, J. R., J. L. Radke, R. A. Dodd, F. J. Worzala and N. C. Tran. 1987. "Plasma Source Ion-Implanted Technique for Surface Modification of Materials," *J. Appl. Phys.*, 62:4591.)

The description above is quite general, and nowhere begins to describe the actual sophisticated design and operation of the equipment used. Such details can be found in (or traced from) the references given at the end of this chapter.

INTERACTION OF IONS AND MATERIAL

The ions have sufficient energy to enter the surface of the solid and to continue to a depth of many Angstroms. Along this path, they collide with many host atoms, and with sufficient energy to cause some of these to be displaced many atom distances from their lattice sites. Some of these in turn have sufficient energy to displace other atoms. An impinging ion finally comes to rest some distance from the surface, leaving in its wake a zone of considerable damage. This effect is shown schematically in Figure 169. Thus the effect of a stream of ions is to produce a volume that has a chemical composition *and* structure different from that of the underlying bulk of the material.

A distribution of implanted ions versus depth is approximately bell-shaped. Of course, too far from the surface, no ions penetrate. Nearer the surface, the concentration of implanted ions increases. Still closer to the surface, however, fewer ions are present, found here only by back-scattering during the many collisions of the ions with the lattice. This gives the distribution shown in Figure 170. Note also that the lattice damage attains a maximum at some distance below the surface. Nearer the surface there is less damage, as the ions must penetrate some distance before interaction begins with the lattice atoms. Thus, damage nearer the

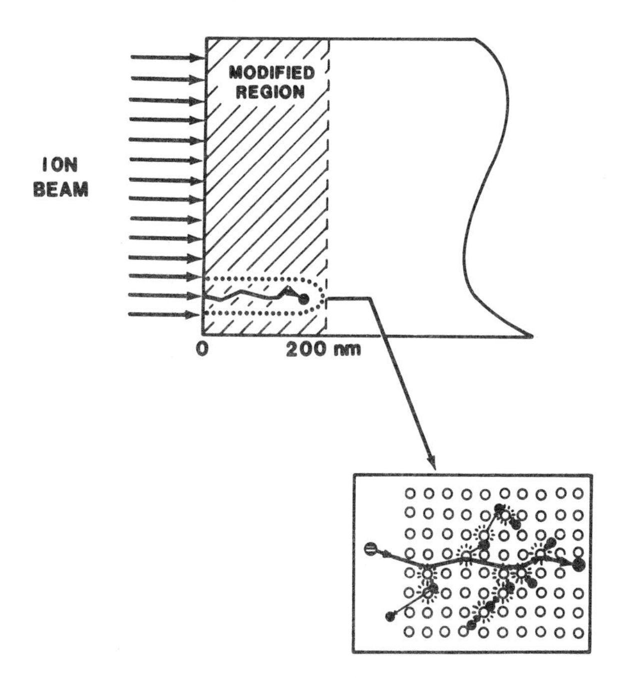

FIGURE 169. Schematic illustration of the collision process of ions with the lattice, and the formation from this of the chemically and structurally modified surface region (adapted with permission from Rehn, L. E., S. T. Picraux and H. Wiedersich. 1987. In *Surface Alloying by Ion, Electron and Laser Beams*, L. E. Rehn, S. T. Picraux and H. Wiedersich, editors, Metals Park, OH: American Society for Metals).

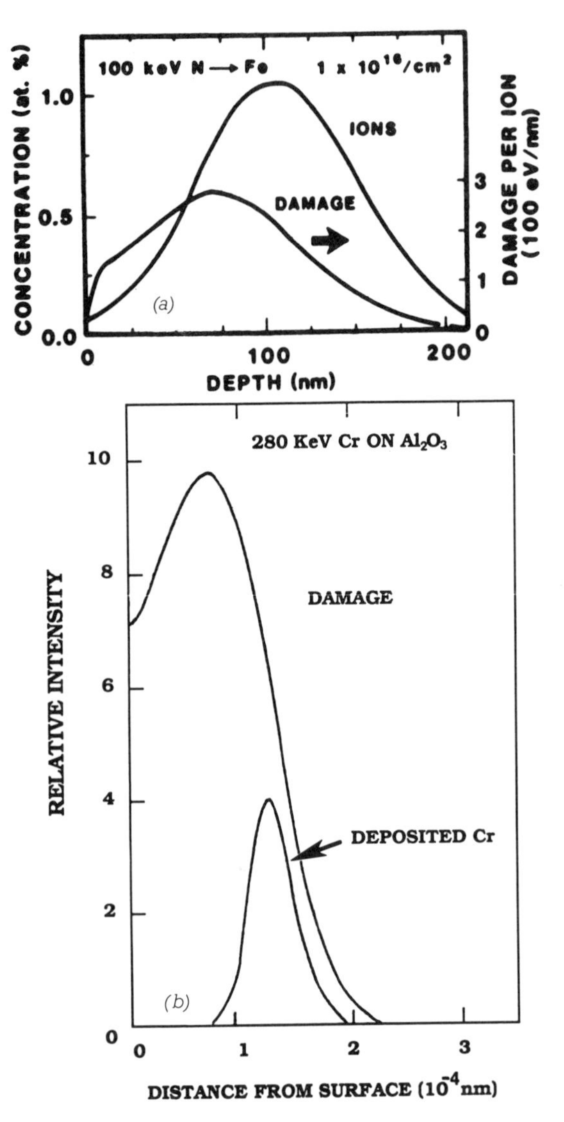

Distribution of Cr ions and displacement defects in Al_2O_3 resulting from bombardment with 280 KeV Cr ions.

FIGURE 170. (a) Schematic illustration of the concentration and lattice damage profile from ion implantation of a medium atomic number target (iron) with 100 keV nitrogen ions, corresponding to the region from 0 to 200 nm in Figure 169. (b) Actual data for Cr implanted in Al_2O_3 [(a) adapted with permission from Rehn, L. E., S. T. Picraux and H. Wiedersich. 1987. In *Surface Alloying by Ion, Electron and Laser Beams*, L. E. Rehn, S. T. Picraux and H. Wiedersich, editors, Metals Park, OH: American Society for Metals; (b) adapted with permission from McHargue, C. J. and C. S. Yust. *J. Am. Chem. Soc.*, 67:117, © 1984, American Chemical Society].

surface is caused by secondary collisions. Figure 170(b) shows actual profiles for Cr ions implanted in Al_2O_3.

The concentration profiles depend upon many factors, such as ion beam energy, type of ions, ion beam flux (number of ions per area), part temperature, material of part to be implanted, etc. Establishment of these parameters in order to obtain a suitable implant profile is important in the production of implanted parts. These important details are not covered here, but can be found in some of the references listed at the end of the chapter.

SPUTTERING

When the high-energy ions impact the surface of a material, they may have sufficient energy to cause some of the atoms of the material to be ejected from the surface. This is called *sputtering*, and it can be utilized to develop a desired fine surface topology by controlling the beam position, such as in lithography. Recognition that this may occur in ion beam implantation may be important because the concentration profile is affected. This is illustrated in Figure 171.

5c. MICROSTRUCTURAL EFFECTS IN METALS AND ALLOYS

The microstructural effects that occur upon ion implantation of metals and alloys depend upon several factors, such as the energy of the impinging ions, flux, temperature of the part, the materials from which the part is made, etc. Thus, prediction of what will occur is difficult. Of course, the chemical composition changes, and this itself affects the structure that develops. For example, the ability to form an amorphous structure may depend upon the concentration of the implanted ion. Some microstructural effects that occur are listed in Table 4.

TABLE 4.
List of Some Microstructural Effects That Can Occur in the Ion-Implanted Layer of Metals and Alloys.

Melting and subsequent solidification
Formation of an amphorous layer
Radiation damage (interstitial atoms and vacancies)
Formation of a metastable solid solution (crystalline)
Resolution of second phases and precipitates
Formation of second phases and precipitates
Other phase transformations

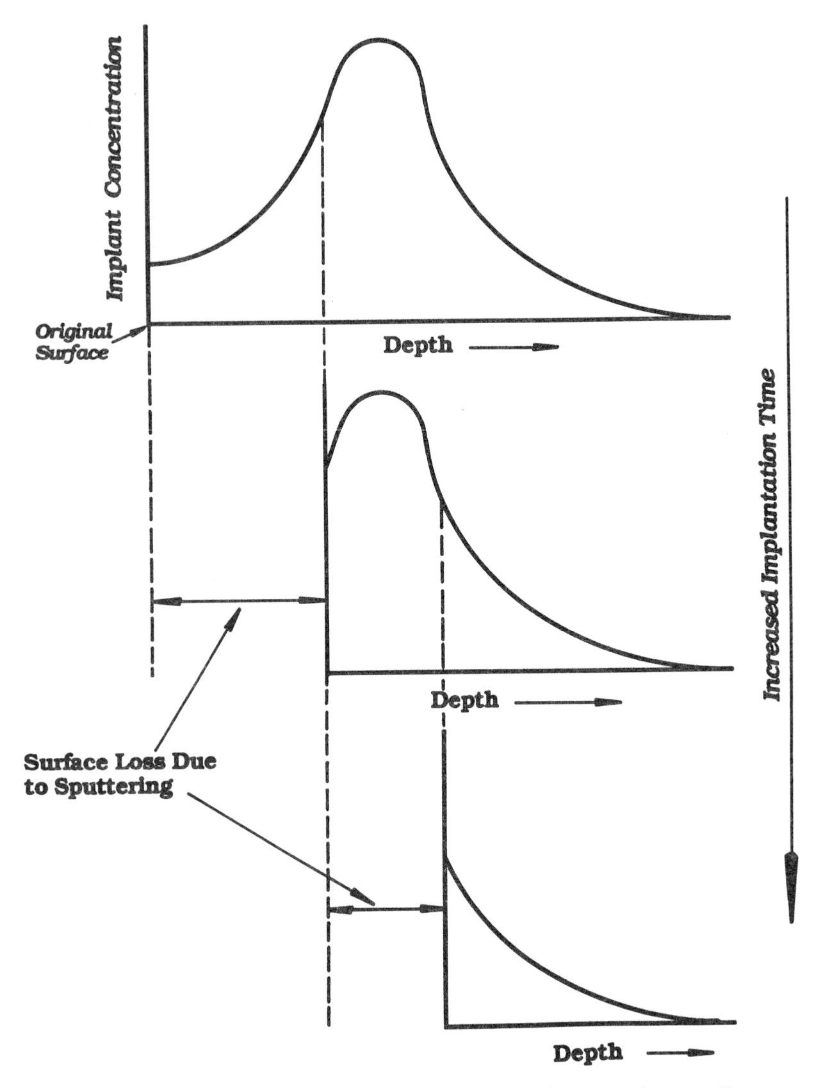

FIGURE 171. Schematic illustration of the effect of sputtering on the concentration profile of the implanted ion.

5d. EFFECT ON WEAR AND CORROSION RESISTANCE

For most metals and alloys, the main use of ion implantation at present is to impart improvements in wear and corrosion resistance. In this section, examples of the improvement in these properties are presented.

WEAR RESISTANCE

Wear is a complicated process, with different types of wear recognized. For example, adhesive wear is caused by the bonding of contact points on the two mating surfaces, such that plastic deformation and fracture of these regions occurs with material adhering to the surfaces. Abrasive wear involves the removal of material from one surface, with debris generated entering into the wear process.

In general, the main factor that improves wear resistance is increased hardness, and thus it would be expected that in order to improve wear resistance ion implantation should increase the surface hardness. That implantation can cause a hardened surface is illustrated by the data in Figure 172. (Note that due to the thinness of the affected layer, conventional microhardness measurements cannot resolve this effect.) Here commercially pure iron was implanted with nitrogen. The surface region was increased in hardness by 80 to 120%. Also, note that the higher flux caused higher hardness. However, note that the depth of the implanted layer was only about 3000 Angstroms, or about 600 atom layers. This emphasizes an important restriction of ion implantation for wear resistance, since such a thin layer is not capable of supporting heavy loads. (It has a low crushing strength.) Thus, the applicability of the improved wear resistance by ion implantation must take such a limitation into consideration.

In many alloys, ion implantation markedly reduces the coefficient of friction, and Figure 173 shows an example. However, as shown in Figure 174, the effectiveness may depend upon the load, probably because of the thinness of the layer (see the preceding paragraph). Not all implanted species reduce the coefficient of friction. In Figure 175 are shown data for Kr, Sn, Mo and Mo + S added to a steel, but only Sn reduced the coefficient of friction.

In measuring actual wear, the problem arises as to how to characterize the process. For example, the volume of wear debris can be measured, the power to move one surface past the other, the depth of the wear track, etc. The results obtained depend markedly upon the type of wear test (ball-on-plate, two cylinders at right angles to each other, etc.) and upon the lubricant. Thus, it is sometimes difficult to compare the behavior of different materials if they are tested differently. However, generally,

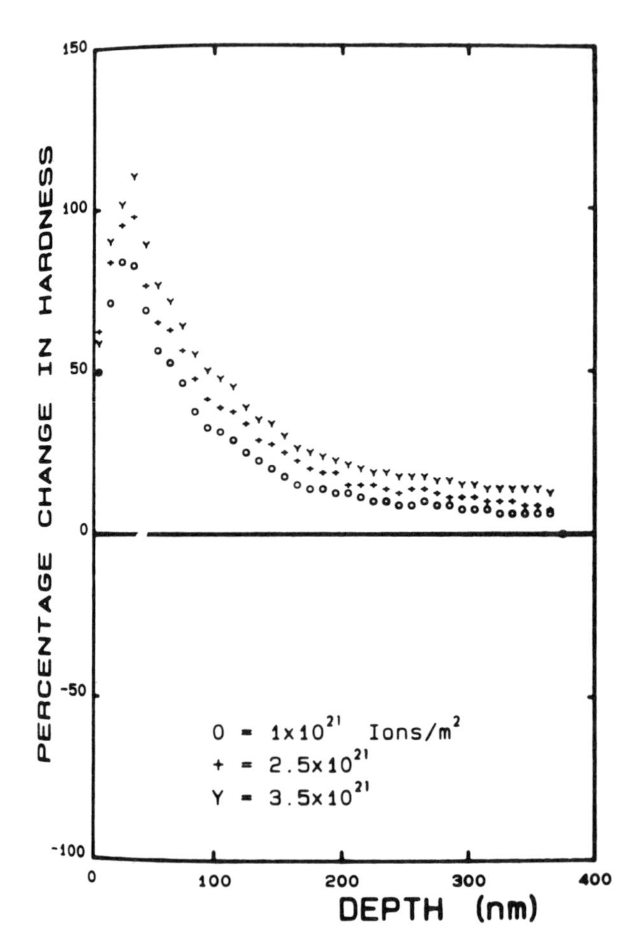

FIGURE 172. Microhardness profiles for Armco iron implanted with nitrogen. Note that the depth of hardening only extends to about 4000 Angstroms (about 1000 atom layers) (reprinted with permission from Pethica, J. B., R. Hutchings and W. C. Oliver. 1983. *Nucl. Inst. and Methods*, 209/210:995, Elsevier Science Publishers).

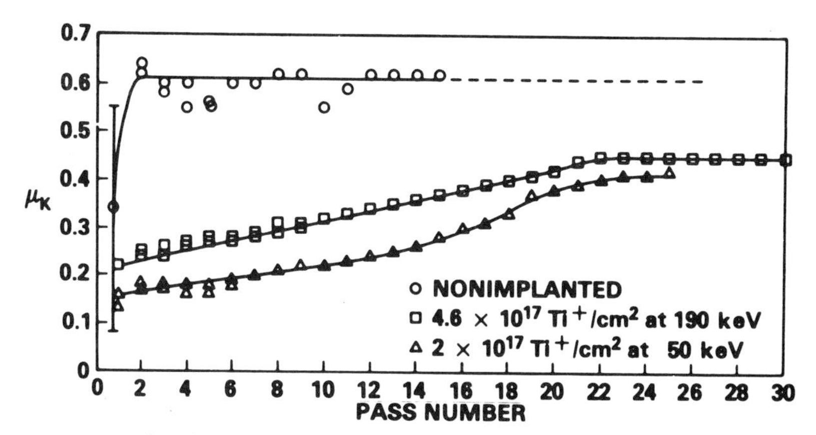

FIGURE 173. Coefficient of friction versus pass number for hardened 52100 bearing steel disks, unimplanted and implanted with titanium, sliding without lubrication against 52100 hardened steel balls (reprinted with permission from Singer, I. L. and R. A. Jeffries. 1984. In *Ion Implantation and Ion Beam Processing of Materials*, G. K. Hubler, O. W. Holland, C. R. Clayton and C. W. White, editors, Pittsburgh, PA: Materials Research Society).

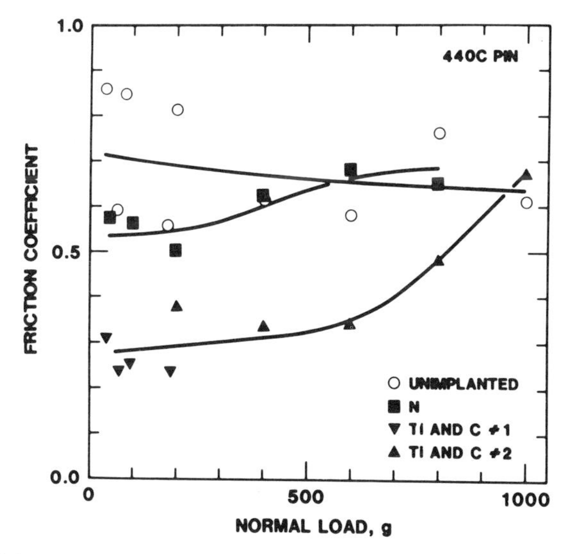

Friction coefficient for unlubricated pin-on-disk test vs normal load for ion-implanted AISI 440C steel. The titanium was implanted at energies between 90 and 180 keV to a total dose of 2 × 10^{17} C$^+$/cm^2, and the nitrogen at 50 keV to a dose of 4 × 10^{17} N$^+$/cm^2.

FIGURE 174. Effect of load on the coefficient of friction of a 440 C stainless steel, implanted and unimplanted (pin-on-disk tests) (reprinted with permission from Pope, L. E., F. G. Yost, D. M. Follstaedt, S. T. Picraux and J. A. Knapp. 1984. In *Ion Implantation and Ion Beam Processing of Materials*, G. K. Hubler, O. W. Holland, C. R. Clayton and C. W. White, editors, Pittsburgh, PA: Materials Research Society).

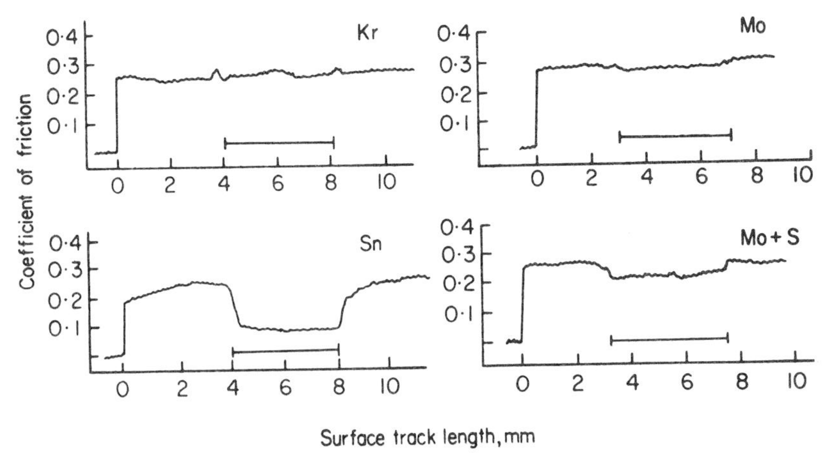

Frictional changes induced in steel as a result of ion implantation. The markers indicate the position of implant. The curves are for Kr^+ at 400 keV, 2.8 × 10^{16} ions cm^{-2}; Mo^+ at 400 keV, 2.8 × 10^{16} ions cm^{-2}; Mo^+ 400 keV (2.8 × 10^{16} ions cm^{-2}) plus S^+ at 150 keV (5.6 × 10^{16} ions cm^{-2}); and Sn^+ at 380 keV, 2.8 × 10^{16} ions cm^{-2}.

FIGURE 175. The effect of implantation of an En 352 steel on the coefficient of friction. The tests used a tungsten carbide ball on the implanted steel surface (adapted from Hartley, H. E. W., G. Dearnaley, J. F. Turner and J. Saunders. 1976. Cited in *Ion Implantation, Sputtering and Their Applications*, P. D. Townsend, J. C. Kelly and N. E. W. Hartley, editors, New York: Academic Press).

results can be obtained by controlling these variables so that the effect of ion implantation on wear can be properly compared to the wear of the same material without implantation. The example in Figure 176 shows a reduction in wear rate with increased ion implantation flux. Note that upon repeated sliding, the wear rate increases as the implant layer is worn off.

The results cited here illustrate that the wear response of ion-implanted surfaces is very complex (as it is in the case of unimplanted surfaces), and that it is difficult even to generalize as to the effect implantation will have on wear. An important factor here is the variety of complex microstructures that are developed in the implanted layer.

CORROSION RESISTANCE

The corrosion resistance of a metal or alloy is related to the ability of the material to form an insoluble, protective coating (usually an oxide) and to the degree of heterogeneity of the surface structure (grain boundaries, inclusions, precipitates, etc.). In most cases the latter is difficult to control, so the common corrosion-resistant alloys have elements added to promote passivation (e.g., Cr in steels). Ion implantation can be used to add such species just to the surface of a part that normally would have a

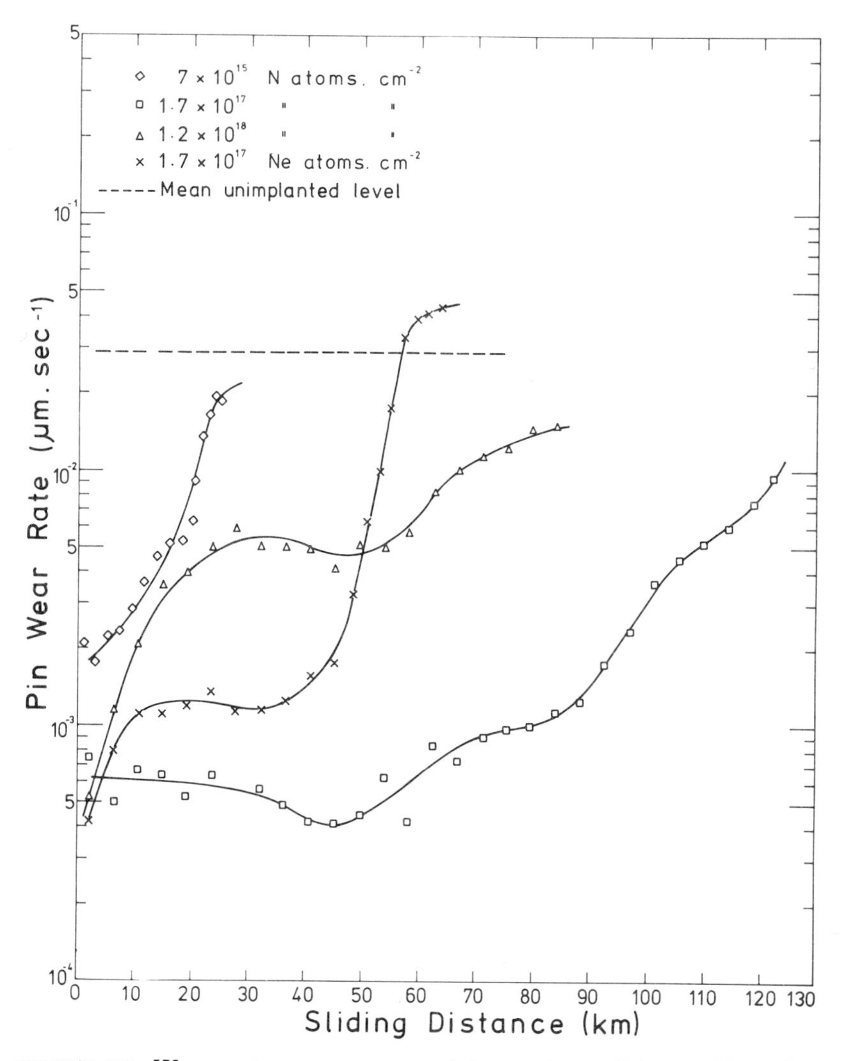

FIGURE 176. Wear rates, as measured in a pin-on-disk machine, for unimplanted and implanted iron (adapted with permission from Goode, P. D., A. T. Peacock and J. Asher. 1983. *Nucl. Inst. Meth.*, 209/210:925, Amsterdam: Elsevier Science Publishers).

low corrosion resistance. But in addition to this, ion implantation may develop a more homogeneous surface layer freer of heterogeneities. In the extreme case, this would involve the formation of an amorphous layer. Alloys that can be processed to be amorphous (by, for example, very rapid cooling) are known to have considerably better corrosion resistance, and in general, this effect is also found in alloys for which an amorphous layer can be formed by ion implantation. An example of such improvement is shown in Figure 177, which shows the effect of implantation of the compound NiTi with N. The passive region is changed to a current density range about four orders of magnitude lower than that of the unimplanted material.

Implantation of the alloy Ti-6% Al-4% V has been studied with the aim of improving the corrosion resistance for better hip joint replacement components. Figure 178 shows the marked improvement in corrosion resistance of this alloy upon implantation with nitrogen.

The reasons for the improvement in corrosion resistance upon implantation (and in some cases the lack of improvement) are not always understood. Part of the difficulty is that the implantation not only changes the

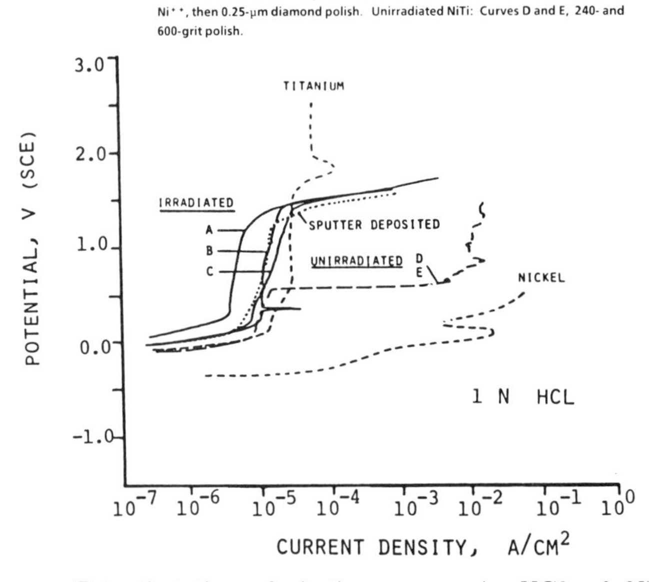

FIGURE 177. Potentiostatic polarization curves in HC1 of NiTi ion-implanted with Ni (from Wang, R. and J. L. Brimhall. 1984. In *Ion Implantation and Ion Beam Processing of Materials*, G. K. Hubler, O. W. Holland, C. R. Clayton and C. W. White, editors, Pittsburgh, PA: Materials Research Society).

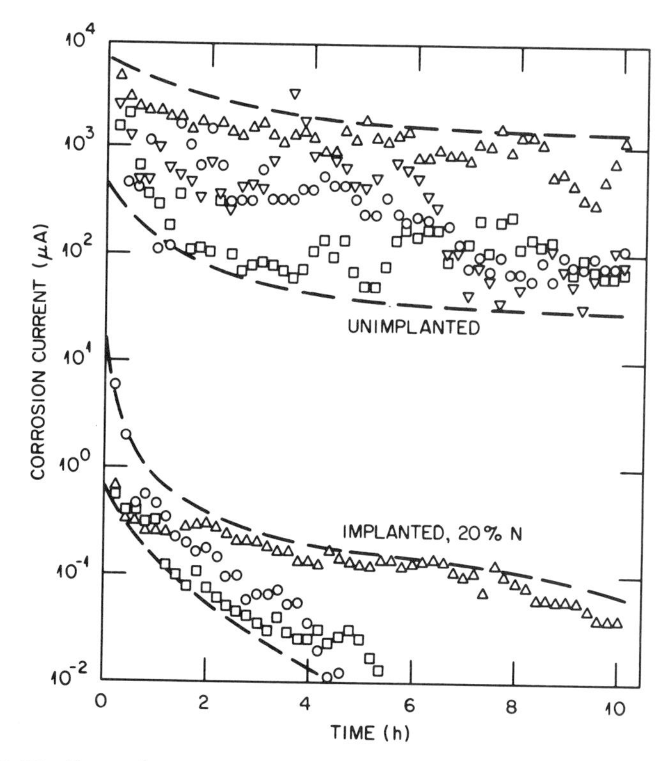

FIGURE 178. Corrosion current versus time during a corrosion-wear process for unimplanted and implanted (with nitrogen) Ti-6Al-4V alloys (reprinted with permission from Williams, J. M. and R. A. Buchanan. 1985. *Mat. Sci. and Engr.*, 69:237, Elsevier Sequoia S. A.).

chemistry and induces radiation damage, but it alters the microstructure as well. Thus, depending upon the implantation variables (e.g., flux), the implanting species and the material, phases may form or disappear, which can have a great effect upon the corrosion response of the surface.

5e. EXAMPLES OF IMPLANTATION OF STEELS AND THE EFFECT ON PROPERTIES

The preceding information was designed to give a general idea of the effect of ion implantation on the microstructures and the wear and corrosion properties of implanted surfaces. The main interest, however, is the ion implantation of steels. Thus, in this section, specific examples of the implantation of steels are given, with emphasis on those that have commercial application.

That ion implantation is receiving attention for use in the surface treatment of steels is illustrated by the listings of applications in Table 5.

NITROGEN IMPLANTATION

The implantation of nitrogen into the surfaces of some steels showed early promise for commercial applications. Since conventional nitriding (see Chapter 4) is used in the surface treatment of steels, it is natural to expect that beneficial effects would accrue from ion implantation of nitrogen. Typical implantation concentration profiles are obtained, as shown in Figure 179. Considerable improvement in wear resistance has been reported, but the results may be quite sensitive to the dose, as illustrated in Figure 180.

The effect of nitrogen on the microstructure (and hence properties) can be complicated. In implanting steels, the microstructure produced by implantation will depend on the prior microstructure (e.g., pearlite, martensite, etc.). Nitrogen is a potent solid solution strengthener of ferrite. Although the solubility is low, implantation might allow concentration in excess of the solubility limit, and hence produce considerable hardening. However, the microstructural effects will depend upon the temperature of implantation, whether due to ion irradiation or external heating. Nitrogen has a relatively high diffusivity (characteristic of interstitial elements) and therefore formation of nitrides and carbonitrides may occur. Due to the low temperature on implantation, these may be very fine (e.g., 100 Angstroms) and therefore provide a significant hardening contribution. Some of the nitrides found in implanted steels are metastable, and their presence depends upon the implanted dose and temperature. Thus, in nitrogen ion implantation of steels, the presence of ferrite, austenite, martensite (its formation induced by the stresses developed), and various nitrides and carbonitrides have been detected. The iron carbide present in the steel prior to implantation may be dissolved by the radiation damage, and then reprecipitated, which may also influence the properties.

It has been reported in some cases that after wear the nitrogen is found to depths beyond that expected from implantation, although some studies have not found this effect. Figure 181 shows the nitrogen concentration profile before wear tests, where the implantation depth is about 0.7 μm. After wear-testing the sample, so that the steel was worn down to 2 μm from the original surface, far beyond the original implant depth, nitrogen was still found. This effect could be caused by radiation-enhanced diffusion, by diffusion enhanced by irradiation heating, and by heating locally due to friction during the wear tests.

Nitrogen implantation has been found to improve the fatigue life of a 1020 steel, as shown in Figure 182. This may be due to residual stresses on the surface.

TABLE 5.
Reported Improvements in Wear Life Obtained by Nitrogen Implantation.

Application/Material	Lifetime Increases, Additional Benefits
Scoring die for aluminum beverage can lids (D2 tool steel)	3×
Forming die for aluminum beverage can bottoms (D2 steel)	Lowered wear 10×, lowered material pickup
Wire guides (hard Cr plate)	3× without significant wear
Finishing rolls for Cu rod (H-13 steel)	Negligible wear after 3× normal lifetime; improved surface finish of product
Paper slitters (1.6% Cr, 1% C steel)	2×
Punches for acetate sheet (Cr-plated steel)	Improved product
Taps for phenolic resin (M2 high-speed steel)	Up to 5×
Tool inserts (4% Ni, 1% Cr steel)	Reduced tool corrosion by 3
Forming tools (12% Cr, 2% C steel)	Greatly reduced adhesive wear
Fuel injectors and metering pump (tool steel)	>100× in engine tests
Fabric slitters (tool steel)	4 to 16×
Plastic cutting (diamond tools)	2 to 4× lifetime
Hip joint prostheses (Ti/6Al/4V)	5 to 100× in laboratory tests
Dental drills (WC-Co)	2 to 3×, significant lower cutting force required
Precision punches for electronic parts (WC-15% Co)	Greater than 2×
Injection molding nozzle, mold screws, gate pads for glass and mineral filled plastics (tool steels and hard Cr-plated steels)	4 to 6×
N into thermally nitrided steel molds (tool steels)	Combination better than either process alone
Plastic calibrator die (nitrided H-13 tool steel)	2×
Profile hot die for plastic extrusion (P-20 tool steel)	4×

Reprinted with permission from J. K. Hirvonen. 1986. In *Surface Alloying by Ion, Electron and Laser Beams*, L. E. Rehn, S. T. Picraux and H. Wiedersich, editors, Metals Park, Ohio: American Society for Metals.

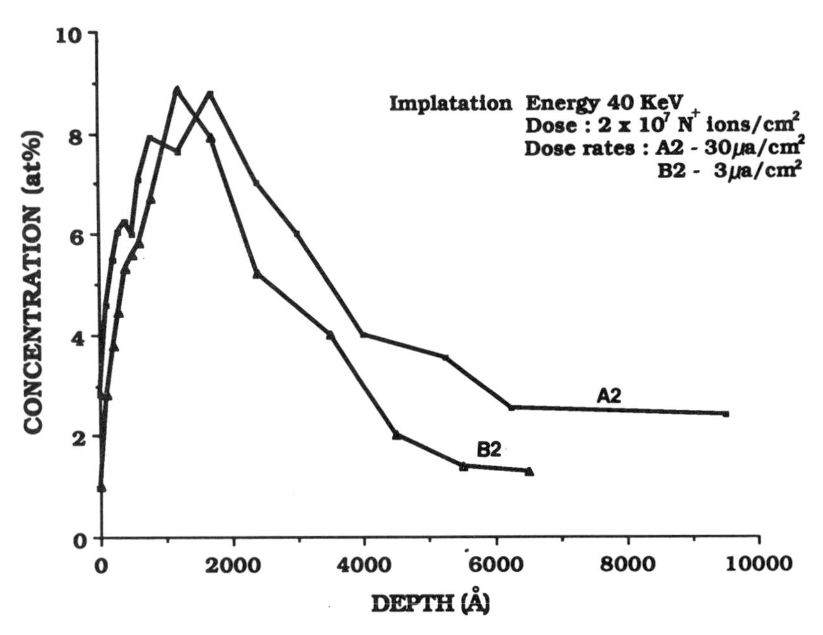

FIGURE 179. Concentration of nitrogen profiles for implantation in mild steel (adapted with permission from Dearnaley, G., P. D. Goode, N. E. W. Hartley, G. W. Proctor, J. F. Turner and R. B. J. Watkins. 1979. In *Proceedings of Conference on Ion Plating and Allied Techniques*, R. Hurley, editor, Edinburgh: CEP Consultants).

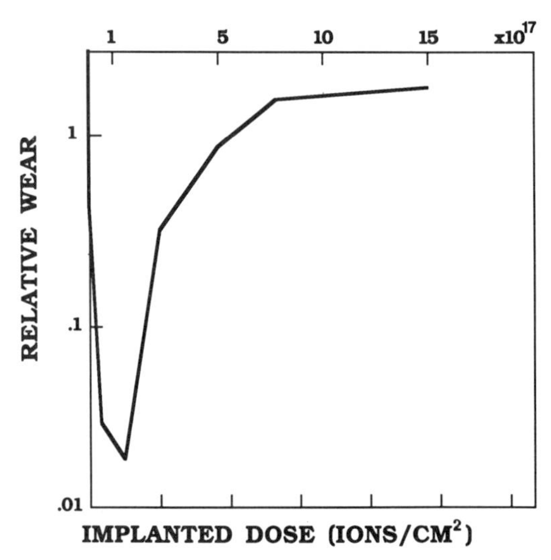

FIGURE 180. The effect of ion implantation dose on the relative wear of a carbon steel ion-implanted with nitrogen (adapted with permission from Varjoranta, T., J. Hirvonen and A. Anttila, 1981. *Thin Sol. Films,* 75:241, Elsevier Sequoia S.A.)

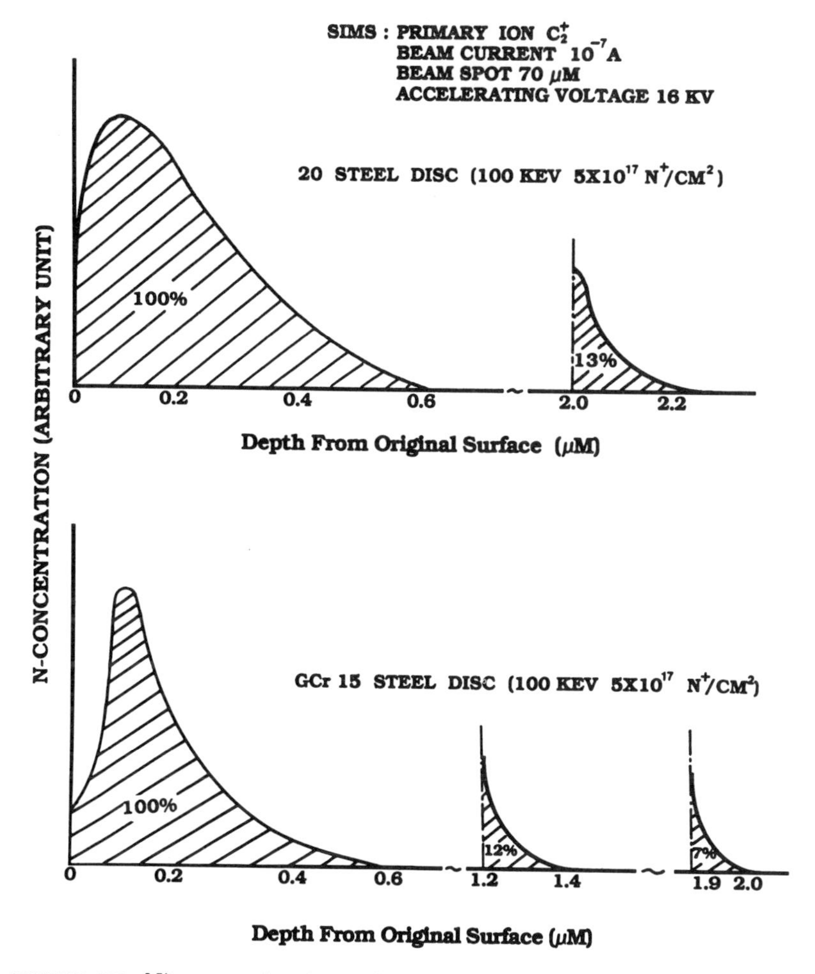

FIGURE 181. Nitrogen depth profiles, before (left) and after wear, in nitrogen-implanted steel. 20 steel is a 0.2% C steel and GCr15 is a 1% C–1.5% Cr steel (adapted with permission from Fu-Zhai, C., L. Heng-De and Z. Xiao-Zhong. 1983. *Nucl. Meth. and Inst. in Physics,* 209/210:881, Amsterdam: Elsevier Science Publishers).

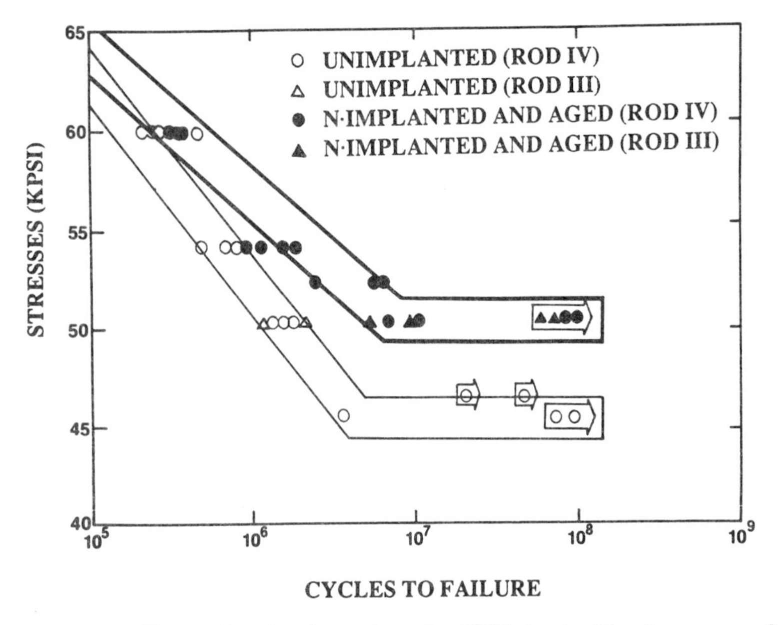

FIGURE 182. Effect of ion implantation of a 1020 steel with nitrogen on the fatigue strength (adapted with permission from Herman, H. 1981. *Nucl. Inst. Meth.*, 182/183:865, Amsterdam: Elsevier Science Publishers).

Figure 183 shows the results of wear tests of a hardened M2 tool steel used for punching holes in mild steel plate. The TiN-coated steel was obtained by vapor deposition, and the result denoted PSII-treated refers to plasma source ion implantation of nitrogen. The wear data are shown in Figure 184. However, as shown in the next subsection, nitrogen implantation in tool steels does not always promote wear improvement.

TITANIUM IMPLANTATION

Bearing steels are a logical potential application of surface implantation. 52100 steel (Fe-1% C-1.5% Cr) is a widely used bearing material, and there have been several studies on the effect of implantation on its properties. It has been found that implantation with nitrogen is not very effective in improving wear resistance, but that implantation with titanium is. The reduction in the coefficient of friction by implantation with titanium is illustrated by the data in Figures 173 and 185. Note that in these tests the reduction required a relatively high dose. Some wear data are shown in Figure 186. Note that implantation with nitrogen or boron did not reduce the wear rate. Also note that simultaneous implantation with titanium and carbon was used. The improvement in wear with titanium implantation

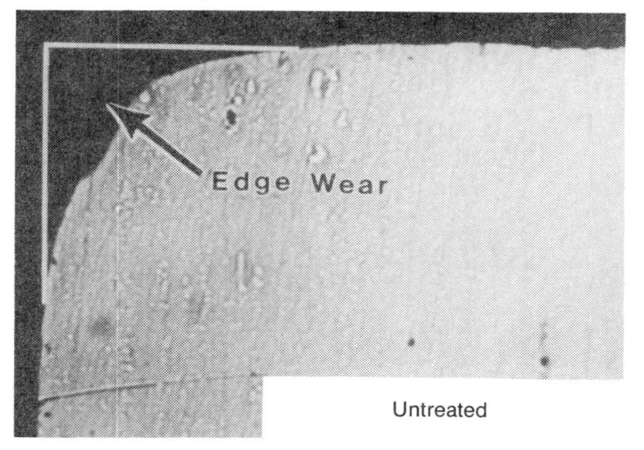

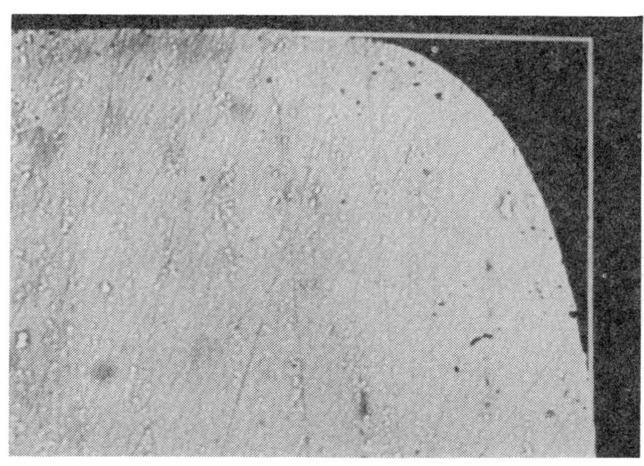

FIGURE 183. Representative cutting edge profiles of M2 tool steel punches sectioned after 20,000 operations (courtesy of Professor John R. Conrad, University of Wisconsin).

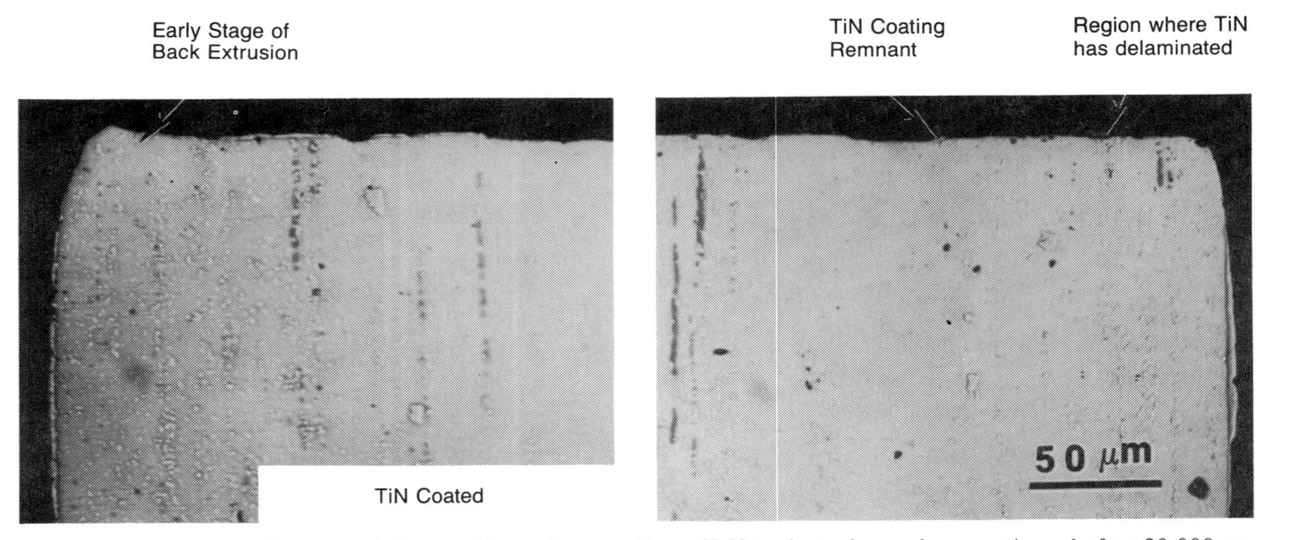

FIGURE 183 (continued). Representative cutting edge profiles of M2 tool steel punches sectioned after 20,000 operations (courtesy of Professor John R. Conrad, University of Wisconsin).

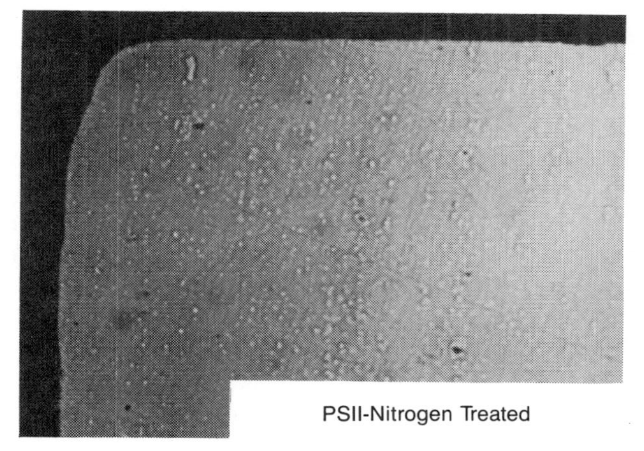

FIGURE 183 (continued). Representative cutting edge profiles of M2 tool steel punches sectioned after 20,000 operations (courtesy of Professor John R. Conrad, University of Wisconsin).

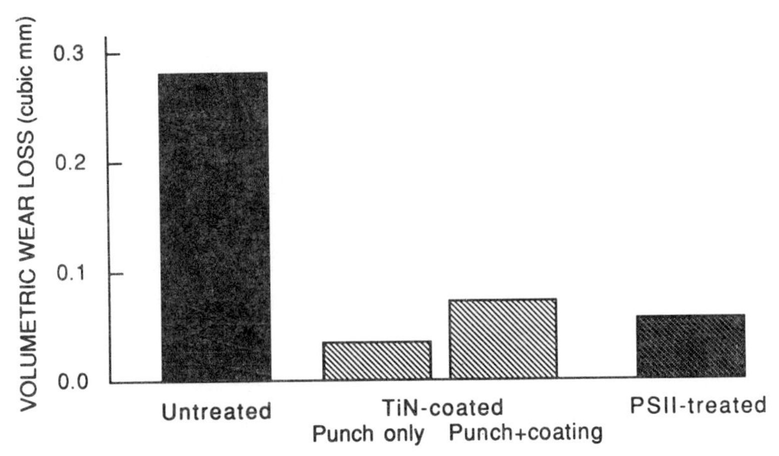

FIGURE 184. Volumetric wear for M2 tool steel punches after 20,000 operations. Only the PSII-treated material involved ion implantation (courtesy of Professor John R. Conrad, University of Wisconsin).

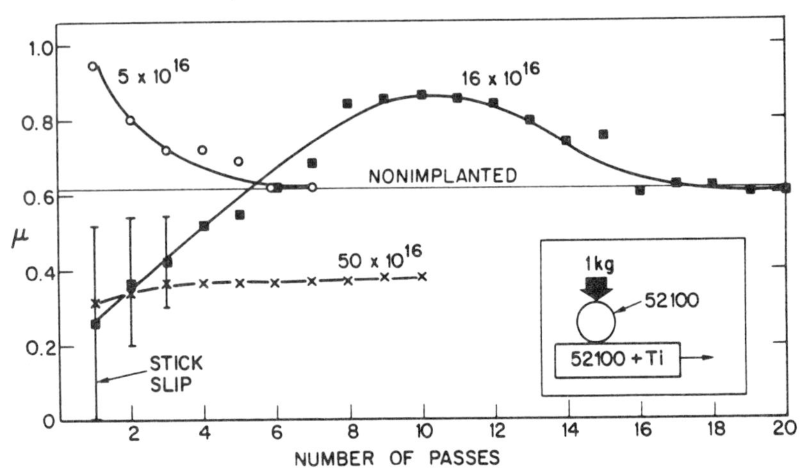

Experimental arrangement (lower right hand corner), and data for the coefficient of friction between an AISI-52100 steel ball and a 52100 flat surface implanted with Ti. Three distinct types of behavior are observed which are dependent on the fluence of Ti ions implanted onto the surface.

FIGURE 185. Coefficient of friction versus pass number for hardened 52100 bearing steel disks implanted with titanium and unimplanted, sliding without lubrication against 52100 hardened steel balls (reprinted with permission from Smidt, F. A., J. K. Hirvonen and S. Ramalingam. 1983. In *Ion Implantation for Materials Processing*, F. A. Smidt, editor, Park Ridge, NJ: Noyes Data Corporation).

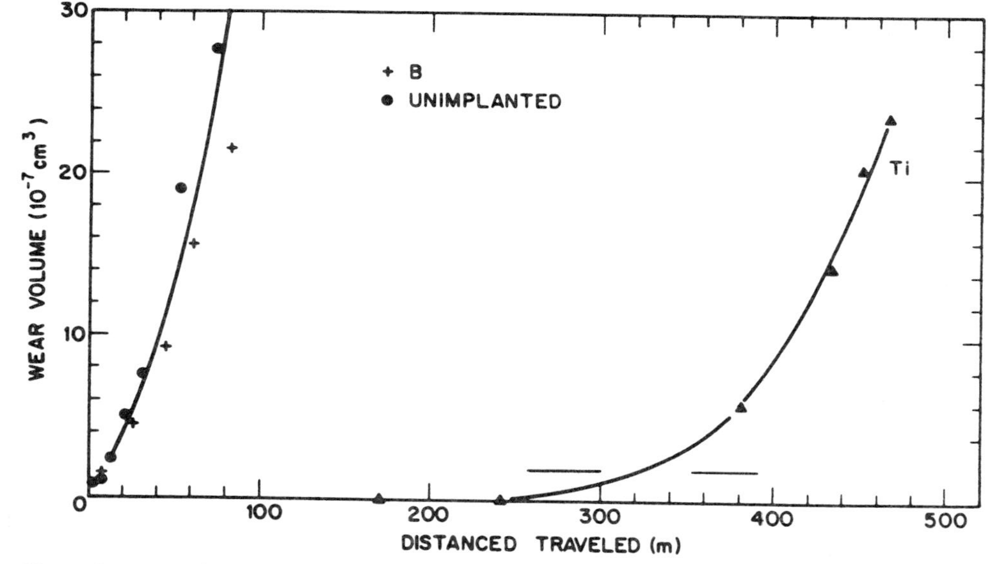

Wear volume for AISE-52100 ball on 52100 disk surfaces unimplanted, implanted with B, and implanted with 5×10^{17} Ti ions/cm². The arrows beside the Ti curve indicate that the onset of severe wear is variable but begins at substantially larger distances than in the unimplanted specimen.

(a)

FIGURE 186. Examples of the wear of ion-implanted 52100 steel (from Smidt, F. A., J. K. Hirvonen and S. Ramalingam. 1983. In *Ion Implantation for Materials Processing*, F. A. Smidt, editor, Park Ridge, NJ: Noyes Data Corporation).

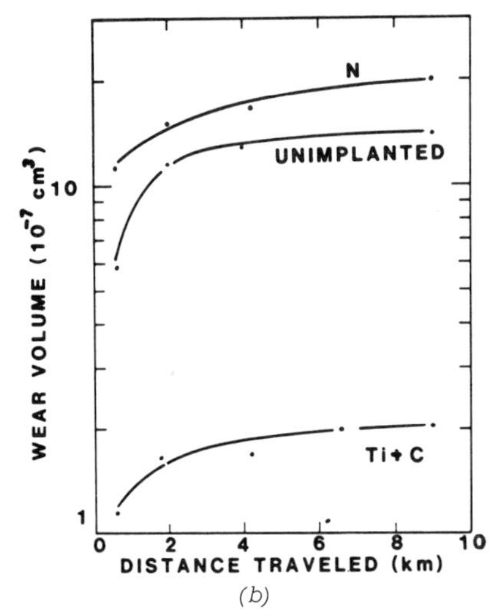

(b)

Wear experiments using a ball on cylinder configuration to investigate the "running-in" portion of the wear regime. Ti implants of 2 × 10^{17} ions/cm^2 show substantially better performance than unimplanted and nitrogen implanted specimens.

FIGURE 186 (continued). Examples of the wear of ion-implanted 52100 steel (from Smidt, F. A., J. K. Hirvonen and S. Ramalingam. 1983. In *Ion Implantation for Materials Processing*, F. A. Smidt, editor, Park Ridge, NJ: Noyes Data Corporation).

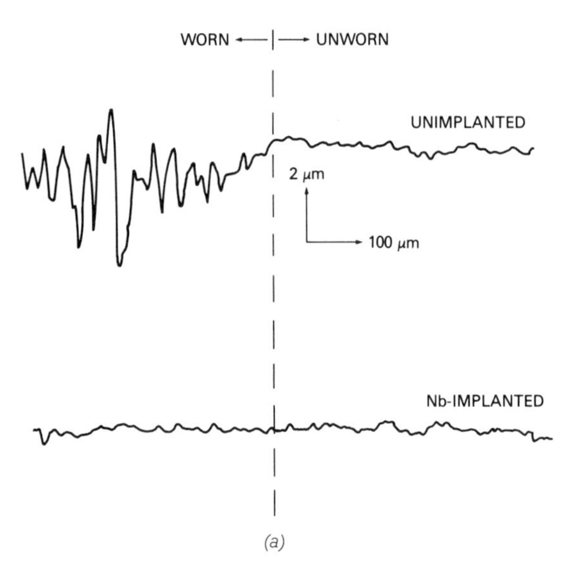

(a)

FIGURE 187. Surface profiles of unimplanted and niobium- and tantalum-implanted 9310 steel after wear-testing (reprinted with permission from Hubler, G. K. and E. T. Hodge. 1986. In *Surface Alloying by Ion, Electron and Laser Beams*, L. E. Rehn, S. T. Picraux and H. Widersich, editors, Metals Park, Ohio: American Society for Metals).

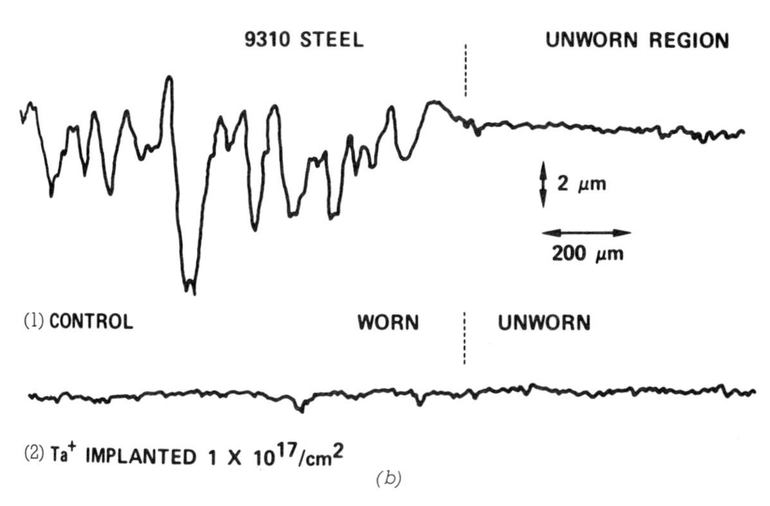

FIGURE 187 (continued). Surface profiles of unimplanted and niobium- and tantalum-implanted 9310 steel after wear-testing (reprinted with permission from Hartley, E. W. and J. K. Hirvonen. 1983. *Nucl. Instrum. and Methods in Phys.*, 209/210:933 Amsterdam: Elsevier Science Publishers).

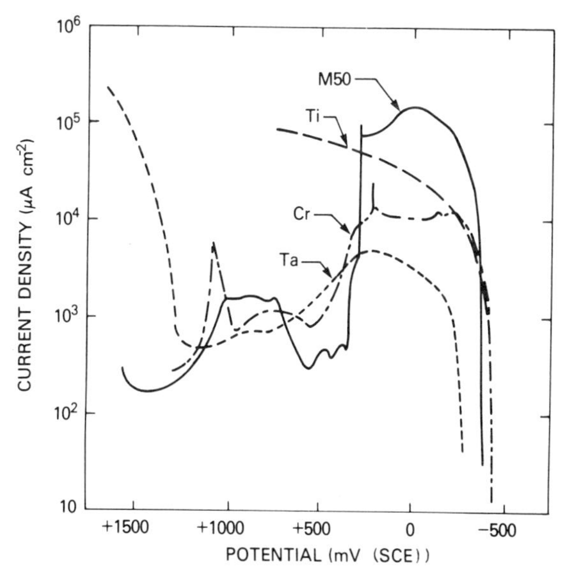

FIGURE 188. The effect of ion implantation of M50 tool steel (0.85% C, 4% Cr, 4% Mo, 1% V) with Ti, Mo and Cr on the corrosion behavior. These are potentiodynamic polarization curves produced in 0.5 M H_2SO_4 (adapted with permission from Wang, Y. F., C. R. Clayton, G. K. Hubler, W. H. Lucke and G. K. Hirvonen. 1979. *Thin Solid Films*, 63:11, Elsevier Sequoia S. A.).

may be associated with the formation of hard TiC carbides, or with the formation of an amorphous layer. This latter effect requires the presence of carbon.

The effect of Ti and C additions on a hardened 440 C stainless steel was shown in Figure 174. Note that nitrogen had no beneficial effect. In this case, the implanted surface layer was amorphous. Also note that the implantation was less effective at higher loads.

OTHER IMPLANTED IONS

Some success has been obtained with implantation of other ions. Figure 187 shows the wear improvement of a 9310 steel by implantation with niobium and tantalum. Improved corrosion resistance of M50 bearing steel was found upon implantation with Cr and Mo, as shown in Figure 188.

5f. CLOSURE

In this chapter, the general ideas behind ion implantation of the surfaces of steels have been presented. No attempt has been made to give a detailed review. This is partly because of the complexity of the subject, and partly because the technology is changing rapidly. The reader is referred to the following sources for additional information.

It is clear that ion implantation of the surfaces of steels is a viable method of affecting useful properties. However, the reasons for the improvement in properties (or the lack of improvement in some cases) are not well understood. This is partly due to the complex microstructures that form, and the sensitivity of their formation to the processing variables.

For further reading on the topic of ion implantation, the following sources are highly recommended.

1. Ryssel, H. and H. Glawischnig, editors. 1983. *Ion Implantation: Equipment and Techniques*. New York: Springer-Verlag.
2. Hubler, G. K., O. W. Holland, C. R. Clayton and C. W. White, editors. 1984. *Ion Implantation and Ion Beam Processing of Materials*. New York: North-Holland.
3. 1985. *Nuclear Instruments and Methods in Physics Research, Section B, Vol. B7, B8.*
4. 1985. *Materials Science and Engineering, Vol. 69.*
5. McHargue, C. J. 1986. "Ion Implantation in Metals and Ceramics," in *International Metals Reviews*, 31:49.

6. Rehn, L. E., S. T. Picraux and H. Wiedersich, editors. 1987. *Surface Alloying by Ion, Electron, and Laser Beams*. Metals Park, OH: American Society for Metals.
7. Destefani, J. D. 1988. "Ion Implantation Update," in *Advanced Materials and Processes*, 123(4).
8. Hochman, R. F., H. Lalnick-Legg and K. O. Legg, editors. 1988. *Ion Implantation and Plasma Assisted Processes*. Metals Park, Ohio: ASM International.

CHAPTER 6 A Comment on References

There is a wealth of literature on the surface treatments of steels. When information is presented in this book, many of these are referenced. However, it is useful to list a few general sources (in English) that cover in some detail aspects of conventional surface treatments reviewed in the book and some that deal with the newer surface treatments. The older sources (many excellent and useful) can be traced from references in the following list.

1. 1981. *Metals Handbook, 9th Edition, Vol. 4, Heat Treating*. Metals Park, OH: American Society for Metals.
2. Thelning, K.-E. 1986. *Steel and Its Heat Treatment, 2nd Edition*. Boston: Butterworths.
3. Semiatin, S. L., D. E. Stutz and I. L. Harry. 1986. *Induction Heat Treatment of Steel*. Metals Park, OH: American Society for Metals.
4. 1977. *Source Book on Nitriding*. Metals Park, OH: American Society for Metals.
5. 1977. *Carburizing and Carbonitriding*. Metals Park, OH: American Society for Metals.
6. Niku-Lari, A., editor. 1984. *Advances in Surface Treatments, Vol. 1*. NY: Pergamon Press; Niku-Lari, A., editor. 1986. *Advances in Surface Treatments, Vol. 2* NY: Pergamon Press.
7. Nemenyi, R. 1984. *Controlled Atmospheres for Heat Treatment*. NY: Pergamon Press.
8. Parrish, G. and G. S. Harper. 1985. *Production Gas Carburizing*. NY: Pergamon Press.
9. Krauss, G., editor. 1988. *Heat Treatment and Surface Engineering*. Metals Park, OH: ASM International.
10. Budinski, K. G. 1988. *Surface Engineering for Wear Resistance*. Englewood Cliffs, NJ: Prentice Hall.
11. Parrish, G. 1980. *The Influence of Microstructure on the Properties of Case-Carburized Components*. Metals Park, OH: American Society for Metals.

12. Diesburg, D. E., editor. 1984. *Case-Hardened Steels: Microstructural and Residual Stress Effects*. Warrendale, PA: Metallurgical Society of AIME.
13. Boyer, H. E., editor. 1987. *Case Hardening of Steel*. Metals Park, OH: ASM International.
14. Krauss, G., editor. 1989. *Carburizing*. Metals Park, OH: ASM International.
15. Sinha, A. K. 1989. *Ferrous Physical Metallurgy*. Boston: Butterworths.
16. Krauss, G. 1990. *Steels: Heat Treatment and Processing Principles*, Materials Park, OH: ASM International.

Three journals are listed that carry articles covering surface treatments.

17. *Metallurgical Transactions*, published by ASM International, Materials Park, Ohio.
18. *Journal of Heat Treating*, published by ASM International, Materials Park, Ohio.
19. *Heat Treatment of Metals,* published by Wolfson Heat Treatment Centre, Aston University, Aston Triangle, Birmingham B4 7ET, England.

APPENDIX 1 **Physical Properties of Pure Metals**

Many physical properties depend on the purity and physical state (annealed, hard drawn, cast, etc.) of the metal. The data refer to metal in the highest state of purity available, and are sufficiently accurate for most purposes. The reader should, however, consult the references before accepting the values quoted as applying to a particular sample.

Metal	*Melting point* °C	*Boiling point* °C	*Density at* 20 °C $g cm^{-3}$‡	*Thermal conductivity* 0–100 °C $W m^{-1} K^{-1}$	*Mean specific heat* 0–100 °C $J kg^{-1} K^{-1}$	*Resistivity at* 20 °C μΩcm	*Temp. coeff. of resistivity* 0–100 °C $10^{-3} K^{-1}$	*Coefficient of expansion* 0–100 °C $10^{-6} K^{-1}$
Aluminium	660.37	2 520	2.70	238	917	2.67	4.5	23.5
Antimony	630.74	1 590	6.68	23.8	209	40.1	5.1	8–11
Arsenic	(817)	616	5.727	—	331	33.3	—	5.6
Barium	729	2 130	3.5	—	285	60 (0 °C)	—	18
Beryllium	1 287	2 470	1.848	194	2052	3.3	9.0	12
Bismuth	271.442	1 564	9.80	9	124.8	117	4.6	13.4
Cadmium	321.108	767	8.64	103	233.2	7.3	4.3	31
Caesium	28.5	670	1.87	36.1 (s)	234	20	4.8	97
Calcium	839	1 484	1.54	125	624	3.7	4.57	22
Cerium	798	3 430	6.75	11.9	188	85.4	8.7	8
Chromium	1 860	2 680	7.1	91.3	461	13.2	2.14	6.5
Cobalt	1 494	2 930	8.9	96	427	6.34	6.6	12.5
Copper	1 084.8	2 560	8.96	397	386.0	1.694	4.3	17.0
Dysprosium	1 500	(2 630)	8.536	10.0	173	91	1.19	8.6
Erbium	1 530	(2 600)	9.051	9.6	166	86	2.01	9.2
Gadolinium	1 350	(3 000)	α7.895 β7.80	8.8	298	134	0.9/1.76	6.4
Gallium	29.7	2 205	5.91	41.0 (s)	377	*	—	18.3
Germanium	937	2 830	5.32	56.4	310	$\sim 89 \times 10^3$	—	5.75
Gold	1 064.43	2 860	19.3[2]	315.5	130	2.20	4.0	14.1
Hafnium	2 227	4 600	13.1	22.9	147	32.2	4.4	6.0
Holmium	1 461	2 600	8.803	—	164	94	1.71	9.5**
Indium	156.4	2 070	7.3	80.0	243	8.8	5.2	24.8
Iridium	2 447	4 390	22.4	146.9	130.6	5.1	4.5	6.8
Iron	1 536	2 860	7.87	78.2	456	10.1	6.5	12.1
Lanthanum	920	(3 420)	α6.174 β6.186 γ5.97	13.8	200	57	2.18	4.9

Metal	*Melting point* °C	*Boiling point* °C	*Density at 20°C* $g\,cm^{-3}$‡	*Thermal conductivity 0–100°C* $W\,m^{-1}\,K^{-1}$	*Mean specific heat 0–100°C* $J\,kg^{-1}\,K^{-1}$	*Resistivity at 20°C* μΩcm	*Temp. coeff. of resistivity 0–100°C* $10^{-3}\,K^{-1}$	*Coefficient of expansion 0–100°C* $10^{-6}\,K^{-1}$
Lead	327.502	1 750	11.68	34.9	129.8	20.6	4.2	29.0
Lithium	181	1 342	0.534	76.1	3 517	9.29	4.35	56
Lutetium	1 652	3 327	9.842	—	154	68	—	125**
Magnesium	649	1 090	1.74	155.5	1 038	4.2	4.25	26.0
Manganese	1 244	2 060	7.4	7.8	486	160(α)	—	23
Mercury	−38.87	357	13.546	8.65	138	95.9	1.0	61
Molybdenum	2 615	4 610	10.2	137	251	5.7	4.35	5.1
Neodymium	1 024	(3 060)	α7.004 β6.80	13.0	209	64	1.64	6.7
Nickel	1 455	2 915	8.9	88.5	452	6.9	6.8	13.3
Niobium	2 467	4 740	8.6	54.1	268	16.0	2.6	7.2
Osmium	3 030	5 000	22.5	86.9	130	8.8	4.1	4.57
Palladium	1 554	2 960	12.0	75.2	247	10.8	4.2	11.0
Platinum	1 769.9	3 830	21.45	73.4	134.4	10.58	3.92	9.0
Plutonium	640	3 235	19.84	8.4	142	146.5	—	55
Polonium	246	965	—	—	—	—	—	—
Potassium	63.2	759	0.86	104(s)	754	6.8	5.7	83
Praeseodymium	932	(3 020)	α6.782 β6.64	11.7	192	68	1.71	4.8
Radium	700	1 500	5	—	—	—	—	—
Rhenium	3 180	5 690	21.0	47.6	138	18.7	4.5	6.6
Rhodium	1 966	3 700	12.4	148	243	4.7	4.4	8.5
Rubidium	38.8	688	1.53	58.3(s)	356	12.1	4.8	9.0
Ruthenium	2 310	4 120	12.2	116.3	234	7.7	4.1	9.6
Samarium	1 072	1 803	7.536 7.40	—	181	92	1.48	—
Scandium	1 538	(2 870)	2.99	—	558	66	—	12
Selenium	220.5	685	4.79	—	339	12	—	37
Silicon	1 412	3 270	2.34	138.5	729	10^3–10^6	—	7.6
Silver	961.93	2 163	10.5	425	234	1.63	4.1	19.1
Sodium	97.8	883	0.97	128	1 227	4.7	5.5	71
Strontium	770	1 375	2.6	—	737	23 (0°C)	—	100
Tantalum	2 980	5 370	16.6	57.55	142	13.5	3.5	6.5
Terbium	1 356	(2 500)	8.272	—	172	116	—	7.0
Tellurium	450	988	6.24	3.8	134	1.6×10^5 (0°C)	— —	1.7 ‖c axis 27.5 ⊥c axis
Thallium	304	1 473	11.85	45.5	130	16.6	5.2	30
Thorium	1 755	4 290	11.5	49.2	100	14	4.0	11.2
Thulium	1 543	1 727	9.322	—	160	90	1.95	11.6**
Tin	231.968	2 625	7.3	73.2	226	12.6	4.6	23.5
Titanium	1 667	3 285	4.5	21.6	528	54	3.8	8.9
Tungsten	3 387	5 555	19.3	174	138	5.4	4.8	4.5
Uranium	1 132	4 400	19.05(α) 18.89(β)	28	117	27	3.4	‡
Vanadium	1 902	3 410	6.1	31.6	498	19.6	3.9	8.3
Ytterbium	824	1 427	6.977 6.54	—	145	28	1.30	25.0

Metal	Melting point °C	Boiling point °C	Density at 20°C $g cm^{-3}$‡	Thermal conductivity 0–100°C $W m^{-1} K^{-1}$	Mean specific heat 0–100°C $J kg^{-1} K^{-1}$	Resistivity at 20°C μΩcm	Temp. coeff. of resistivity 0–100°C $10^{-3} K^{-1}$	Coefficient of expansion 0–100°C $10^{-6} K^{-1}$
Yttrium	1520	3300	4.478 4.25	10.2	309	53	2.71	10.8**
Zinc	419.58	911	7.14	119.5	394	5.96	4.2	31
Zirconium	1852	4400	6.49	22.6	289	44	4.4	5.9

(s) = solid   * *See* 'Electrical properties.'   ‡ α-Uranium 23 ∥ *a* axis, −3.5 ∥ *b* axis, 17 ∥ *c* axis } 25–300°C   β-Uranium 4.6 ∥ *c* axis, 23 ⊥ *c* axis } 20–720°C

** At 400°C.   ‡ Densities of higher allotropes not at 20°C.

Rare Earths and Rare Metals ( ).

Melting and boiling points (1) *see also* 'Thermochemical data' p. 8–1. Electrical resistivity (2,3) *see also* 'Electrical properties, p. 19–1. Specific heat (4,5) Thermal conductivity (6).

APPENDIX 2 **Metric Conversion Factors**

To convert from	To	Multiply by
angstrom	m	1.0000×10^{-10}(a)
atm	Pa	1.0133×10^{5}
Btu(b)	J	1.054×10^{3}
Btu(b)/ft²·h	W/m²	3.1525
Btu(b)/ft²·h·°F	W/m²·K	5.6745
Btu(b)·ft/h·ft²·°F	W/m·K	1.7296
Btu(b)/ft²·s	W/m²	1.135×10^{4}
Btu(b)·in./ft²·h·°F	W/m·K	1.4413×10^{-1}
Btu(b)·in./s·ft²·°F	W/m·K	5.1887×10^{2}
Btu(b)/lbm·°F	J/kg·K	4.1840×10^{3}
cal(b)	J	4.1840 (a)
cal(b)/cm·s·°C	W/m·K	4.1840×10^{2}(a)
cal(b)/g	J/kg	4.1840×10^{3}(a)
cal(b)/g·°C	J/kg·K	4.1840×10^{3}(a)
circ mil	m²	5.0671×10^{-10}
°C	K	$t_K = t_{°C} + 273.15$
degree	rad	1.7453×10^{-2}
dyne/cm²	Pa	1.0000×10^{-1}(a)
°F	°C	$t_{°C} = (t_{°F} - 32)/1.8$
°F	K	$t_K = (t_{°F} + 459.67)/1.8$
ft	m	3.0480×10^{-1}
ft²	m²	9.2903×10^{-2}
ft³	m³	2.8317×10^{-2}
ft of water(c)	Pa	2.9890×10^{3}
ft²/h (thermal diffusivity)	m²/s	2.58064×10^{-5}(a)
ft·lbf	J	1.3558
ft·lbf/s	W	1.3558
ft/s	m/s	3.0480×10^{-1}
gauss	T	1.0000×10^{-4}(a)
gallon(d)	m³	3.7854×10^{-3}
g/cm³	kg/m³	1.0000×10^{3}(a)
g/cm³	Mg/m³	1.0000(a)
hp(e)	W	7.4570×10^{2}
hp(f)	W	7.4600×10^{2}
in.	m	2.5400×10^{-2}
in.²	m²	6.4516×10^{-4}
in.³	m³	1.6387×10^{-5}
in. of Hg(g)	Pa	3.3864×10^{3}
in. of water(c)	Pa	2.4908×10^{2}
K	°C	$t_{°C} = t_K - 273.15$

Reprinted with permission from Boyer, H. E. and T. L. Gall, editors. 1985. *Metals Handbook, Desk Edition*. Metals Park, Ohio: American Society for Metals.

To convert from	To	Multiply by
kgf	N	9.80665(a)
kgf/mm^2	Pa	9.80665×10^6(a)
ksi	MPa	6.8948
ksi	Pa	6.8948×10^6
ksi$\sqrt{\text{in.}}$	MPa$\sqrt{\text{m}}$	1.089
lb(h)	kg	4.5359×10^{-1}
lb/in.^{3}	kg/m^3	2.7680×10^4
lbf	N	4.4482
lbf·in.	N·m	1.1298×10^{-1}
lbf·ft	N·m	1.3558
MPa$\sqrt{\text{m}}$	MNm$^{-3/2}$	1.0000(a)
μin.	m	2.5400×10^8(a)
mil	m	2.5400×10^{-5}(a)
N/m^2	Pa	1.0000(a)
oersted	A/m	79.578
oz/ft^2	kg/m^2	3.0515×10^{-1}
psi	Pa	6.8948×10^3
°R	K	$t_K = t_{°R}/1.8$
ton(j)	kg	9.0718×10^2
ton(k)	kg	1.0160×10^3
ton/in.^{2}	Pa	1.3786×10^4
tonne	kg	1.0000×10^3(a)
torr	Pa	1.3332×10^2
Ω/circ mil·ft	Ω·m	1.6624×10^{-9}

(a) Exactly. (b) Thermochemical. (c) At 4 °C (39.2 °F). (d) U.S. liquid. (e) Mechanical (1 hp = 550 ft·lbf/s). (f) Electrical. (g) At 0 °C (32 °F). (g) Avoirdupois. (j) Short; equal to 2000 lbm. (k) Long; 2240 lbm.

APPENDIX 3 **Common Units for Converting from the English to the Metric (SI) System**

The International System of Units (SI for short) is a modernized version of the metric system. It is built upon seven base units and two supplementary units. Derived units are related to base and supplementary units by formulas in the right-hand column. Symbols for units with specific names are given in parentheses. The information in this Data Sheet, adapted from the revised "Metric Practice Guide," Standard E380 ASTM, includes a selected list of factors for converting U. S. customary units to SI units.

Metric Units and Conversion Factors

Quantity	Unit	Formula
	Base Units	
length	metre (m)	
mass	kilogram (kg)	
time	second (s)	
electric current	ampere (A)	
thermodynamic temperature	kelvin (K)	
amount of substance	mole (mol)	
luminous intensity	candela (cd)	
	Supplementary Units	
plane angle	radian (rad)	
solid angle	steradian (sr)	
	Derived Units	
acceleration	metre per second squared	m/s^2
activity (of a radioactive source)	disintegration per second	(disintegration)/s
angular acceleration	radian per second squared	rad/s^2
angular velocity	radian per second	rad/s
area	square metre	m^2
density	kilogram per cubic metre	kg/m^3
electric capacitance	farad (F)	A • s/V
electric conductance	siemens (S)	A/V
electric field strength	volt per metre	V/m
electric inductance	henry (H)	V • s/A
electric potential difference	volt (V)	W/A
electric resistance	ohm (Ω)	V/A
electromotive force	volt (V)	W/A
energy	joule (J)	N • m
entropy	joule per kelvin	J/K
force	newton (N)	$kg \cdot m/s^2$
frequency	hertz (Hz)	(cycle)/s
illuminance	lux (lx)	lm/m^2
luminance	candela per square metre	cd/m^2
luminous flux	lumen (lm)	cd • sr
magnetic field strength	ampere per metre	A/m
magnetic flux	weber (Wb)	V • s
magnetic flux density	tesla (T)	Wb/m^2
magnetomotive force	ampere (A)	—
power	watt (W)	J/s
pressure	pascal (Pa)	N/m^2

Quantity	Unit	Formula
quantity of electricity	coulomb (C)	A • s
quantity of heat	joule (J)	N • m
radiant intensity	watt per steradian	W/sr
specific heat	joule per kilogram-kelvin	J/kg • K
stress	pascal (Pa)	N/m^2
thermal conductivity	watt per metre-kelvin	W/m • K
velocity	metre per second	m/s
viscosity, dynamic	pascal-second	Pa • s
viscosity, kinematic	square metre per second	m^2/s
voltage	volt (V)	W/A
volume	cubic metre	m^3
wavenumber	reciprocal metre	(wave)/m
work	joule (J)	N • m

Metric Conversion Factors

To convert from	To	Multiply by
atmosphere (760 mm Hg)	Pa	$1.013\,25 \times 10^{5}$
Btu (International Table)	J	$1.055\,056 \times 10^{3}$
Btu (International Table)/hour	W	$2.930\,711 \times 10^{-1}$
calorie (International Table)	J	4.186 800*
centipoise	Pa • s	$1.000\,000^* \times 10^{-3}$
centistoke	m^2/s	$1.000\,000^* \times 10^{-6}$
circular mil	m^2	$5.067\,075 \times 10^{-10}$
degree Fahrenheit	°C	$t_C = (t_F - 32)/1.8$
foot	m	$3.048\,000^* \times 10^{-1}$
$foot^2$	m^2	$9.290\,304^* \times 10^{-2}$
$foot^3$	m^3	$2.831\,685 \times 10^{-2}$
foot-pound-force	J	1.355 818
foot-pound-force/minute	W	$2.259\,697 \times 10^{-2}$
$foot/second^2$	m/s^2	$3.048\,000^* \times 10^{-1}$
gallon (U.S. liquid)	m^3	$3.785\,412 \times 10^{-3}$
horsepower (electric)	W	$7.460\,000^* \times 10^{2}$
inch	m	$2.540\,000^* \times 10^{-2}$
$inch^2$	m^2	$6.451\,600^* \times 10^{-4}$
$inch^3$	m^3	$1.638\,706 \times 10^{-5}$
inch of mercury (60 F)	Pa	$3.376\,85 \times 10^{3}$
inch of water (60 F)	Pa	$2.488\,4 \times 10^{2}$
$kilogram\text{-}force/centimetre^2$	Pa	$9.806\,650^* \times 10^{4}$
kip (1000 lbf)	N	$4.448\,222 \times 10^{3}$
$kip/inch^2$ (ksi)	Pa	$6.894\,757 \times 10^{6}$
ounce (U.S. fluid)	m^3	$2.957\,353 \times 10^{-5}$
ounce-force (avoirdupois)	N	$2.780\,139 \times 10^{-1}$
ounce-mass (avoirdupois)	kg	$2.834\,952 \times 10^{-2}$
$ounce\text{-}mass/ft^2$	kg/m^2	$3.051\,52 \times 10^{-1}$
$ounce\text{-}mass/yard^2$	kg/m^2	$3.390\,575 \times 10^{-2}$
pint (U.S. liquid)	m^3	$4.731\,765 \times 10^{-4}$
pound-force (lbf avoirdupois)	N	4.448 222
pound-mass (lbm avoirdupois)	kg	$4.535\,924 \times 10^{-1}$
$pound\text{-}force/inch^2$ (psi)	Pa	$6.894\,757 \times 10^{3}$
$pound\text{-}mass/inch^3$	kg/m^3	$2.767\,990 \times 10^{4}$
$pound\text{-}mass/foot^3$	kg/m^3	$1.601\,846 \times 10$
quart (U.S. liquid)	m^3	$9.463\,529 \times 10^{-4}$
ton (short, 2000 lbm)	kg	$9.071\,847 \times 10^{2}$
torr (mm-Hg)	Pa	$1.333\,22 \times 10^{2}$
watt-hour	J	$3.600\,000^* \times 10^{3}$
yard	m	$9.144\,000^* \times 10^{-1}$
$yard^2$	m^2	$8.361\,274 \times 10^{-1}$
$yard^3$	m^3	$7.645\,549 \times 10^{-1}$

*Exact

Multiplication Factors	Prefix	SI Symbol
$1\,000\,000\,000\,000 = 10^{12}$	tera	T
$1\,000\,000\,000 = 10^{9}$	giga	G
$1\,000\,000 = 10^{6}$	mega	M
$1\,000 = 10^{3}$	kilo	k
$100 = 10^{2}$	hecto*	h
$10 = 10^{1}$	deka*	da
$0.1 = 10^{-1}$	deci*	d
$0.01 = 10^{-2}$	centi*	c
$0.001 = 10^{-3}$	milli	m
$0.000\,001 = 10^{-6}$	micro	μ
$0.000\,000\,001 = 10^{-9}$	nano	n
$0.000\,000\,000\,001 = 10^{-12}$	pico	p
$0.000\,000\,000\,000\,001 = 10^{-15}$	femto	f
$0.000\,000\,000\,000\,000\,001 = 10^{-18}$	atto	a

*To be avoided where possible

APPENDIX 4 **Metric Energy Conversions**

The middle column of figures (in bold-faced type) contains the reading (in J or ft·lb) to be converted. If converting from ft·lb to J, read the J equivalent in the column headed "J". If converting from J to ft·lb, read the equivalent in the column headed "ft·lb". 1 ft·lb = 1.355818 J.

ft·lb		J	ft·lb		J	ft·lb		J
0.7376	**1**	1.3558	28.7649	**39**	52.8769	56.7923	**77**	104.3980
1.4751	**2**	2.7116	29.5025	**40**	54.2327	57.5298	**78**	105.7538
2.2127	**3**	4.0675	30.2400	**41**	55.5885	58.2674	**79**	107.1096
2.9502	**4**	5.4233	30.9776	**42**	56.9444	59.0050	**80**	108.4654
3.6878	**5**	6.7791	31.7152	**43**	58.3002	59.7425	**81**	109.8212
4.4254	**6**	8.1349	32.4527	**44**	59.6560	60.4801	**82**	111.1771
5.1629	**7**	9.4907	33.1903	**45**	61.0118	61.2177	**83**	112.5329
5.9005	**8**	10.8465	33.9279	**46**	62.3676	61.9552	**84**	113.8887
6.6381	**9**	12.2024	34.6654	**47**	63.7234	62.6928	**85**	115.2445
7.3756	**10**	13.5582	35.4030	**48**	65.0793	63.4303	**86**	116.6003
8.1132	**11**	14.9140	36.1405	**49**	66.4351	64.1679	**87**	117.9562
8.8507	**12**	16.2698	36.8781	**50**	67.7909	64.9055	**88**	119.3120
9.5883	**13**	17.6256	37.6157	**51**	69.1467	65.6430	**89**	120.6678
10.3259	**14**	18.9815	38.3532	**52**	70.5025	66.3806	**90**	122.0236
11.0634	**15**	20.3373	39.0908	**53**	71.8583	67.1182	**91**	123.3794
11.8010	**16**	21.6931	39.8284	**54**	73.2142	67.8557	**92**	124.7452
12.5386	**17**	23.0489	40.5659	**55**	74.5700	68.5933	**93**	126.0911
13.2761	**18**	24.4047	41.3035	**56**	75.9258	69.3308	**94**	127.4469
14.0137	**19**	25.7605	42.0410	**57**	77.2816	70.0684	**95**	128.8027
14.7512	**20**	27.1164	42.7786	**58**	78.6374	70.8060	**96**	130.1585
15.4888	**21**	28.4722	43.5162	**59**	79.9933	71.5435	**97**	131.5143
16.2264	**22**	29.8280	44.2537	**60**	81.3491	72.2811	**98**	132.8702
16.9639	**23**	31.1838	44.9913	**61**	82.7049	73.0186	**99**	134.2260
17.7015	**24**	32.5396	45.7288	**62**	84.0607	73.7562	**100**	135.5818
18.4390	**25**	33.8954	46.4664	**63**	85.4165	77.4440	**105**	142.3609
19.1766	**26**	35.2513	47.2040	**64**	86.7723	81.1318	**110**	149.1400
19.9142	**27**	36.6071	47.9415	**65**	88.1282	84.8196	**115**	155.9191
20.6517	**28**	37.9629	48.6791	**66**	89.4840	88.5075	**120**	162.6982
21.3893	**29**	39.3187	49.4167	**67**	90.8398	92.1953	**125**	169.4772
22.1269	**30**	40.6745	50.1542	**68**	92.1956	95.8831	**130**	176.2563
22.8644	**31**	42.0304	50.8918	**69**	93.5514	99.5709	**135**	183.0354
23.6020	**32**	43.3862	51.6293	**70**	94.9073	103.2587	**140**	189.8145
24.3395	**33**	44.7420	52.3669	**71**	96.2631	106.9465	**145**	196.5936
25.0771	**34**	46.0978	53.1045	**72**	97.6189	110.6343	**150**	203.3727
25.8147	**35**	47.4536	53.8420	**73**	98.9747	114.3221	**155**	210.1518
26.5522	**36**	48.8094	54.5796	**74**	100.3305	118.0099	**160**	216.9308
27.2898	**37**	50.1653	55.3172	**75**	101.6863	121.6977	**165**	223.7099
28.0274	**38**	51.5211	56.0547	**76**	103.0422	125.3856	**170**	230.4890

Reprinted with permission from Boyer, H. E. and T. L. Gall, editors. 1985. *Metals Handbook, Desk Edition*. Metals Park, Ohio: American Society for Metals.

ft·lb		J	ft·lb		J	ft·lb		J
129.0734	**175**	237.2681	199.1418	**270**	366.0708	287.6492	**390**	528.7690
132.7612	**180**	244.0472	206.5174	**280**	379.6290	295.0248	**400**	542.3272
136.4490	**185**	250.8263	213.8930	**290**	393.1872	302.4005	**410**	555.8854
140.1368	**190**	257.6054	221.2686	**300**	406.7454	309.7761	**420**	569.4435
143.8246	**195**	264.3845	228.6442	**310**	420.3036	317.1517	**430**	583.0017
147.5124	**200**	271.1636	236.0199	**320**	433.8617	324.5273	**440**	596.5599
154.8880	**210**	284.7218	243.3955	**330**	447.4199	331.9029	**450**	610.1181
162.2637	**220**	298.2799	250.7711	**340**	460.9781	339.2786	**460**	623.6762
169.6393	**230**	311.8381	258.1467	**350**	474.5363	346.6542	**470**	637.2344
177.0149	**240**	325.3963	265.5224	**360**	488.0944	354.0298	**480**	650.7926
184.3905	**250**	338.9545	272.8980	**370**	501.6526	361.4054	**490**	664.3508
191.7661	**260**	352.5126	280.2736	**380**	515.2108	368.7811	**500**	677.9090

APPENDIX 5 **Metric Stress or Pressure Conversions**

The middle column of figures (in bold-faced type) contains the reading (in MPa or ksi) to be converted. If converting from ksi to MPa, read the MPa equivalent in the column headed "MPa." If converting from MPa to ksi, read the ksi equivalent in the column headed "ksi." 1 ksi = 6.894757 MPa. 1 psi = 6.894757 kPa.

ksi		MPa	ksi		MPa	ksi		MPa
0.14504	**1**	6.895	6.0916	**42**	289.58	12.038	**83**	572.26
0.29008	**2**	13.790	6.2366	**43**	296.47	12.183	**84**	579.16
0.43511	**3**	20.684	6.3817	**44**	303.37	12.328	**85**	586.05
0.58015	**4**	27.579	6.5267	**45**	310.26	12.473	**86**	592.95
0.72519	**5**	34.474	6.6717	**46**	317.16	12.618	**87**	599.84
0.87023	**6**	41.369	6.8168	**47**	324.05	12.763	**88**	606.74
1.0153	**7**	48.263	6.9618	**48**	330.95	12.909	**89**	613.63
1.1603	**8**	55.158	7.1068	**49**	337.84	13.053	**90**	620.53
1.3053	**9**	62.053	7.2519	**50**	344.74	13.198	**91**	627.42
1.4504	**10**	68.948	7.3969	**51**	351.63	13.343	**92**	634.32
1.5954	**11**	75.842	7.5420	**52**	358.53	13.489	**93**	641.21
1.7405	**12**	82.737	7.6870	**53**	365.42	13.634	**94**	648.11
1.8855	**13**	89.632	7.8320	**54**	372.32	13.779	**95**	655.00
2.0305	**14**	96.527	7.9771	**55**	379.21	13.924	**96**	661.90
2.1756	**15**	103.42	8.1221	**56**	386.11	14.069	**97**	668.79
2.3206	**16**	110.32	8.2672	**57**	393.00	14.214	**98**	675.69
2.4656	**17**	117.21	8.4122	**58**	399.90	14.359	**99**	682.58
2.6107	**18**	124.11	8.5572	**59**	406.79	14.504	**100**	689.48
2.7557	**19**	131.00	8.7023	**60**	413.69	15.954	**110**	758.42
2.9008	**20**	137.90	8.8473	**61**	420.58	17.405	**120**	827.37
3.0458	**21**	144.79	8.9923	**62**	427.47	18.855	**130**	896.32
3.1908	**22**	151.68	9.1374	**63**	434.37	20.305	**140**	965.27
3.3359	**23**	158.58	9.2824	**64**	441.26	21.756	**150**	1034.2
3.4809	**24**	165.47	9.4275	**65**	448.16	23.206	**160**	1103.2
3.6259	**25**	172.37	9.5725	**66**	455.05	24.656	**170**	1172.1
3.7710	**26**	179.26	9.7175	**67**	461.95	26.107	**180**	1241.1
3.9160	**27**	186.16	9.8626	**68**	468.84	27.557	**190**	1310.0
4.0611	**28**	193.05	10.008	**69**	475.74	29.008	**200**	1379.0
4.2061	**29**	199.95	10.153	**70**	482.63	30.458	**210**	1447.9
4.3511	**30**	206.84	10.298	**71**	489.53	31.908	**220**	1516.8
4.4962	**31**	213.74	10.443	**72**	496.42	33.359	**230**	1585.8
4.6412	**32**	220.63	10.588	**73**	503.32	34.809	**240**	1654.7
4.7862	**33**	227.53	10.733	**74**	510.21	36.259	**250**	1723.7
4.9313	**34**	234.42	10.878	**75**	517.11	37.710	**260**	1792.6
5.0763	**35**	241.32	11.023	**76**	524.00	39.160	**270**	1861.6
5.2214	**36**	248.21	11.168	**77**	530.90	40.611	**280**	1930.5
5.3664	**37**	255.11	11.313	**78**	537.79	42.061	**290**	1999.5
5.5114	**38**	262.00	11.458	**79**	544.69	43.511	**300**	2068.4
5.6565	**39**	268.90	11.603	**80**	551.58	44.962	**310**	2137.4
5.8015	**40**	275.79	11.748	**81**	558.48	46.412	**320**	2206.3
5.9465	**41**	282.69	11.893	**82**	565.37	47.862	**330**	2275.3

Reprinted with permission from Boyer, H. E. and T. L. Gall, editors. 1985. *Metals Handbook, Desk Edition*. Metals Park, Ohio: American Society for Metals.

ksi		MPa
49.313	**340**	2344.2
50.763	**350**	2413.2
52.214	**360**	2482.1
53.664	**370**	2551.1
55.114	**380**	2620.0
56.565	**390**	2689.0
58.015	**400**	2757.9
59.465	**410**	2826.9
60.916	**420**	2895.8
62.366	**430**	2964.7
63.817	**440**	3033.7
65.267	**450**	3102.6
66.717	**460**	3171.6
66.168	**470**	3240.5
69.618	**480**	3309.5
71.068	**490**	3378.4
72.519	**500**	3447.4
73.969	**510**	...
75.420	**520**	...
76.870	**530**	...
78.320	**540**	...
79.771	**550**	...
81.221	**560**	...
82.672	**570**	...
84.122	**580**	...
85.572	**590**	...
87.023	**600**	...
88.473	**610**	...
89.923	**620**	...
91.374	**630**	...
92.824	**640**	...
94.275	**650**	...
95.725	**660**	...
97.175	**670**	...
98.626	**680**	...
100.08	**690**	...
101.53	**700**	...
102.98	**710**	...
104.43	**720**	...
105.88	**730**	...
107.33	**740**	...
108.78	**750**	...
110.23	**760**	...
111.68	**770**	...
113.13	**780**	...
114.58	**790**	...
116.03	**800**	...
117.48	**810**	...

ksi		MPa
118.93	**820**	...
120.38	**830**	...
121.83	**840**	...
123.28	**850**	...
124.73	**860**	...
126.18	**870**	...
127.63	**880**	...
129.08	**890**	...
130.53	**900**	...
131.98	**910**	...
133.43	**920**	...
134.89	**930**	...
136.34	**940**	...
137.79	**950**	...
139.24	**960**	...
140.69	**970**	...
142.14	**980**	...
143.59	**990**	...
145.04	**1000**	...
147.94	**1020**	...
150.84	**1040**	...
153.74	**1060**	...
156.64	**1080**	...
159.54	**1100**	...
162.44	**1120**	...
165.34	**1140**	...
168.24	**1160**	...
171.14	**1180**	...
174.05	**1200**	...
176.95	**1220**	...
179.85	**1240**	...
182.75	**1260**	...
185.65	**1280**	...
188.55	**1300**	...
191.45	**1320**	...
194.35	**1340**	...
197.25	**1360**	...
200.15	**1380**	...
203.05	**1400**	...
205.95	**1420**	...
208.85	**1440**	...
211.76	**1460**	...
214.66	**1480**	...
217.56	**1500**	...
220.46	**1520**	...
223.36	**1540**	...
226.26	**1560**	...

ksi		MPa
229.16	**1580**	...
232.06	**1600**	...
234.96	**1620**	...
237.86	**1640**	...
240.76	**1660**	...
243.66	**1680**	...
246.56	**1700**	...
249.46	**1720**	...
252.37	**1740**	...
255.27	**1760**	...
258.17	**1780**	...
261.07	**1800**	...
263.97	**1820**	...
266.87	**1840**	...
269.77	**1860**	...
272.67	**1880**	...
275.57	**1900**	...
278.47	**1920**	...
281.37	**1940**	...
284.27	**1960**	...
287.17	**1980**	...
290.08	**2000**	...
292.98	**2020**	...
295.88	**2040**	...
298.78	**2060**	...
301.68	**2080**	...
304.58	**2100**	...
307.48	**2120**	...
310.38	**2140**	...
313.28	**2160**	...
316.18	**2180**	...
319.08	**2200**	...
321.98	**2220**	...
324.88	**2240**	...
327.79	**2260**	...
330.69	**2280**	...
333.59	**2300**	...
336.49	**2320**	...
339.39	**2340**	...
342.29	**2360**	...
345.19	**2380**	...
348.09	**2400**	...
350.99	**2420**	...
353.89	**2440**	...
356.79	**2460**	...
359.69	**2480**	...
362.59	**2500**	...

APPENDIX 6 **Temperature Conversions**

The general arrangement of this conversion table was devised by Sauveur and Boylston. The middle columns of numbers (in **boldface** type) contain the temperature readings (°F or °C) to be converted. When converting from degrees Fahrenheit to degrees Celsius, read the Celsius equivalent in the column headed "C." When converting from Celsius to Fahrenheit, read the Fahrenheit equivalent in the column headed "F."

F		C	F		C	F		C
.....	**−458**	−272.22	.....	**−358**	−216.67	−432.4	**−258**	−161.11
.....	**−456**	−271.11	.....	**−356**	−215.56	−428.8	**−256**	−160.00
.....	**−454**	−270.00	.....	**−354**	−214.44	−425.2	**−254**	−158.89
.....	**−452**	−268.89	.....	**−352**	−213.33	−421.6	**−252**	−157.78
.....	**−450**	−267.78	.....	**−350**	−212.22	−418.0	**−250**	−156.67
.....	**−448**	−266.67	.....	**−348**	−211.11	−414.4	**−248**	−155.56
.....	**−446**	−265.56	.....	**−346**	−210.00	−410.8	**−246**	−154.44
.....	**−444**	−264.44	.....	**−344**	−208.89	−407.2	**−244**	−153.33
.....	**−442**	−263.33	.....	**−342**	−207.78	−403.6	**−242**	−152.22
.....	**−440**	−262.22	.....	**−340**	−206.67	−400.0	**−240**	−151.11
.....	**−438**	−261.11	.....	**−338**	−205.56	−396.4	**−238**	−150.00
.....	**−436**	−260.00	.....	**−336**	−204.44	−392.8	**−236**	−148.89
.....	**−434**	−258.89	.....	**−334**	−203.33	−389.2	**−234**	−147.78
.....	**−432**	−257.78	.....	**−332**	−202.22	−385.6	**−232**	−146.67
.....	**−430**	−256.67	.....	**−330**	−201.11	−382.0	**−230**	−145.56
.....	**−428**	−255.56	.....	**−328**	−200.00	−378.4	**−228**	−144.44
.....	**−426**	−254.44	.....	**−326**	−198.89	−374.8	**−226**	−143.33
.....	**−424**	−253.33	.....	**−324**	−197.78	−371.2	**−224**	−142.22
.....	**−422**	−252.22	.....	**−322**	−196.67	−367.6	**−222**	−141.11
.....	**−420**	−251.11	.....	**−320**	−195.56	−364.0	**−220**	−140.00
.....	**−418**	−250.00	.....	**−318**	−194.44	−360.4	**−218**	−138.89
.....	**−416**	−248.89	.....	**−316**	−193.33	−356.8	**−216**	−137.78
.....	**−414**	−247.78	.....	**−314**	−192.22	−353.2	**−214**	−136.67
.....	**−412**	−246.67	.....	**−312**	−191.11	−349.6	**−212**	−135.56
.....	**−410**	−245.56	.....	**−310**	−190.00	−346.0	**−210**	−134.44
.....	**−408**	−244.44	.....	**−308**	−188.89	−342.4	**−208**	−133.33
.....	**−406**	−243.33	.....	**−306**	−187.78	−338.8	**−206**	−132.22
.....	**−404**	−242.22	.....	**−304**	−186.67	−335.2	**−204**	−131.11
.....	**−402**	−241.11	.....	**−302**	−185.56	−331.6	**−202**	−130.00
.....	**−400**	−240.00	.....	**−300**	−184.44	−328.0	**−200**	−128.89
.....	**−398**	−238.89	.....	**−298**	−183.33	−324.4	**−198**	−127.78
.....	**−396**	−237.78	.....	**−296**	−182.22	−320.8	**−196**	−126.67
.....	**−394**	−236.67	.....	**−294**	−181.11	−317.2	**−194**	−125.56
.....	**−392**	−235.56	.....	**−292**	−180.00	−313.6	**−192**	−124.44
.....	**−390**	−234.44	.....	**−290**	−178.89	−310.0	**−190**	−123.33
.....	**−388**	−233.33	.....	**−288**	−177.78	−306.4	**−188**	−122.22
.....	**−386**	−232.22	.....	**−286**	−176.67	−302.8	**−186**	−121.11
.....	**−384**	−231.11	.....	**−284**	−175.56	−299.2	**−184**	−120.00
.....	**−382**	−230.00	.....	**−282**	−174.44	−295.6	**−182**	−118.89
.....	**−380**	−228.89	.....	**−280**	−173.33	−292.0	**−180**	−117.78
.....	**−378**	−227.78	.....	**−278**	−172.22	−288.4	**−178**	−116.67
.....	**−376**	−226.67	.....	**−276**	−171.11	−284.8	**−176**	−115.56
.....	**−374**	−225.56	.....	**−274**	−170.00	−281.2	**−174**	−114.44
.....	**−372**	−224.44	−457.6	**−272**	−168.89	−277.6	**−172**	−113.33
.....	**−370**	−223.33	−454.0	**−270**	−167.78	−274.0	**−170**	−112.22
.....	**−368**	−222.22	−450.4	**−268**	−166.67	−270.4	**−168**	−111.11
.....	**−366**	−221.11	−446.8	**−266**	−165.56	−266.8	**−166**	−110.00
.....	**−364**	−220.00	−443.2	**−264**	−164.44	−263.2	**−164**	−108.89
.....	**−362**	−218.89	−439.6	**−262**	−163.33	−259.6	**−162**	−107.78
.....	**−360**	−217.78	−436.0	**−260**	−162.22	−256.0	**−160**	−106.67

Reprinted with permission from Boyer, H. E. and T. L. Gall, editors. 1985. *Metals Handbook, Desk Edition.* Metals Park, Ohio: American Society for Metals.

F		C	F		C	F		C
−252.4	−158	−105.56	+35.6	+2	−16.67	323.6	162	72.22
−248.8	−156	−104.44	+39.2	+4	−15.56	327.2	164	73.33
−245.2	−154	−103.33	+42.8	+6	−14.44	330.8	166	74.44
−241.6	−152	−102.22	+46.4	+8	−13.33	334.4	168	75.56
−238.0	−150	−101.11	+50.0	+10	−12.22	338.0	170	76.67
−234.4	−148	−100.00	+53.6	+12	−11.11	341.6	172	77.78
−230.8	−146	−98.89	+57.2	+14	−10.00	345.2	174	78.89
−227.2	−144	−97.78	+60.8	+16	−8.89	348.8	176	80.00
−223.6	−142	−96.67	+64.4	+18	−7.78	352.4	178	81.11
−220.0	−140	−95.56	+68.0	+20	−6.67	356.0	180	82.22
−216.4	−138	−94.44	+71.6	+22	−5.56	359.6	182	83.33
−212.8	−136	−93.33	+75.2	+24	−4.44	363.2	184	84.44
−209.2	−134	−92.22	+78.8	+26	−3.33	366.8	186	85.56
−205.6	−132	−91.11	+82.4	+28	−2.22	370.4	188	86.67
−202.0	−130	−90.00	+86.0	+30	−1.11	374.0	190	87.78
−198.4	−128	−88.89	+89.6	+32	±0.00	377.6	192	88.89
−194.8	−126	−87.78	+93.2	+34	+1.11	381.2	194	90.00
−191.2	−124	−86.67	+96.8	+36	+2.22	384.8	196	91.11
−187.6	−122	−85.56	+100.4	+38	+3.33	388.4	198	92.22
−184.0	−120	−84.44	+104.0	+40	+4.44	392.0	200	93.33
−180.4	−118	−83.33	107.6	42	5.56	395.6	202	94.44
−176.8	−116	−82.22	111.2	44	6.67	399.2	204	95.56
−173.2	−114	−81.11	114.8	46	7.78	402.8	206	96.67
−169.6	−112	−80.00	118.4	48	8.89	406.4	208	97.78
−166.0	−110	−78.89	122.0	50	10.00	410.0	210	98.89
−162.4	−108	−77.78	125.6	52	11.11	413.6	212	100.00
−158.8	−106	−76.67	129.2	54	12.22	417.2	214	101.11
−155.2	−104	−75.56	132.8	56	13.33	420.8	216	102.22
−151.6	−102	−74.44	136.4	58	14.44	424.4	218	103.33
−148.0	−100	−73.33	140.0	60	15.56	428.0	220	104.44
−144.4	−98	−72.22	143.6	62	16.67	431.6	222	105.56
−140.8	−96	−71.11	147.2	64	17.78	435.2	224	106.67
−137.2	−94	−70.00	150.8	66	18.89	438.8	226	107.78
−133.6	−92	−68.89	154.4	68	20.00	442.4	228	108.89
−130.0	−90	−67.78	158.0	70	21.11	446.0	230	110.00
−126.4	−88	−66.67	161.6	72	22.22	449.6	232	111.11
−122.8	−86	−65.56	165.2	74	23.33	453.2	234	112.22
−119.2	−84	−64.44	168.8	76	24.44	456.8	236	113.33
−115.6	−82	−63.33	172.4	78	25.56	460.4	238	114.44
−112.0	−80	−62.22	176.0	80	26.67	464.0	240	115.56
−108.4	−78	−61.11	179.6	82	27.78	467.6	242	116.67
−104.8	−76	−60.00	183.2	84	28.89	471.2	244	117.78
−101.2	−74	−58.89	186.8	86	30.00	474.8	246	118.89
−97.6	−72	−57.78	190.4	88	31.11	478.4	248	120.00
−94.0	−70	−56.67	194.0	90	32.22	482.0	250	121.11
−90.4	−68	−55.56	197.6	92	33.33	485.6	252	122.22
−86.8	−66	−54.44	201.2	94	34.44	489.2	254	123.33
−83.2	−64	−53.33	204.8	96	35.56	492.8	256	124.44
−79.6	−62	−52.22	208.4	98	36.67	496.4	258	125.56
−76.0	−60	−51.11	212.0	100	37.78	500.0	260	126.67
−72.4	−58	−50.00	215.6	102	38.89	503.6	262	127.78
−68.8	−56	−48.89	219.2	104	40.00	507.2	264	128.89
−65.2	−54	−47.78	222.8	106	41.11	510.8	266	130.00
−61.6	−52	−46.67	226.4	108	42.22	514.4	268	131.11
−58.0	−50	−45.56	230.0	110	43.33	518.0	270	132.22
−54.4	−48	−44.44	233.6	112	44.44	521.6	272	133.33
−50.8	−46	−43.33	237.2	114	45.56	525.2	274	134.44
−47.2	−44	−42.22	240.8	116	46.67	528.8	276	135.56
−43.6	−42	−41.11	244.4	118	47.78	532.4	278	136.67
−40.0	−40	−40.00	248.0	120	48.89	536.0	280	137.78
−36.4	−38	−38.89	251.6	122	50.00	539.6	282	138.89
−32.8	−36	−37.78	255.2	124	51.11	543.2	284	140.00
−29.2	−34	−36.67	258.8	126	52.22	546.8	286	141.11
−25.6	−32	−35.56	262.4	128	53.33	550.4	288	142.22
−22.0	−30	−34.44	266.0	130	54.44	554.0	290	143.33
−18.4	−28	−33.33	269.6	132	55.56	557.6	292	144.44
−14.8	−26	−32.22	273.2	134	56.67	561.2	294	145.56
−11.2	−24	−31.11	276.8	136	57.78	564.8	296	146.67
−7.6	−22	−30.00	280.4	138	58.89	568.4	298	147.78
−4.0	−20	−28.89	284.0	140	60.00	572.0	300	148.89
−0.4	−18	−27.78	287.6	142	61.11	575.6	302	150.00
+3.2	−16	−26.67	291.2	144	62.22	579.2	304	151.11
+6.8	−14	−25.56	294.8	146	63.33	582.8	306	152.22
+10.4	−12	−24.44	298.4	148	64.44	586.4	308	153.33
+14.0	−10	−23.33	302.0	150	65.56	590.0	310	154.44
+17.6	−8	−22.22	305.6	152	66.67	593.6	312	155.56
+21.2	−6	−21.11	309.2	154	67.78	597.2	314	156.67
+24.8	−4	−20.00	312.8	156	68.89	600.8	316	157.78
+28.4	−2	−18.89	316.4	158	70.00	604.4	318	158.89
+32.0	±0	−17.78	320.0	160	71.11	608.0	320	160.00

F		C
611.6	**322**	161.11
615.2	**324**	162.22
618.8	**326**	163.33
622.4	**328**	164.44
626.0	**330**	165.56
629.6	**332**	166.67
633.2	**334**	167.78
636.8	**336**	168.89
640.4	**338**	170.00
644.0	**340**	171.11
647.6	**342**	172.22
651.2	**344**	173.33
654.8	**346**	174.44
658.4	**348**	175.56
662.0	**350**	176.67
665.6	**352**	177.78
669.2	**354**	178.89
672.8	**356**	180.00
676.4	**358**	181.11
680.0	**360**	182.22
683.6	**362**	183.33
687.2	**364**	184.44
690.8	**366**	185.56
694.4	**368**	186.67
698.0	**370**	187.78
701.6	**372**	188.89
705.2	**374**	190.00
708.8	**376**	191.11
712.4	**378**	192.22
716.0	**380**	193.33
719.6	**382**	194.44
723.2	**384**	195.56
726.8	**386**	196.67
730.4	**388**	197.78
734.0	**390**	198.89
737.6	**392**	200.00
741.2	**394**	201.11
744.8	**396**	202.22
748.4	**398**	203.33
752.0	**400**	204.44
755.6	**402**	205.56
759.2	**404**	206.67
762.8	**406**	207.78
766.4	**408**	208.89
770.0	**410**	210.00
773.6	**412**	211.11
777.2	**414**	212.22
780.8	**416**	213.33
784.4	**418**	214.44
788.0	**420**	215.56
791.6	**422**	216.67
795.2	**424**	217.78
798.8	**426**	218.89
802.4	**428**	220.00
806.0	**430**	221.11
809.6	**432**	222.22
813.2	**434**	223.33
816.8	**436**	224.44
820.4	**438**	225.56
824.0	**440**	226.67
827.6	**442**	227.78
831.2	**444**	228.89
834.8	**446**	230.00
838.4	**448**	231.11
842.0	**450**	232.22
845.6	**452**	233.33
849.2	**454**	234.44
852.8	**456**	235.56
856.4	**458**	236.67
860.0	**460**	237.78
863.6	**462**	238.89
867.2	**464**	240.00

F		C
870.8	**466**	241.11
874.4	**468**	242.22
878.0	**470**	243.33
881.6	**472**	244.44
885.2	**474**	245.56
888.8	**476**	246.67
892.4	**478**	247.78
896.0	**480**	248.89
899.6	**482**	250.00
903.2	**484**	251.11
906.8	**486**	252.22
910.4	**488**	253.33
914.0	**490**	254.44
917.6	**492**	255.56
921.2	**494**	256.67
924.8	**496**	257.78
928.4	**498**	258.89
932.0	**500**	260.00
935.6	**502**	261.11
939.2	**504**	262.22
942.8	**506**	263.33
946.4	**508**	264.44
950.0	**510**	265.56
953.6	**512**	266.67
957.2	**514**	267.78
960.8	**516**	268.89
964.4	**518**	270.00
968.0	**520**	271.11
971.6	**522**	272.22
975.2	**524**	273.33
978.8	**526**	274.44
982.4	**528**	275.56
986.0	**530**	276.67
989.6	**532**	277.78
993.2	**534**	278.89
996.8	**536**	280.00
1000.4	**538**	281.11
1004.0	**540**	282.22
1007.6	**542**	283.33
1011.2	**544**	284.44
1014.8	**546**	285.56
1018.4	**548**	286.67
1022.0	**550**	287.78
1040.0	**560**	293.33
1058.0	**570**	298.89
1076.0	**580**	304.44
1094.0	**590**	310.00
1112.0	**600**	315.56
1130.0	**610**	321.11
1148.0	**620**	326.67
1166.0	**630**	332.22
1184.0	**640**	337.78
1202.0	**650**	343.33
1220.0	**660**	348.89
1238.0	**670**	354.44
1256.0	**680**	360.00
1274.0	**690**	365.56
1292.0	**700**	371.11
1310.0	**710**	376.67
1328.0	**720**	382.22
1346.0	**730**	387.78
1364.0	**740**	393.33
1382.0	**750**	398.89
1400.0	**760**	404.44
1418.0	**770**	410.00
1436.0	**780**	415.56
1454.0	**790**	421.11
1472.0	**800**	426.67
1490.0	**810**	432.22
1508.0	**820**	437.78
1526.0	**830**	443.33
1544.0	**840**	448.89
1562.0	**850**	454.44
1580.0	**860**	460.00
1598.0	**870**	465.56

F		C
1616.0	**880**	471.11
1634.0	**890**	476.67
1652.0	**900**	482.22
1670.0	**910**	487.78
1688.0	**920**	493.33
1706.0	**930**	498.89
1724.0	**940**	504.44
1742.0	**950**	510.00
1760.0	**960**	515.56
1778.0	**970**	521.11
1796.0	**980**	526.67
1814.0	**990**	532.22
1832.0	**1000**	537.78
1850.0	**1010**	543.33
1868.0	**1020**	548.89
1886.0	**1030**	554.44
1904.0	**1040**	560.00
1922.0	**1050**	565.56
1940.0	**1060**	571.11
1958.0	**1070**	576.67
1976.0	**1080**	582.22
1994.0	**1090**	587.78
2012.0	**1100**	593.33
2030.0	**1110**	598.89
2048.0	**1120**	604.44
2066.0	**1130**	610.00
2084.0	**1140**	615.56
2102.0	**1150**	621.11
2120.0	**1160**	626.67
2138.0	**1170**	632.22
2156.0	**1180**	637.78
2174.0	**1190**	643.33
2192.0	**1200**	648.89
2210.0	**1210**	654.44
2228.0	**1220**	660.00
2246.0	**1230**	665.56
2264.0	**1240**	671.11
2282.0	**1250**	676.67
2300.0	**1260**	682.22
2318.0	**1270**	687.78
2336.0	**1280**	693.33
2354.0	**1290**	698.89
2372.0	**1300**	704.44
2390.0	**1310**	710.00
2408.0	**1320**	715.56
2426.0	**1330**	721.11
2444.0	**1340**	726.67
2462.0	**1350**	732.22
2480.0	**1360**	737.78
2498.0	**1370**	743.33
2516.0	**1380**	748.89
2534.0	**1390**	754.44
2552.0	**1400**	760.00
2570.0	**1410**	765.56
2588.0	**1420**	771.11
2606.0	**1430**	776.67
2624.0	**1440**	782.22
2642.0	**1450**	787.78
2660.0	**1460**	793.33
2678.0	**1470**	798.89
2696.0	**1480**	804.44
2714.0	**1490**	810.00
2732.0	**1500**	815.56
2750.0	**1510**	821.11
2768.0	**1520**	826.67
2786.0	**1530**	832.22
2804.0	**1540**	837.78
2822.0	**1550**	843.33
2840.0	**1560**	848.89
2858.0	**1570**	854.44
2876.0	**1580**	860.00
2894.0	**1590**	865.56

F		C
2912.0	1600	871.11
2930.0	1610	876.67
2948.0	1620	882.22
2966.0	1630	887.78
2984.0	1640	893.33
3002.0	1650	898.89
3020.0	1660	904.44
3038.0	1670	910.00
3056.0	1680	915.56
3074.0	1690	921.11
3092.0	1700	926.67
3110.0	1710	932.22
3128.0	1720	937.78
3146.0	1730	943.33
3164.0	1740	948.89
3182.0	1750	954.44
3200.0	1760	960.00
3218.0	1770	965.56
3236.0	1780	971.11
3254.0	1790	976.67
3272.0	1800	982.22
3290.0	1810	987.78
3308.0	1820	993.33
3326.0	1830	998.89
3344.0	1840	1004.4
3362.0	1850	1010.0
3380.0	1860	1015.6
3398.0	1870	1021.1
3416.0	1880	1026.7
3434.0	1890	1032.2
3452.0	1900	1037.8
3470.0	1910	1043.3
3488.0	1920	1048.9
3506.0	1930	1054.4
3524.0	1940	1060.0
3542.0	1950	1065.6
3560.0	1960	1071.1
3578.0	1970	1076.7
3596.0	1980	1082.2
3614.0	1990	1087.8
3632.0	2000	1093.3
3650.0	2010	1098.9
3668.0	2020	1104.4
3686.0	2030	1110.0
3704.0	2040	1115.6
3722.0	2050	1121.1
3740.0	2060	1126.7
3758.0	2070	1132.2
3776.0	2080	1137.8
3794.0	2090	1143.3
3812.0	2100	1148.9
3830.0	2110	1154.4
3848.0	2120	1160.0
3866.0	2130	1165.6
3884.0	2140	1171.1
3902.0	2150	1176.7
3920.0	2160	1182.2
3938.0	2170	1187.8
3956.0	2180	1193.3
3974.0	2190	1198.9
3992.0	2200	1204.4
4010.0	2210	1210.0
4028.0	2220	1215.6
4046.0	2230	1221.1
4064.0	2240	1226.7
4082.0	2250	1232.2
4100.0	2260	1237.8

F		C
4118.0	2270	1243.3
4136.0	2280	1248.9
4154.0	2290	1254.4
4172.0	2300	1260.0
4190.0	2310	1265.6
4208.0	2320	1271.1
4226.0	2330	1276.7
4244.0	2340	1282.2
4262.0	2350	1287.8
4280.0	2360	1293.3
4298.0	2370	1298.9
4316.0	2380	1304.4
4334.0	2390	1310.0
4352.0	2400	1315.6
4370.0	2410	1321.1
4388.0	2420	1326.7
4406.0	2430	1332.2
4424.0	2440	1337.8
4442.0	2450	1343.3
4460.0	2460	1348.9
4478.0	2470	1354.4
4496.0	2480	1360.0
4514.0	2490	1365.6
4532.0	2500	1371.1
4550.0	2510	1376.7
4568.0	2520	1382.2
4586.0	2530	1387.8
4604.0	2540	1393.3
4622.0	2550	1398.9
4640.0	2560	1404.4
4658.0	2570	1410.0
4676.0	2580	1415.6
4694.0	2590	1421.1
4712.0	2600	1426.7
4730.0	2610	1432.2
4748.0	2620	1437.8
4766.0	2630	1443.3
4784.0	2640	1448.9
4802.0	2650	1454.4
4820.0	2660	1460.0
4838.0	2670	1465.6
4856.0	2680	1471.1
4874.0	2690	1476.7
4892.0	2700	1482.2
4910.0	2710	1487.8
4928.0	2720	1493.3
4946.0	2730	1498.9
4964.0	2740	1504.4
4982.0	2750	1510.0
5000.0	2760	1515.6
5018.0	2770	1521.1
5036.0	2780	1526.7
5054.0	2790	1532.2
5072.0	2800	1537.8
5090.0	2810	1543.3
5108.0	2820	1548.9
5126.0	2830	1554.4
5144.0	2840	1560.0
5162.0	2850	1565.6
5180.0	2860	1571.1
5198.0	2870	1576.7
5216.0	2880	1582.2
5234.0	2890	1587.8
5252.0	2900	1593.3
5270.0	2910	1598.9
5288.0	2920	1604.4
5306.0	2930	1610.0

F		C
5324.0	2940	1615.6
5342.0	2950	1621.1
5360.0	2960	1626.7
5378.0	2970	1632.2
5396.0	2980	1637.8
5414.0	2990	1643.3
5432.0	3000	1648.9
5450.0	3010	1654.4
5468.0	3020	1660.0
5486.0	3030	1665.6
5504.0	3040	1671.1
5522.0	3050	1676.7
5540.0	3060	1682.2
5558.0	3070	1687.8
5576.0	3080	1693.3
5594.0	3090	1698.9
5612.0	3100	1704.4
5702.0	3150	1732.2
5792.0	3200	1760.0
5882.0	3250	1787.7
5972.0	3300	1815.5
6062.0	3350	1843.3
6152.0	3400	1871.1
6242.0	3450	1898.8
6332.0	3500	1926.6
6422.0	3550	1954.4
6512.0	3600	1982.2
6602.0	3650	2010.0
6692.0	3700	2037.7
6782.0	3750	2065.5
6872.0	3800	2093.3
6962.0	3850	2121.1
7052.0	3900	2148.8
7142.0	3950	2176.6
7232.0	4000	2204.4
7322.0	4050	2232.2
7412.0	4100	2260.0
7502.0	4150	2287.7
7592.0	4200	2315.5
7682.0	4250	2343.3
7772.0	4300	2371.1
7862.0	4350	2398.8
7952.0	4400	2426.6
8042.0	4450	2454.4
8132.0	4500	2482.2
8222.0	4550	2510.0
8312.0	4600	2537.7
8402.0	4650	2565.5
8492.0	4700	2593.3
8582.0	4750	2621.1
8672.0	4800	2648.8
8762.0	4850	2676.6
8852.0	4900	2704.4
8942.0	4950	2732.2
9032.0	5000	2760.0
9122.0	5050	2787.7
9212.0	5100	2815.5
9302.0	5150	2843.3
9392.0	5200	2871.1
9482.0	5250	2898.8
9572.0	5300	2926.6
9662.0	5350	2954.4
9752.0	5400	2982.2
9842.0	5450	3010.0
9932.0	5500	3037.7
10022.0	5550	3065.5
10112.0	5600	3093.3

APPENDIX 7 **Iron-Carbon Equilibrium Diagram**

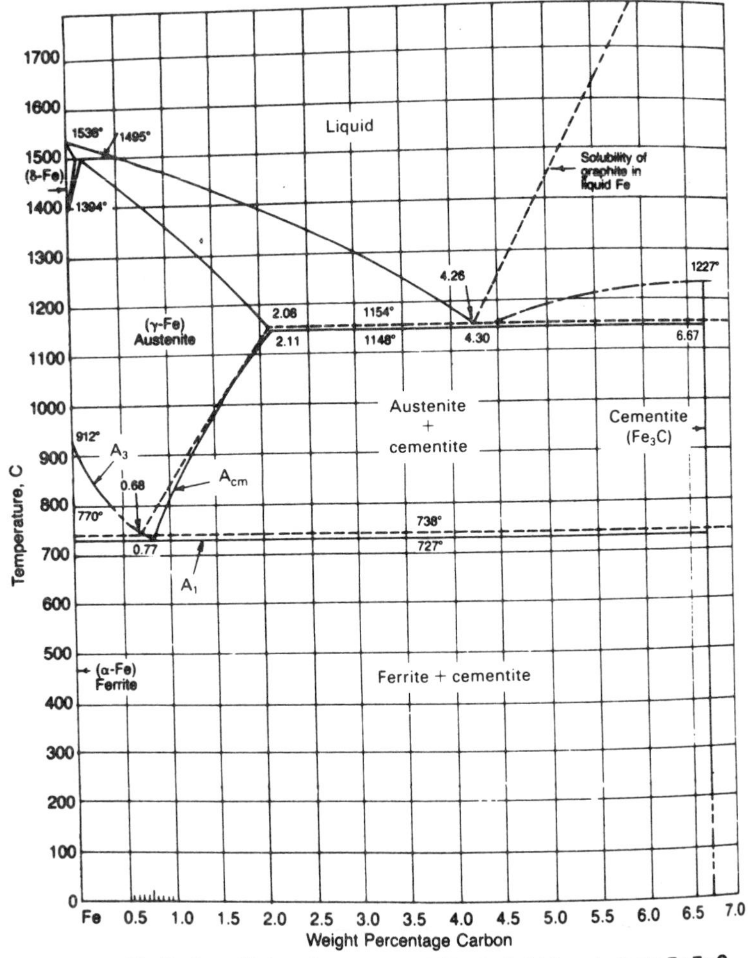

The Fe-C equilibrium diagram up to 6.67% C. Solid lines indicate Fe-Fe_3C diagram; dashed lines indicate Fe-graphite diagram.

Adapted from 1973. *Metals Handbook, 8th Ed., Vol. 8.* Metals Park, Ohio: American Society for Metals.

APPENDIX 8 **Description of Hardness Code and Specification of Hardness Numbers for Metals**

Specify hardness according to the code described below.
This code is in agreement with the method of designation used by the following standards organizations:

1. American Society for Testing and Materials (ASTM)
2. American National Standards Institute (ANSI)
3. International Standards Organization (ISO)

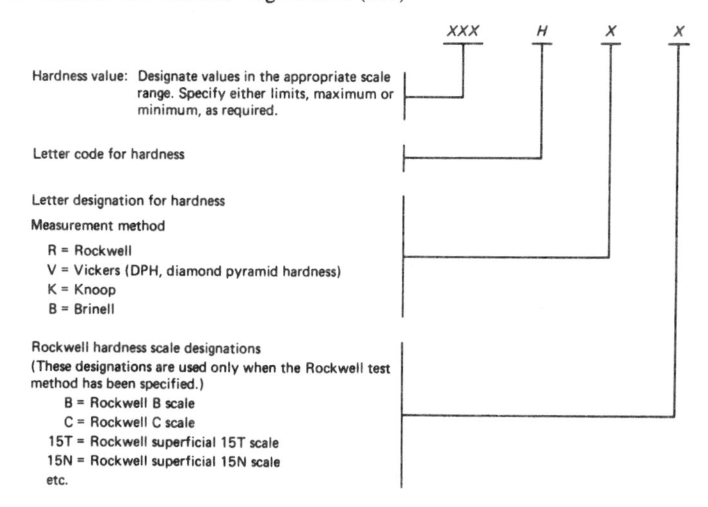

Example

1. 50-60 HRC — means: a hardness value of 50 to 60 using the Rockwell C scale.
2. 85 HR15T MAX — means: a maximum hardness value of 85 using the Rockwell Superficial 15T scale.
3. 185-240 HV 1Kgf — means: a hardness value of 185-240 using the Vickers hardness tester and a test load of 1 kilogram-force.
4. 500 HK MIN 200gf — means: a minimum hardness value of 500 using the Knoop hardness tester and a test load of 200 grams-force.

APPENDIX 9 **Comparison of Hardness Tests**

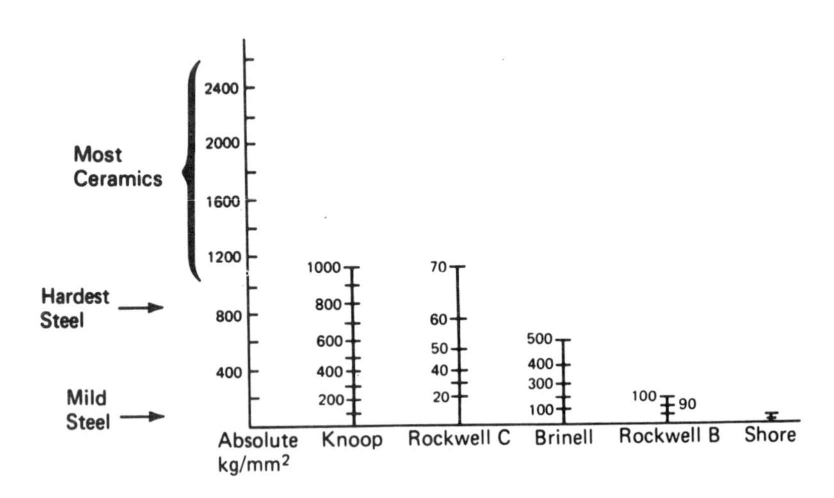

Hardness Test	Indentor	Load	Application
Knoop or Vickers	Diamond	1 g to 2000 g	Microhardness of soft steels to ceramics
Brinell	Ball	500 & 3000 kg	Soft steels & metals to 40 HRC
Rockwell B	Ball	100 kg	Soft steels & nonferrous metals
Rockwell T	Ball	15, 30 & 45 kg	Thin soft metals
Rockwell N	Diamond	15, 30 & 45 kg	Hard thin sheet metals
Rockwell A	Diamond	50 kg	Cemented carbides
Rockwell R	Ball	10 kg	Polymers
Shore Durometer	Needle	Spring	Elastomers
Rockwell C	Diamond	150 kg	Hardened metals (thick)

APPENDIX 10 **Approximate Equivalent Hardness Numbers and Tensile Strengths for Rockwell C and B Hardness Numbers for Steel**[(a)]

Rockwell C-scale hardness No.	Vickers hardness No.	Brinell hardness No., 3000-kg load, 10-mm ball: Standard ball	Brinell hardness No., 3000-kg load, 10-mm ball: Tungsten carbide ball	Rockwell hardness No.: A scale, 60-kg load, Brale indenter	Rockwell hardness No.: B scale, 100-kg load, 1/16-in.-diam ball	Rockwell hardness No.: D scale, 100-kg load, Brale indenter	Rockwell superficial hardness No., superficial Brale indenter: 15N scale, 15-kg load	30N scale, 30-kg load	45N scale, 45-kg load	Knoop hardness No., 500-g load and greater	Shore Scleroscope hardness No.	Tensile strength (approx), 1000 psi	Rockwell C-scale hardness No.
Rockwell C-Scale Hardness Numbers													
68	940	...	...	85.6	...	76.9	93.2	84.4	75.4	920	97	...	68
67	900	...	...	85.0	...	76.1	92.9	83.6	74.2	895	95	...	67
66	865	...	...	84.5	...	75.4	92.5	82.8	73.3	870	92	...	66
65	832	...	(739)	83.9	...	74.5	92.2	81.9	72.0	846	91	...	65
64	800	...	(722)	83.4	...	73.8	91.8	81.1	71.0	822	88	...	64
63	772	...	(705)	82.8	...	73.0	91.4	80.1	69.9	799	87	...	63
62	746	...	(688)	82.3	...	72.2	91.1	79.3	68.8	776	85	...	62
61	720	...	(670)	81.8	...	71.5	90.7	78.4	67.7	754	83	...	61
60	697	...	(654)	81.2	...	70.7	90.2	77.5	66.6	732	81	...	60
59	674	...	(634)	80.7	...	69.9	89.8	76.6	65.5	710	80	351	59
58	653	...	615	80.1	...	69.2	89.3	75.7	64.3	690	78	338	58
57	633	...	595	79.6	...	68.5	88.9	74.8	63.2	670	76	325	57
56	613	...	577	79.0	...	67.7	88.3	73.9	62.0	650	75	313	56
55	595	...	560	78.5	...	66.9	87.9	73.0	60.9	630	74	301	55
54	577	...	543	78.0	...	66.1	87.4	72.0	59.8	612	72	292	54
53	560	...	525	77.4	...	65.4	86.9	71.2	58.6	594	71	283	53
52	544	(500)	512	76.8	...	64.6	86.4	70.2	57.4	576	69	273	52
51	528	(487)	496	76.3	...	63.8	85.9	69.4	56.1	558	68	264	51
50	513	(475)	481	75.9	...	63.1	85.5	68.5	55.0	542	67	255	50
49	498	(464)	469	75.2	...	62.1	85.0	67.6	53.8	526	66	246	49
48	484	(451)	455	74.7	...	61.4	84.5	66.7	52.5	510	64	238	48
47	471	442	443	74.1	...	60.8	83.9	65.8	51.4	495	63	229	47
46	458	432	432	73.6	...	60.0	83.5	64.8	50.3	480	62	221	46
45	446	421	421	73.1	...	59.2	83.0	64.0	49.0	466	60	215	45
44	434	409	409	72.5	...	58.5	82.5	63.1	47.8	452	58	208	44
43	423	400	400	72.0	...	57.7	82.0	62.2	46.7	438	57	201	43
42	412	390	390	71.5	...	56.9	81.5	61.3	45.5	426	56	194	42
41	402	381	381	70.9	...	56.2	80.9	60.4	44.3	414	55	188	41
40	392	371	371	70.4	...	55.4	80.4	59.5	43.1	402	54	182	40
39	382	362	362	69.9	...	54.6	79.9	58.6	41.9	391	52	177	39
38	372	353	353	69.4	...	53.8	79.4	57.7	40.8	380	51	171	38
37	363	344	344	68.9	...	53.1	78.8	56.8	39.6	370	50	166	37
36	354	336	336	68.4	(109.0)	52.3	78.3	55.9	38.4	360	49	161	36
35	345	327	327	67.9	(108.5)	51.5	77.7	55.0	37.2	351	48	157	35
34	336	319	319	67.4	(108.0)	50.8	77.2	54.2	36.1	342	47	153	34
33	327	311	311	66.8	(107.5)	50.0	76.6	53.3	34.9	334	46	149	33
32	318	301	301	66.3	(107.0)	49.2	76.1	52.1	33.7	326	44	145	32
31	310	294	294	65.8	(106.0)	48.4	75.6	51.3	32.5	318	43	141	31

(a) For carbon and alloy steels in the annealed, normalized, and quenched-and-tempered conditions; less accurate for cold worked condition and for austenitic steels. The values in **boldface type** correspond to the values in the joint SAE-ASM-ASTM hardness conversions as printed in ASTM E140, Table 2. The values in parentheses are beyond normal range and are given for information only.

From Boyer, H. E. and T. L. Gall, editors. 1985. *Metals Handbook, Desk Edition*. Metals Park, Ohio: American Society for Metals.

Rockwell C-scale hardness No.	Vickers hardness No.	Brinell hardness No., 3000-kg load, 10-mm ball: Standard ball	Brinell hardness No., 3000-kg load, 10-mm ball: Tungsten carbide ball	Rockwell hardness No.: A scale, 60-kg load Brale indenter	Rockwell hardness No.: B scale, 100-kg load, 1/16-in.-diam ball	Rockwell hardness No.: D scale, 100-kg load, Brale indenter	Rockwell superficial hardness No., superficial Brale indenter: 15N scale, 15-kg load	Rockwell superficial hardness No., superficial Brale indenter: 30N scale, 30-kg load	Rockwell superficial hardness No., superficial Brale indenter: 45N scale, 45-kg load	Knoop hardness No., 500-g load and greater	Shore Scleroscope hardness No.	Tensile strength (approx), 1000 psi	Rockwell C-scale hardness No.
30	302	286	286	65.3	(105.5)	47.7	75.0	50.4	31.3	311	42	138	30
29	294	279	279	64.7	(104.5)	47.0	74.5	49.5	30.1	304	41	135	29
28	286	271	271	64.3	(104.0)	46.1	73.9	48.6	28.9	297	40	131	28
27	279	264	264	63.8	(103.0)	45.2	73.3	47.7	27.8	290	39	128	27
26	272	258	258	63.3	(102.5)	44.6	72.8	46.8	26.7	284	38	125	26
25	266	253	253	62.8	(101.5)	43.8	72.2	45.9	25.5	278	38	122	25
24	260	247	247	62.4	(101.0)	43.1	71.6	45.0	24.3	272	37	119	24
23	254	243	243	62.0	100.0	42.1	71.0	44.0	23.1	266	36	117	23
22	248	237	237	61.5	99.0	41.6	70.5	43.2	22.0	261	35	114	22
21	243	231	231	61.0	98.5	40.9	69.9	42.3	20.7	256	35	112	21

Rockwell B-Scale Hardness Numbers

Rockwell B-scale hardness No.	Vickers hardness No.	Brinell hardness No., 10-mm-diam ball: 500-kg load	Brinell hardness No., 10-mm-diam ball: 3000-kg load	Rockwell hardness No.: A scale, 60-kg load, Brale indenter	Rockwell hardness No.: C scale, 150-kg load, Brale indenter	Rockwell hardness No.: F scale, 60-kg load, 1/16-in.-diam ball	Rockwell superficial hardness No., 1/16-in.-diam ball: 15T scale, 15-kg load	Rockwell superficial hardness No., 1/16-in.-diam ball: 30T scale, 30-kg load	Rockwell superficial hardness No., 1/16-in.-diam ball: 45T scale, 45-kg load	Knoop hardness No., 500-g load and greater	Shore Scleroscope hardness No.	Tensile strength (approx), 1000 psi	Rockwell B-scale hardness No.
98	228	189	228	60.2	(19.9)	...	92.5	81.8	70.9	241	34	107	98
97	222	184	222	59.5	(18.6)	...	92.1	81.1	69.9	236	33	104	97
96	216	179	216	58.9	(17.2)	...	91.8	80.4	68.9	231	32	102	96
95	210	175	210	58.3	(15.7)	...	91.5	79.8	67.9	226	...	99	95
94	205	171	205	57.6	(14.3)	...	91.2	79.1	66.9	221	31	97	94
93	200	167	200	57.0	(13.0)	...	90.8	78.4	65.9	216	30	94	93
92	195	163	195	56.4	(11.7)	...	90.5	77.8	64.8	211	...	92	92
91	190	160	190	55.8	(10.4)	...	90.2	77.1	63.8	206	29	90	91
90	185	157	185	55.2	(9.2)	...	89.9	76.4	62.8	201	28	88	90
89	180	154	180	54.6	(8.0)	...	89.5	75.8	61.8	196	27	86	89
88	176	151	176	54.0	(6.9)	...	89.2	75.1	60.8	192	...	84	88
87	172	148	172	53.4	(5.8)	...	88.9	74.4	59.8	188	26	82	87
86	169	145	169	52.8	(4.7)	...	88.6	73.8	58.8	184	26	81	86
85	165	142	165	52.3	(3.6)	...	88.2	73.1	57.8	180	25	79	85
84	162	140	162	51.7	(2.5)	...	87.9	72.4	56.8	176	...	78	84
83	159	137	159	51.1	(1.4)	...	87.6	71.8	55.8	173	24	76	83
82	156	135	156	50.6	(0.3)	...	87.3	71.1	54.8	170	24	75	82
81	153	133	153	50.0	...	...	86.9	70.4	53.8	167	...	73	81
80	150	130	150	49.5	...	...	86.6	69.7	52.8	164	23	72	80
79	147	128	147	48.9	...	...	86.3	69.1	51.8	161	...	70	79
78	144	126	144	48.4	...	...	86.0	68.4	50.8	158	22	69	78
77	141	124	141	47.9	...	...	85.6	67.7	49.8	155	22	68	77
76	139	122	139	47.3	...	...	85.3	67.1	48.8	152	...	67	76
75	137	120	137	46.8	...	99.6	85.0	66.4	47.8	150	21	66	75
74	135	118	135	46.3	...	99.1	84.7	65.7	46.8	148	21	65	74
73	132	116	132	45.8	...	98.5	84.3	65.1	45.8	145	...	64	73
72	130	114	130	45.3	...	98.0	84.0	64.4	44.8	143	20	63	72
71	127	112	127	44.8	...	97.4	83.7	63.7	43.8	141	20	62	71
70	125	110	125	44.3	...	96.8	83.4	63.1	42.8	139	...	61	70
69	123	109	123	43.8	...	96.2	83.0	62.4	41.8	137	19	60	69
68	121	107	121	43.3	...	95.6	82.7	61.7	40.8	135	19	59	68
67	119	106	119	42.8	...	95.1	82.4	61.0	39.8	133	19	58	67
66	117	104	117	42.3	...	94.5	82.1	60.4	38.7	131	...	57	66
65	116	102	116	41.8	...	93.9	81.8	59.7	37.7	129	18	56	65
64	114	101	114	41.4	...	93.4	81.4	59.0	36.7	127	18	...	64
63	112	99	112	40.9	...	92.8	81.1	58.4	35.7	125	18	...	63
62	110	98	110	40.4	...	92.2	80.8	57.7	34.7	124	...	...	62
61	108	96	108	40.0	...	91.7	80.5	57.0	33.7	122	17	...	61
60	107	95	107	39.5	...	91.1	80.1	56.4	32.7	120	...	...	60
59	106	94	106	39.0	...	90.5	79.8	55.7	31.7	118	...	...	59
58	104	92	104	38.6	...	90.0	79.5	55.0	30.7	117	...	...	58
57	103	91	103	38.1	...	89.4	79.2	54.4	29.7	115	...	...	57
56	101	90	101	37.7	...	88.8	78.8	53.7	28.7	114	...	...	56
55	100	89	100	37.2	...	88.2	78.5	53.0	27.7	112	...	...	55

APPENDIX 11 **Summary of Hardness of Martensite as a Function of Carbon Content in Fe–C Steels**

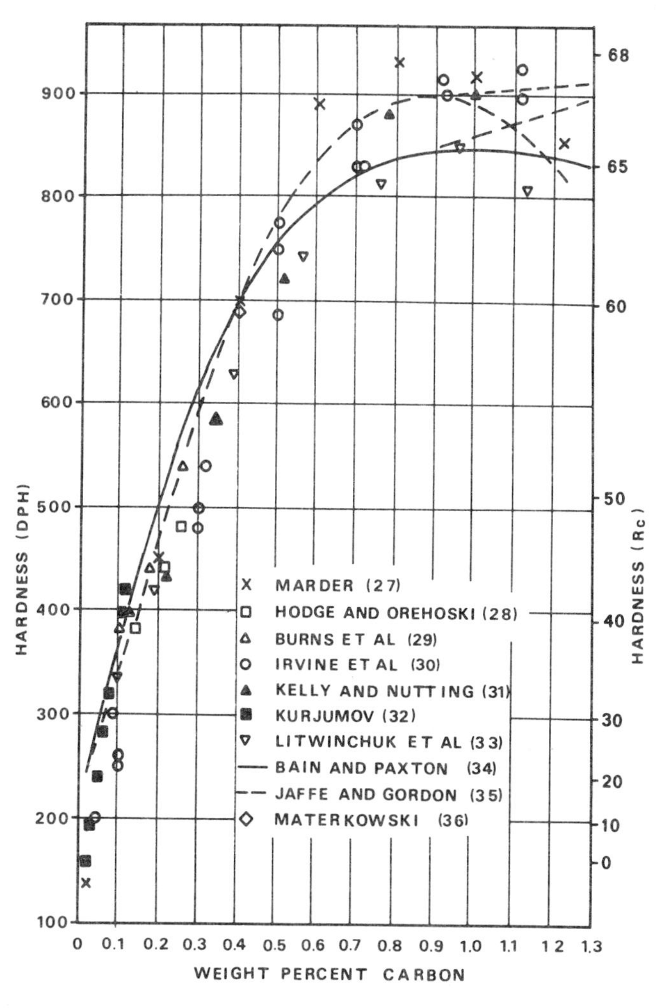

Reprinted with permission from Krauss, G. 1978. In *Hardenability Concepts with Applications to Steel,* D. V. Doane and J. S. Kirkaldy, editors, Warrendale, PA: AIME, p. 229.

APPENDIX 12 **AISI-SAE System of Designations for Steels**

Numerals and digits(a)	Type of steel and/or nominal alloy content
Carbon steels	
10xx	Plain carbon (Mn 1.00% max)
11xx	Resulfurized
12xx	Resulfurized and rephosphorized
15xx	Plain carbon (max Mn range—1.00 to 1.65%)
Manganese steels	
13xx	Mn 1.75
Nickel steels	
23xx	Ni 3.50
25xx	Ni 5.00
Nickel-chromium steels	
31xx	Ni 1.25; Cr 0.65 and 0.80
32xx	Ni 1.75; Cr 1.07
33xx	Ni 3.50; Cr 1.50 and 1.57
34xx	Ni 3.00; Cr 0.77
Molybdenum steels	
40xx	Mo 0.20 and 0.25
44xx	Mo 0.40 and 0.52
Chromium-molybdenum steels	
41xx	Cr 0.50, 0.80 and 0.95; Mo 0.12, 0.20, 0.25 and 0.30
Nickel-chromium-molybdenum steels	
43xx	Ni 1.82; Cr 0.50 and 0.80; Mo 0.25
43BVxx	Ni 1.82; Cr 0.50; Mo 0.12 and 0.25; V 0.03 min
47xx	Ni 1.05; Cr 0.45; Mo 0.20 and 0.35
81xx	Ni 0.30; Cr 0.40; Mo 0.12
86xx	Ni 0.55; Cr 0.50; Mo 0.20
87xx	Ni 0.55; Cr 0.50; Mo 0.25
88xx	Ni 0.55; Cr 0.50; Mo 0.35
93xx	Ni 3.25; Cr 1.20; Mo 0.12
94xx	Ni 0.45; Cr 0.40; Mo 0.12
97xx	Ni 0.55; Cr 0.20; Mo 0.20
98xx	Ni 1.00; Cr 0.80; Mo 0.25

(a) "xx" in the last two (or three) digits of these designations indicates that the carbon content (in hundredths of a percent) is to be inserted.

Numerals and digits(a)	Type of steel and/or nominal alloy content
Nickel-molybdenum steels	
46xx	Ni 0.85 and 1.82; Mo 0.20 and 0.25
48xx	Ni 3.50; Mo 0.25
Chromium steels	
50xx	Cr 0.27, 0.40, 0.50 and 0.65
51xx	Cr 0.80, 0.87, 0.92, 0.95, 1.00 and 1.05
Chromium steels	
50xxx	Cr 0.50 } C 1.00 min
51xxx	Cr 1.02 } C 1.00 min
52xxx	Cr 1.45 } C 1.00 min
Chromium-vanadium steels	
61xx	Cr 0.60, 0.80 and 0.95; V 0.10 and 0.15 min
Tungsten-chromium steel	
72xx	W 1.75; Cr 0.75
Silicon-manganese steels	
92xx	Si 1.40 and 2.00; Mn 0.65, 0.82 and 0.85; Cr 0.00 and 0.65
High-strength low-alloy steels	
9xx	Various SAE grades
Boron steels	
xxBxx	B denotes boron steel
Leaded steels	
xxLxx	L denotes leaded steel

APPENDIX 13 Composition and Heat Treating Temperatures of Commonly Nitrided Steels

Trade name	SAE	AMS	Nominal composition, % C	Mn	Si	Cr	Ni	Mo	Al	Other	Austenitize °C	°F	Temper °C	°F
Nitralloy G	···	···	0.35	0.55	0.30	1.2	···	0.20	1.0	···	950	1750	565-700	1050-1300
135M	7140	6470	0.42	0.55	0.30	1.6	···	0.38	1.0	···	950	1750	565-700	1050-1300
N	···	6475	0.24	0.55	0.30	1.15	3.5	0.25	1.0	···	900	1650	650-675	1200-1250
EZ	···	···	0.35	0.80	0.30	1.25	···	0.20	1.0	0.2 Se	950	1750	565-700	1050-1300
H 11 mod	···	6485	0.40	0.30	0.90	5.0	···	1.3	···	0.5 V	1010	1850	540-625	1000-1150(a)
H 13	···	···	0.40	0.40	1.10	5.00	···	1.35	···	1.10 V	1010	1850	540-625	1000-1150(a)
18 Ni 250	···	···	0.015	0.05	0.05	···	18.0	4.8	0.10	7.5 Co, 0.40 Ti	···	···	(b)	(b)
18 Ni 300	···	···	0.015	0.05	0.05	···	18.0	4.8	0.10	9.0 Co, 0.60 Ti	···	···	(b)	(b)
···	4140	6382	0.40	0.90	0.30	0.95	···	0.20	···	···	850	1550	575-625	1075-1150
···	4340	6415	0.40	0.70	0.30	0.80	1.85	0.25	···	···	815	1500	575-625	1075-1150
European Grades														
31 CrMoV 9 (W Germany)	···	···	0.30	0.55	0.25	2.5	···	0.2	···	0.15 V	860	1580	575-625	1075-1165
22 40 (Sweden)	···	···	0.32	0.70	0.27	2.5	0.5	0.3	···	0.25 V	915	1675	575-675	1070-1245
30 CD 12 (France)	···	···	0.32	0.55	0.25	3.0	···	0.4	···	···	900	1650	575-650	1070-1200

(a) Double temper. (b) Age 3 h at 480 °C (900 °F).

APPENDIX 14 Typical Heat Treatments for Carburizing Grades of Carbon and Alloy Steels

AISI-SAE designations (a)	Pretreatment (b)	Cooling after carburizing (c)	Reheat temperature °C	Reheat temperature °F	Quench after reheating (c)	Maximum tempering temperature °C	Maximum tempering temperature °F
10XX(e)	...	WQ	785	1450	WQ	200	400
11XX(e)	...	WQ, OQ	785-900	1450-1650	WQ, OQ	200	400
15XX(e)	...	OQ	785	1450	OQ	200	400
40XX, 4118	N	OQ(f)	...	...	...	175	350
4320	N, C	OQ(f)	...	...	...	175	350
		C(g)	830-845	1525-1550	OQ	175	350
44XX	N, C	OQ(f)	...	...	...	175	350
46XX	N, C	OQ(f)	...	...	...	175	350
		C(g)	815-845	1500-1550	OQ	175	350
		OQ(h)	815-845	1500-1550	OQ	175	350
4720	N, C	OQ(h)	815-845	1500-1550	OQ	175	350
48XX	NT, C	OQ(f)	...	...	...	160	325
		C(g)	795-830	1475-1525	OQ	160	325
		OQ(h)	795-830	1475-1525	OQ	160	325
5015, 51XX	N	OQ(f)	...	...	...	175	350
6118	N	OQ(f)	...	...	...	160	325
8115, 86XX	N, C	OQ(f)	...	...	...	175	350
8720, 8822	...	C(g)	845-870	1550-1600	OQ	175	350
		OQ(h)	845-870	1550-1600	OQ	175	350
9310	NT	C(g)	790-830	1450-1525	OQ	160	325
		OQ(h)	790-830	1450-1525	OQ	160	325
94BXX	N	OQ(f)	...	...	...	175	350

Note: Carburizing temperature typically 900 to 925 °C (1650 to 1700 °F). Applicable to fine grain steels only. (a) XX indicates that there are two or more grades having carbon contents of 0.30% or less that are commonly used for carburizing applications. (b) N indicates that the steel should be normalized at a temperature at least as high as the carburizing temperature, followed by air cooling; NT indicates normalizing and tempering at 600 to 650 °C (1100 to 1200 °F); C indicates cycle annealing, consisting of austenitizing at a temperature as high as the carburizing temperature, cooling rapidly to 535 to 675 °C (1000 to 1250 °F) and holding for 1 to 3 h, then air cooling. (c) WQ, water quench; OQ, oil quench; C, slow cooling. (d) Minimum tempering temperature is 120 °C (250 °F). (e) This grade is sometimes carbonitrided at 790 to 900 °C (1450 to 1650 °F), quenched in oil and tempered at 120 to 200 °C (250 to 400 °F). (f) This treatment is most commonly used; it generally minimizes distortion of the part. Parts are normally cooled from carburizing temperature to 850 °C (1550 °F) before quenching. (g) This treatment is used if machining after carburizing is required, or if facilities for direct quench from carburizing furnace are not available. (h) This treatment is used to refine grain size of the steel; it produces a desirable combination of case and core properties. It is rather likely to produce distortion in the parts.

Reprinted with permission from *Metals Handbook, 9th Ed., Vol. 1.* 1978. Metals Park, Ohio: American Society for Metals.

APPENDIX 15 **The Relationship between the ASTM Grain-Size and Average "Diameter" of the Grain**

ASTM micro-grain size number	Calculated "diameter" of average grain		ASTM micro-grain size number	Calculated "diameter" of average grain	
	mm	in.		mm	in.
		X10^{-3}			X10^{-3}
00	0.508	20.0	7.5	0.027	1.05
0	0.359	14.1	-	0.025	0.984
0.5	0.302	11.9	8.0	0.0224	0.884
1.0	0.254	10.0	-	0.0200	0.787
-	0.250	9.84	8.5	0.0189	0.743
1.5	0.214	8.41	9.0	0.0159	0.625
-	0.200	7.87	-	0.0150	0.591
-	0.180	7.09	9.5	0.0134	0.526
2.0	0.179	7.07	10.0	0.0112	0.442
2.5	0.151	5.95	-	0.0100	0.394
-	0.150	5.91	10.5	0.00944	0.372
3.0	0.127	5.00	-	0.00900	0.354
-	0.120	4.72	-	0.00800	0.315
3.5	0.107	4.20	11.0	0.00794	0.313
-	0.099	3.90	-	0.00700	0.276
4.0	0.0898	3.54	11.5	0.00667	0.263
4.5	0.076	2.97	-	0.00600	0.236
-	0.070	2.76	12.0	0.00561	0.221
5.0	0.064	2.50	-	0.00500	0.197
-	0.060	2.36	12.5	0.00472	0.186
5.5	0.053	2.10	-	0.00400	0.158
-	0.050	1.97	13.0	0.00397	0.156
6.0	0.045	1.77	13.5	0.00334	0.131
-	0.040	1.58	-	0.00300	0.118
6.5	0.038	1.49	14.0	0.00281	0.111
-	0.035	1.38	-	0.00250	0.098
7.0	0.032	1.25			
-	0.030	1.18			

Adapted from *1966 Book of ASTM Standards, Part 31*. 1966. Philadelphia: American Society for Testing and Materials, © ASTM.

APPENDIX 16 **A Comment on Magnification Markers**

It is customary to indicate the magnification of a photograph of a microstructure (micrograph), but when the photograph size is changed (e.g., enlarged), the original magnification must be corrected. A method of avoiding this is to place on the micrograph a bar or marker that has the correct dimension for the magnification of the photograph. Then if the magnification of the photograph is changed, the marker also changes dimensions to maintain the correct dimension at the new magnification.

For example, consider a micrograph at $100\times$. Then a marker 1 cm long would correspond to a dimension of $1/100 = 0.01$ cm. It is common to express this in micrometers (microns). Since a micron is equal to 10^{-4} cm, then the 0.01 cm dimension could also be labeled as 100 μm ($0.01 \times 10{,}000$). If the micrograph is enlarged 2.5 times, then the marker becomes $2.5 \times 1 = 2.5$ cm long, and is still labeled as 100 μm.

In general,

$$M = x/(y \times 10^{-4})$$

where M is the magnification, x is the marker length in cm, and y is the marker value in μm on the micrograph.

At very high magnification, the micron marker value may be less than 1. For example, if the magnification is $40{,}000\times$, then 1 cm would correspond to 0.25 μm. In such cases, the marker corresponding to a certain value in Angstroms (Å) may be used. An Angstrom is equal to 10^{-8} cm, so in this case the marker dimension y in Å would be obtained from the equation above to be $40{,}000 = 1/(y \times 10^{-8})$, so $y = 2500$ Å.

For the SI system, the nanometer is used. One nanometer equals 10 Å, so for the case above, the marker would be labeled 250 nm.

The use of such markers is very useful because the actual sizes of microstructural features can be determined. For example, if the micron marker is labeled 2 μm and is 1 cm long, and a particle is present which is about 0.5 cm long in the micrograph, then its *actual* dimension in the sample is about 1 μm.

APPENDIX 17 **A Glossary of Terms**

A_1 temperature—The eutectoid temperature of a steel.

A_3 temperature—The temperature at which proeutectoid ferrite begins to separate from austenite under conditions of slow cooling.

A_{cm} temperature—The temperature at which proeutectoid cementite begins to separate from austenite under conditions of slow cooling.

abrasion—The process of grinding or wearing away using abrasives.

abrasive—A substance capable of grinding away another material.

age-hardening—Hardening by aging, usually after rapid cooling or cold working.

aging—A change in properties that occurs at ambient or moderately elevated temperatures after hot working or a heat-treating operation (quench aging in ferrous alloys), or after a cold working operation (strain aging). The change in properties is often, but not always, due to a phase change (precipitation), but does not involve a change in chemical composition.

allotriomorph—A particle of a phase that has no regular external shape.

allotropy—The property whereby certain elements may exist in more than one crystal structure. *See* POLYMORPHISM.

alloy—A substance having metallic properties and composed of two or more chemical elements of which at least one is a metal.

alloy steel—Steel containing significant quantities of alloying elements (other than carbon and the commonly accepted amounts of manganese, silicon, sulfur and phosphorus) added to effect changes in mechanical or physical properties. Those containing less than 5% total metallic alloying elements tend to be termed low-alloy steels, and those containing more than 5% tend to be termed high-alloy steels.

alloying element—An element added to a metal, and remaining in the metal, that effects changes in structure and properties.

angstrom unit (Å)—A unit of linear measure equal to 10^{10} m, or 0.1 nm; not an accepted SI unit, but still sometimes used for small distances such as interatomic distances and some wavelengths.

annealing—A general term denoting treatment of metal that involves heating to a suitable temperature followed by cooling at a suitable rate to produce dis-

Reprinted with permission from Samuels, L. E. 1980. *Optical Microscopy of Carbon Steels.* Metals Park, Ohio: American Society for Metals.

crete changes in microstructure and properties. *See* FULL ANNEALING, HOMOGENIZING ANNEALING, SPHEROIDIZING ANNEALING *and* SUBCRITICAL ANNEALING. Full annealing is implied when the term *annealing* is used without qualification.

annealing twin—A twin formed in a metal during an annealing heat treatment.

arc welding—A group of welding processes wherein the metal or metals being joined are coalesced by heating with an arc, with or without the application of pressure and with or without the use of filler metal.

artifact—In microscopy, a false structure introduced during preparation of a specimen.

austempering—Cooling an austenitized steel at a rate high enough to suppress formation of high-temperature transformation products, then holding the steel at a temperature below that for pearlite formation and above that for martensite formation until transformation to an essentially bainitic structure is complete.

austenite—Generally, a solid solution of one or more alloying elements in the fcc polymorph of iron (γ-Fe). Specifically, in carbon steels, the interstitial solid solution of carbon in γ-Fe.

austenitic grain size—The size of the grains in steel heated into the austenitic region.

austenitizing—Forming austenite by heating a steel to between A_1 and A_3 or between A_1 and A_{cm} (partial austenitizing), or above A_1 or A_{cm} (complete austenitizing). When used without qualification, the term implies complete austenitizing.

autoradiograph—A radiograph recorded photographically by radiation spontaneously emitted by radioisotopes that are produced in, or added to, the material. This technique identifies the locations of the radioisotopes.

autotempering—Tempering occurring immediately after martensite has formed, either as the martensite cools from the M_s temperature to room temperature or at room temperature. Also known as *self-tempering*.

bainite—A eutectoid transformation product of ferrite and a fine dispersion of carbide, generally formed at temperatures below 450 to 500°C: upper bainite is an aggregate containing parallel lath-shaped units of ferrite, produces the so-called "feathery" appearance in optical microscopy, and is formed at temperatures above about 350°C; lower bainite consists of individual plate-shape units and is formed at temperatures below about 350°C.

bamboo grain structure—A structure in wire or sheet in which the boundaries of the grains tend to be aligned normal to the long axis and to extend completely through the thickness.

banding—Inhomogeneous distribution of alloying elements or phases aligned in filaments or plates parallel to the direction of working. *See* FERRITE-PEARLITE BANDING *and* SEGREGATION BANDING.

billet—A solid, semifinished steel round or square product that has been hot worked by forging, rolling or extrusion; usually smaller than a bloom.

blister steel—High-carbon steel produced by carburizing wrought iron. The bar, originally smooth, is covered with small blisters when removed from the cementation (carburizing) furnace.

bloom—(1) Ancient: iron produced in a solid condition directly by the reduction of ore in a primitive furnace. The carbon content is variable but usually low. Also known as *bloomery iron*. The earliest ironmaking process, but still used in underdeveloped countries. (2) Modern: a semifinished hot rolled steel product, rectangular in section, usually produced on a blooming mill but sometimes made by forging.

bloomery—A primitive furnace used for direct reduction of ore to iron.

box annealing—Annealing a metal or alloy in a sealed container under conditions that minimize oxidation.

braze welding—A group of welding processes in which metals are joined by a filler metal that has a liquidus temperature below the solidus of the parent metal, but above 450°C. The filler metal is not distributed by capillary action. *Compare* BRAZING.

brazing—A welding process in which a third metal or alloy of comparatively low melting point is melted and caused, sometimes with the assistance of a flux, to wet the surfaces to be joined.

brittle fracture—Fracture preceded by little or no plastic deformation.

brittleness—The tendency of a material to fracture without first undergoing significant plastic deformation.

buffer—A substance added to aqueous solutions to maintain a constant hydrogen-ion concentration, even in the presence of acids or alkalis.

burning—(1) During austenitizing, permanent damage of a metal or alloy by heating to cause incipient melting; *see* OVERHEATING. (2) During subcritical annealing, particularly in continuous annealing, production of a severely decarburized and grain-coarsened surface layer that results from heating to an excessively high temperature for an excessively long time.

capped steel—Semikilled steel cast in a bottletop mold and covered with a cap fitting into the top of the mold to cause the top metal to solidify. Pressure increases in the sealed-in molten metal, resulting in a surface condition in the ingot similar to that of rimmed steel.

carbide—A compound of carbon with one or more metallic elements.

carbon equivalent (for rating of weldability)—A value that takes into account the equivalent additive effects of carbon and other alloying elements on a particular characteristic of a steel. For rating of weldability, a formula commonly used is:

$$CE = C + (Mn/6) + [(Cr + Mo + V)/5] + [(Ni + Cu)/15]$$

carbon potential—A measure of the capacity of an environment containing active carbon to alter or maintain, under prescribed conditions, the carbon concentration in a steel.

carbon restoration—Replacing the carbon lost in the surface layer during previous processing by carburizing this layer to substantially the original carbon level.

carbon steel—A steel containing only residual quantities of elements other than carbon, except those added for deoxidation or to counter the deleterious effects of residual sulfur. Silicon is usually limited to about 0.60%, and manganese to about 1.65%. Also termed *plain carbon steel, ordinary steel, straight carbon steel.*

carbonitriding—A case-hardening process in which a suitable ferrous material is heated above the lower transformation temperature in a gaseous atmosphere having a composition that results in simultaneous absorption of carbon and nitrogen by the surface and, by diffusion, creates a concentration gradient. The process is completed by cooling at a rate that produces the desired properties in the workpiece.

carburizing—A process in which an austenitized ferrous material is brought into contact with a carbonaceous atmosphere having sufficient carbon potential to cause absorption of carbon at the surface and, by diffusion, create a concentration gradient.

case—In a ferrous alloy, the outer portion that has been made harder than the inner portion, or core. *See* CASE-HARDENING.

case-hardening—A generic term covering several processes applicable to steel that change the chemical composition of the surface layer by absorption of carbon or nitrogen, or a mixture of the two, and, by diffusion, create a concentration gradient.

cast iron—Iron containing more carbon than the solubility limit in austenite (about 2%).

cast steel—Steel in the form of castings.

casting—(1) An object at or near finished shape obtained by solidification of a substance in a mold. (2) Pouring molten metal into a mold to produce an object of desired shape.

cementation—(1) Introduction of one or more elements into the outer layer of a metal object by means of diffusion at high temperature. (2) An obsolete process used to convert wrought iron to blister steel by carburizing. Wrought iron bars were packed in sealed chests with charcoal and heated at about 1100°C for 6 to 8 days. Cementation was the predominant method of manufacturing steels, particularly high-carbon tool steels, prior to the introduction of the bessemer and open-hearth methods.

cementite—A metastable carbide, with composition Fe_3C and orthorhombic crystal structure, having limited substitutional solubility for the carbide-forming elements, notably manganese.

chafery—A charcoal-fired furnace used in early ironmaking processes to reheat a bloom of wrought iron for forging to consolidate the iron and expel entrapped slag.

chemical polishing—Improving the specular reflectivity of a metal surface by chemical treatment.

cleavage—Fracture of a crystal by crack propagation across a crystallographic plane of low index.

cleavage fracture—Fracture of a grain, or most of the grains, in a polycrystalline metal by cleavage, resulting in bright reflecting facets.

cleavage plane—A characteristic crystallographic plane or set of planes in a crystal on which cleavage fracture occurs easily.

columnar structure—A structure consisting of elongated grains whose long axes are parallel.

constituent—A phase, or combination of phases, that occurs in a characteristic configuration in a microstructure.

constitutional diagram—A graphical representation of the temperature and composition limits of phase fields in an alloy system as they actually exist under specific conditions of heating and cooling (synonymous with PHASE DIAGRAM). A constitutional diagram may be, or may approximate, an equilibrium diagram, or may represent metastable conditions or phases. *Compare* EQUILIBRIUM DIAGRAM.

continuous phase—In an alloy or portion of an alloy containing more than one phase, the phase that forms the background or matrix in which the other phase or phases are present as isolated volumes.

controlled rolling—A hot rolling process in which the temperature of the steel is closely controlled, particularly during the final rolling passes, to produce a fine-grain microstructure.

coring—A variation of composition between the center and surface of a unit of structure (such as a DENDRITE, a GRAIN or a CARBIDE particle) resulting from nonequilibrium growth over a range of temperature.

corrosion—Deterioration of a metal by chemical or electrochemical reaction with its environment.

creep—Time-dependent strain occurring under stress.

critical cooling rate—The limiting rate at which austenite must be cooled to ensure that a particular type of transformation product is formed.

critical point—(1) The temperature or pressure at which a change in crystal structure, phase or physical properties occurs; same as TRANSFORMATION TEMPERATURE. (2) In an equilibrium diagram, that specific combination of composition, temperature and pressure at which the phases of an inhomogeneous system are in equilibrium.

critical strain—That strain which results in the formation of very large grains during recrystallization.

critical temperature—Synonymous with CRITICAL POINT if pressure is constant.

cross rolling—A (hot) rolling process in which rolling reduction is carried out in a direction perpendicular to, as well as a direction parallel to, the length of the original slab.

crucible steel—High-carbon steel produced by melting blister steel in a covered crucible. Crucible steel was developed by Benjamin Huntsman in about 1750 and remained in use until the late 1940s.

crystalline fracture—A fracture of a polycrystalline metal characterized by a grainy appearance. *Compare* FIBROUS FRACTURE.

decarburization—Loss of carbon from the surface of a ferrous alloy as a result of heating in a medium that reacts with carbon.

decoration (of dislocations)—Segregation of solute atoms to the line of a dislocation in a crystal. In ferrite, the dislocations may be decorated with carbon or nitrogen atoms.

deformation bands—Generally, bands in which deformation has been concentrated inhomogeneously. *See* KINK BAND, MICROBANDS *and* SHEAR BANDS for specific types.

degenerate structure—Usually refers to pearlite that does not have an ideally lamellar structure. The degree of degeneracy may vary from slight perturbations in the lamellar arrangement to structures that are not recognizably lamellar.

dendrite—A crystal that has grown in a treelike branching mode.

dendritic segregation—Inhomogeneous distribution of alloying elements through the arms of dendrites.

deoxidation—(1) Removal of oxygen from molten metals by use of suitable chemical agents. (2) Sometimes refers to removal of undesirable elements other than oxygen by the introduction of elements or compounds that readily react with them.

diffusion—(1) Spreading of a constituent in a gas, liquid or solid, tending to make the composition of all parts uniform. (2) The spontaneous movement of atoms or molecules to new sites within a material.

dilatometer—An instrument for measuring the expansion or contraction of a solid metal resulting from heating, cooling, polymorphic changes, etc.

dislocation—A linear defect in the structure of a crystal.

ductility—The capacity of a material to deform plastically without fracturing.

elastic limit—The maximum stress to which a material may be subjected without any permanent strain remaining upon complete release of the stress.

elastic strain—Dimensional changes accompanying stress where the original dimensions are restored upon release of the stress.

electron beam microprobe analyzer—An instrument for selective chemical analysis of a small volume of material. An electron beam bombards the area of interest and x-radiation thereby emitted is analyzed in a spectrometer.

electropolishing—Improving the specular reflectivity of a metal surface by electrochemical dissolution.

elongation after fracture—In tensile testing, the increase in the gauge length measured after fracture of the specimen within the gauge length and usually expressed as a percentage of the original gauge length.

epitaxy—Induced orientation of the lattice of a crystal of a surface deposit by the lattice of the substrate crystal.

equiaxed structure—A structure in which the grains have approximately the same dimensions in all directions.

equilibrium diagram—A graphical representation of the temperature, pressure and composition limits of phase fields in an alloy system as they exist under conditions of thermodynamical equilibrium. In condensed systems, pressure is usually considered constant.

etchant—A chemical solution used to etch a metal to reveal structural details.

etching—Subjecting the surface of a metal to preferential chemical or electrolytic attack to reveal structural details.

eutectoid—(1) An isothermal reversible transformation in which a solid solution is converted into two or more intimately mixed solids, the number of solids formed being the same as the number of components in the system. (2) An alloy having the composition indicated by the eutectoid point on an equilibrium diagram. (3) An alloy structure of intermixed solid constituents formed by a eutectoid transformation.

fatigue—The phenomenon leading to fracture under repeated or fluctuating stresses (having a maximum value less than the tensile strength of the material).

ferrite—Generally, a solid solution of one or more alloying elements in the bcc polymorph of iron (α-Fe). Specifically, in carbon steels, the interstitial solid solution of carbon in α-Fe.

ferrite-pearlite banding—Inhomogeneous distribution of ferrite and pearlite aligned in filaments or plates parallel to the direction of working.

ferritic grain size—The grain size of the ferritic matrix of a steel.

fiber—(1) The characteristic of wrought metal that indicates directional properties. It is revealed by etching a longitudinal section or manifested by the fibrous appearance of a fracture. It is caused chiefly by extension of the constituents of the metal, both metallic and nonmetallic, in the direction of working. (2) The pattern of preferred orientation of metal crystals after a given deformation process. *See* PREFERRED ORIENTATION.

fibrous fracture—A fracture whose surface is characterized by a dull gray or silky appearance. *Compare* CRYSTALLINE FRACTURE.

filler metal—A third material that is melted concurrently with the parent metals during fusion or braze welding. It is usually, but not necessarily, of different composition from the parent metals.

finery—A charcoal-fueled hearth furnace used in early processes for converting cast iron to wrought iron by melting and oxidizing it in an air blast, then repeatedly oxidizing the product in the presence of a slag. The carbon oxidizes more rapidly than the iron so that a wrought iron of low-carbon content is produced.

finishing temperature—The temperature at which hot working is completed.

flash—(1) In forging, the excess metal forced between the upper and lower dies. (2) In resistance butt welding, a fin formed perpendicular to the direction of applied pressure.

flow lines—(1) Texture showing the direction of metal flow during hot or cold working. Flow lines often can be revealed by etching the surface or a section of a metal part. (2) In mechanical metallurgy, paths followed by volume elements of metal during deformation.

flow stress—The uniaxial true stress required to cause plastic deformation at a specified value of strain.

flux—(1) In refining, a material used to remove undesirable substances as a molten mixture. It may also be used as a protective covering for molten metal. (2) In welding, a material used to prevent the formation of, or to dissolve and facilitate the removal of, oxides and other undesirable substances.

forge welding—Welding hot metal by applying pressure or blows.

forging—Plastically deforming metal, usually hot, into desired shapes with compressive force, with or without dies.

fractography—Descriptive treatment of fracture, especially in metals, with specific reference to photography of the fracture surface.

fragmentation—The subdivision of a grain into small discrete crystallites outlined by a heavily deformed network of intersecting slip bands as a result of cold working. These small crystals or fragments differ from one another in orientation and tend to rotate to a stable orientation determined by the slip systems.

free machining—Pertains to the machining characteristics of an alloy to which one or more ingredients have been introduced to produce small broken chips, low power consumption, better surface finish or longer tool life.

full annealing (ferrous materials)—An annealing treatment in which a steel is austenitized by heating to a temperature above the upper critical temperature (A_3 or A_{cm}) and then cooled slowly to room temperature. A typical cooling rate would be 100°C/h. *Compare* NORMALIZING. Use of the term "annealing" without qualification implies full annealing.

fusion welding—Any welding process in which fusion is employed to complete the weld.

gas welding—Welding with a gas flame.

grain—An individual crystal in a polycrystalline metal or alloy, including twinned regions or subgrains if present.

grain-boundary liquation—An advanced stage of overheating in which material in the region of austenitic grain boundaries melts. Also known as BURNING.

grain-boundary sulfide precipitation—An intermediate stage of overheating in which sulfide inclusions are redistributed to the austenitic grain boundaries by partial solution at the overheating temperature and reprecipitation during subsequent cooling.

grain coarsening—A heat treatment that produces excessively large austenitic grains.

grain flow—Fiberlike lines appearing on polished and etched sections of forgings, caused by orientation of the constituents of the metal in the direction of working during forging.

grain growth—An increase in the average size of the grains in polycrystalline metal, usually a result of heating at elevated temperature.

grain size—A measure of the areas or volumes of grains in a polycrystalline metal or alloy, usually expressed as an average when the individual sizes are fairly uniform. In metals containing two or more phases, the grain size refers to that of the matrix unless otherwise specified. Grain size is reported in terms of number of grains per unit area or volume, average diameter, or as a number derived from area measurements.

granular fracture—A type of irregular surface produced when metal fractures, characterized by a rough, grainlike appearance as differentiated from a smooth silky, or fibrous, type. It can be subclassified into transgranular and intergranular forms. This type of fracture is frequently called crystalline fracture, but the implication that the metal has crystallized is completely misleading.

graphite—The polymorph of carbon with a hexagonal crystal structure.

graphitization—Formation of graphite in iron or steel. Primary graphitization refers to formation of graphite during solidification. Secondary graphitization refers to later formation during heat treatment.

grinding—Removing material from a workpiece with a grinding wheel or abrasive belt.

hardenability—In ferrous alloys, the property that determines the depth and distribution of hardness induced by quenching.

hardening—Increasing hardness by suitable treatment, usually involving heating and cooling. When applicable, the following more specific terms should be used: AGE-HARDENING, CASE-HARDENING, *flame-hardening, induction-hardening, precipitation-hardening,* QUENCH-HARDENING.

hardness (indentation)—Resistance of a metal to plastic deformation by indentation. Various hardness tests such as Brinell, Rockwell and Vickers may be used. In the Vickers test, a diamond pyramid with an included face angle of 136°C is used as the indenter.

heat-affected zone—That portion of the base metal which was not melted during brazing, cutting or welding, but within which microstructure and physical properties were altered by the treatment.

heat tinting—Coloration of a metal surface through oxidation by heating to reveal details of structure.

heat treatment—Heating and cooling a solid metal or alloy in such a way that desired structures, conditions or properties are attained. Heating for the sole purpose of hot working is excluded from the meaning of this term.

hematite—The oxide of iron of highest valency which has a composition close to the stoichiometric composition Fe_2O_3.

homogenizing annealing—An annealing treatment carried out at a high temperature, approaching the solidus temperature, for a sufficiently long time that inhomogeneous distributions of alloying elements are reduced by diffusional processes.

hot working—Deformation under conditions that result in recrystallization.

hypereutectoid alloy—In a eutectoid system, any alloy containing more than the eutectoid concentration of solute.

hypoeutectic alloy—In a eutectic system, any alloy containing less than the eutectic concentration of solute.

idiomorph—A particle of a phase that has a regular external shape.

impact test—A test for determining the behavior of materials when subjected to high rates of loading under conditions designed to promote fracture, usually in bending, tension or torsion. The quantity measured is the energy absorbed when the specimen is broken by a single blow.

impurities—Elements or compounds whose presence in a material is undesired.

inclusion—A nonmetallic material in a solid metallic matrix.

ingot—A casting suitable for hot working or remelting.

ingot iron—Commercially pure iron.

intercrystalline—Between crystals, or between grains. Same as *intergranular.*

internal oxidation—Formation of oxides beneath the surface of a metal.

interstitial solid solution—A solid solution in which the solute atoms occupy (interstitial) positions between the atoms in the structure of the solvent. *See also* SUBSTITUTIONAL SOLID SOLUTION.

intracrystalline—Within or across crystals or grains. Same as TRANSCRYSTALLINE and TRANSGRANULAR.

iron—An element that has an average atomic number of 55.85 and that always, in engineering practice, contains small but significant amounts of carbon. Thus iron-carbon alloys containing less than about 0.1% C may be referred to as irons. Alloys with higher carbon contents are always termed steels.

isothermal transformation (IT) diagram—A diagram that shows the isothermal time required for transformation of austenite to commence and to finish as a function of temperature. Same as TIME-TEMPERATURE-TRANSFORMATION (TTT) DIAGRAM or *S-curve*.

killed steel—Steel deoxidized with a strong deoxidizing agent, such as silicon or aluminum, to reduce the oxygen content to such a level that no reaction occurs between carbon and oxygen during solidification.

kink band (deformation)—In polycrystalline materials, a volume of crystal that has rotated physically to accommodate differential deformation between adjoining parts of a grain while the band itself has deformed homogeneously. This occurs by regular bending of the slip lamellae along the boundaries of the band.

lamellar tear—A system of cracks or discontinuities aligned generally parallel to the worked surface of a plate. Usually associated with a fusion weld in thick plate.

lamination—An abnormal structure resulting in a separation or weakness aligned generally parallel to the worked surface of the metal.

lath martensite—Martensite formed, partly in steels containing less than about 1.0% C and solely in steels containing less than about 0.5% C, as parallel arrays or packets of lath-shape units about 0.1 to 0.3 μm thick, and having a habit plane that is close to $\{111\}_A$.

liquation—Partial melting of an alloy.

liquidus—In a constitutional diagram, the locus of points representing the temperatures at which various components commence freezing on cooling or finish melting on heating.

Lüders lines or bands—Elongated surface markings or depressions caused by localized plastic deformation that results from discontinuous (inhomogeneous) yielding

M_f temperature—The temperature at which martensitic transformation is essentially complete during cooling after austenitization.

***M_s* temperature**—The temperature at which a martensitic transformation starts during cooling after austenitization.

machinability—The capacity of a material to be machined easily.

macroetching—Etching of a metal surface with the objective of accentuating gross structural details, for observation by the unaided eye or at magnifications not exceeding ten diameters.

macrograph—A graphic reproduction of a prepared surface of a specimen at a magnification not exceeding ten diameters. When photographed, the reproduction is known as a PHOTOMACROGRAPH (not a macrophotograph).

macrostructure—The structure of a metal as revealed by examination of the etched surface at a magnification not exceeding ten diameters.

magnetite—The oxide of iron of intermediate valence which has a composition close to the stoichiometric composition Fe_3O_4.

manual welding—Welding wherein the entire welding operation is performed and controlled by hand.

martempering—(1) A hardening procedure in which an austenitized ferrous material is quenched into an appropriate medium at a temperature just above the M_s temperature of the material, held in the medium until the temperature is uniform throughout—but not long enough for bainite to form—and then cooled in air. The treatment is frequently followed by tempering. (2) When the process is applied to carburized material, the controlling M_s temperature is that of the case. This variation of the process is frequently called *marquenching*.

martensite—In steel, a metastable transition phase with a body-centered-tetragonal crystal structure formed by diffusionless transformation of austenite generally during cooling between the M_s and M_f temperatures.

martensite range—The interval between the M_s and M_f temperatures.

matrix—The principal phase or aggregate in which another constituent is embedded.

mechanical twin—A twin formed in a metal during plastic deformation by simple shear of the structure.

mechanical polishing—A method of producing a specularly reflecting surface by use of abrasives.

melting point—Thye temperature at which a pure metal, compound or eutectic changes from solid to liquid; the temperature at which the liquid and the solid are in equilibrium.

metal—An opaque, lustrous, elemental substance that is a good conductor of heat and electricity and, when polished, a good reflector of light. Most metals are malleable and ductile and are, in general, denser than other substances.

metallograph—An optical instrument designed for both visual observation and photomicrography of prepared surfaces of opaque materials at magnifications ranging from about 25 to about 1500 diameters.

metastable—Possessing a state of pseudoequilibrium that has a free energy higher than that of the true equilibrium state but from which a system does not change spontaneously.

microbands (deformation)—Thin sheetlike volumes of constant thickness in which cooperative slip occurs on a fine scale. They are an instability which carry exclusively the deformation at medium strains when normal homogeneous slip is precluded. The sheets are aligned at $\pm 55°$ to the compression direction and are confined to individual grains, which usually contain two sets of bands. *Compare* SHEAR BANDS.

microcrack—A crack of microscopic size.

micrograph—A graphic reproduction of the prepared surface of a specimen at a magnification greater than ten diameters. When photographed, the reproduction is known as a PHOTOMICROGRAPH (not a microphotograph).

microsegregation—Segregation within a grain, crystal or small particle. *See* CORING.

microstructure—The structure of a prepared surface of a metal as revealed by a microscope at a magnification greater than ten diameters.

mild steel—Carbon steel containing a maximum of about 0.25% C.

natural aging—Spontaneous aging of a supersaturated solid solution at room temperature. *See* AGING.

necking—Local reduction of the cross-sectional area of metal by stretching.

Neumann band—A mechanical (deformation) twin in ferrite.

nodular pearlite—Pearlite that has grown as a colony with an approximately spherical morphology.

normalizing—Heating a ferrous alloy to a suitable temperature above A_3 or A_{cm} and then cooling in air to a temperature substantially below A_1. The cooling rate usually is in the range 500 to 1000°C/h.

notch brittleness—A measure of the susceptibility of a material to brittle fracture at locations of stress concentration. For example, in a notch tensile test a material is said to be "notch brittle" if its notch strength is less than its tensile strength; otherwise, it is said to be "notch ductile."

notch ductility. *See* NOTCH BRITTLENESS.

notch sensitivity—A measure of the reduction in strength of a metal caused by the presence of stress concentration. Values can be obtained from static, impact or fatigue tests.

nucleation—Initiation of a phase transformation at discrete sites, the new phase growing from nuclei. *See* NUCLEUS (1).

nucleus—(1) The first structurally stable particle capable of initiating recrystallization of a phase or the growth of a new phase, and separated from the matrix by an interface. (2) The heavy central core of an atom, in which most of the mass and the total positive electrical charge are concentrated.

orientation (crystal)—Directions in space of the axes of the lattice of a crystal with respect to a chosen reference or coordinate system. *See also* PREFERRED ORIENTATION.

overaging—Aging under conditions of time and temperature greater than those required to obtain maximum change in a certain property. *See* AGING.

overheating—Heating a metal or alloy to such a high temperature that its properties are impaired. When the original properties cannot be restored by further heat treating, by mechanical working or by a combination of working and heat treating, the overheating is known as BURNING.

oxidation—(1) A reaction in which there is an increase in valence resulting from a loss of electrons. (2) Chemical combination with oxygen to form an oxide.

oxidized surface—A surface having a thin, tightly adhering oxidized skin.

pack rolling—Hot rolling a pack of two or more sheets of metal; scale prevents the sheets from being welded together.

pancake grain structure—A structure in which the lengths and widths of individual grains are large compared to their thicknesses.

pass—(1) A single transfer of metal through a stand of rolls. (2) The open space between two grooved rolls through which metal is processed. (3) The weld metal deposited in one run along the axis of a weld.

patenting—A heat treatment applied to medium- and high-carbon steel prior to cold drawing to wire. The treatment involves austenitization followed by isothermal transformation at a temperature that produces a microstructure of very fine pearlite.

pattern welding—A process in which strips or other small sections of iron or steel are twisted together and then forge welded. Homogeneity and toughness are thereby improved. A regular decorative pattern can be developed in the final product. Commonly used for making swords as early as the 3rd century A.D.

pearlite—A eutectoid transformation product of ferrite and cementite that ideally has a lamellar structure but that is always degenerate to some extent.

peritectic—An isothermal reversible reaction in which a liquid phase reacts with a solid phase to produce another solid phase.

phase—A physically homogeneous and distinct portion of a material system.

phase diagram—Synonymous with CONSTITUTIONAL DIAGRAM.

photomacrograph—*See* MACROGRAPH.

photomicrograph—*See* MICROGRAPH.

physical properties—Properties, other than mechanical properties, that pertain to the physical nature of a material; e.g., density, electrical conductivity, thermal expansion, reflectivity, magnetic susceptibility, etc.

pig iron—(1) High-carbon iron made by reduction of iron ore in the blast furnace. (2) Cast iron in the form of pigs.

piling—A process in which several bars are stacked and hot rolled together, with the objective of improving the homogeneity of the final product. Used in primitive ironmaking.

plastic deformation—Deformation that remains, or will remain, permanent after release of the stress that caused it.

plasticity—The capacity of a metal to deform nonelastically without rupturing.

plate—A flat-rolled metal product of some minimum thickness and width arbitrarily dependent on the type of metal.

plate martensite—Martensite formed, partly in steels containing more than about 0.5% C and solely in steels containing more than about 1.0% C, as lenticular-shape plates on irrational habit planes that are near $\{225\}_A$, or $\{259\}_A$ in very-high-carbon steels.

polishing—Producing a specularly reflecting surface.

polycrystalline—Comprising an aggregate of more than one crystal, and usually a large number of crystals.

polymorphism—The property whereby certain substances may exist in more than one crystalline form, the particular form depending on the conditions of crystallization—e.g., temperature and pressure. Among elements, this phenomenon is also called ALLOTROPY.

preferred orientation—A condition of a polycrystalline aggregate in which the crystal orientations are not random.

primary crystal—The first type of crystal that separates from a melt during solidification.

proeutectoid (phase)—Particles of a phase that precipitate during cooling after austenitizing but before the eutectoid transformation takes place.

proof stress—*See* YIELD STRENGTH.

puddling process—A process for making wrought iron in which cast iron is melted in a hearth furnace and rabbled with slag and oxide until a pasty mass is obtained. This process was developed by Henry Cort about 1784 and remained in use until 1957, although on a very small scale during the present century.

quench aging—Aging that occurs after quenching following solution heat treatment.

quench-hardening—Hardening by austenitizing and then cooling at a rate such that a substantial amount of austenite is transformed to martensite.

quenching—Rapid cooling.

recarburizing—(1) Increasing the carbon content of molten cast iron or steel by adding carbonaceous material, high-carbon pig iron or a high-carbon alloy. (2) Carburizing a metal part to return surface carbon lost in processing.

recovery—Reduction or removal of work-hardening effects, without motion of large-angle grain boundaries.

recrystallization—(1) A change from one crystal structure to another, such as that occurring on heating or cooling through a critical temperature. (2) Formation of a new, strain-free grain structure from the structure existing in cold worked metal.

recrystallization annealing—Annealing cold worked metal to produce a new grain structure without a phase change.

recrystallization temperature—The approximate minimum temperature at which complete recrystallization of a cold worked metal occurs within a specified time.

residual elements—Small quantities of elements unintentionally present in an alloy.

residual stress—Stress present in a body that is free of external forces or thermal gradients.

resistance welding—Welding with electrical-resistance heating and pressure, the work being part of an electrical circuit.

resolution—The capacity of an optical or radiation system to separate closely spaced forms or entities; also, the degree to which such forms or entities can be discriminated.

resulfurized steel—Steel to which sulfur has been added in controlled amounts after refining. The sulfur is added to improve machinability.

rimmed steel—Low-carbon steel containing sufficient iron oxide to produce continuous evolution of carbon monoxide during ingot solidification, resulting in a case or rim of metal virtually free of voids.

rolling—Reducing the cross-sectional area of metal stock, or otherwise shaping metal products, through the use of rotating rolls.

scale—A layer of oxidation products formed on a metal at high temperature.

scarf joint—A butt joint in which the plane of the joint is inclined with respect to the main axes of the members.

segregation—Nonuniform distribution of alloying elements, impurities or phases.

segregation banding—Inhomogeneous distribution of alloying elements aligned in filaments or plates parallel to the direction of working.

self diffusion—The spontaneous movement of an atom to a new site in a crystal of its own species.

semikilled steel—Steel that is incompletely deoxidized and contains sufficient dissolved oxygen to react with the carbon to form carbon monoxide and thus offset solidification shrinkage.

shear—That type of force that causes or tends to cause two contiguous parts of the same body to slide relative to each other in a direction parallel to their plane of contact.

shear bands (deformation)—Bands in which deformation has been concentrated inhomogeneously in sheets that extend across regional groups of grains. Usually only one system is present in each regional group of grains, different systems being present in adjoining groups. The bands are noncrystallographic and form on planes of maximum shear stress (55° to the compression direction). They carry most of the deformation at large strains. *Compare* MICROBANDS.

shear steel—Steel produced by forge welding together several bars of blister steel, providing a more homogeneous product.

sheet—A flat-rolled metal product of some maximum thickness and minimum width arbitrarily dependent on the type of metal. Sheet is thinner than plate.

shortness—A form of brittleness in metal. It is designated as "cold," "hot," and "red," to indicate the temperature range in which the brittleness occurs.

sintering—Bonding of adjacent surfaces of particles in a mass of metal powders, or in a compact, by heating.

slab—A piece of metal, intermediate between ingot and plate, at least twice as wide as it is thick.

slag—A nonmetallic product resulting from mutual dissolution of flux and nonmetallic impurities in smelting and refining operations.

slip—Plastic deformation by irreversible shear displacement of one part of a crystal relative to another in a definite crystallographic direction and a definite crystallographic plane.

slip direction—The crystallographic direction in which translation of slip takes place.

slip line—Trace of a slip plane on a viewing surface.

slip plane—The crystallographic plane on which slip occurs in a crystal.

solid solution—A solid crystalline phase containing two or more chemical species in concentrations that may vary between limits imposed by phase equilibrium.

solidus—In a constitutional diagram, the locus of points representing the temperatures at which various components finish freezing on cooling or begin to melt on heating.

solute—The component of either a liquid or solid solution that is present to the lesser or minor extent; the component that is dissolved in the solvent.

solution heat treatment—A heat treatment in which an alloy is heated to a suitable temperature, held at that temperature long enough to cause one or more constituents to enter into solid solution, and then cooled rapidly enough to hold these constituents in solution.

solvent—The component of either a liquid or solid solution that is present to the greater or major extent; the component that dissolves the solute.

solvus—In a phase or equilibrium diagram, the locus of points representing the temperature at which solid phases with various compositions coexist with other solid phases; that is, the limits of solid solubility.

spheroidized structure—A microstructure consisting of a matrix containing spheroidal particles of another constituent.

spheroidizing—Heating and cooling to produce a spheroidal or globular form of carbide in steel.

spheroidizing annealing—A subcritical annealing treatment intended to produce spheroidization of cementite or other carbide phases.

steel—An iron-base alloy usually containing carbon and other alloying elements. In carbon steel and low-alloy steel, the maximum carbon content is about 2.0%; in high-alloy steel, about 2.5%. The dividing line between low-alloy and high-alloy steels is generally regarded as the 5% level of total metallic alloying elements. Steel is differentiated from two general classes of "iron"—namely, cast irons, which have high carbon concentrations, and relatively pure irons, which have low carbon concentrations.

strain—A measure of the relative change in the size of a body. Linear strain is the change per unit length of a linear dimension. True (or natural) strain is the natural logarithm of the ratio of the length at the moment of observation to the original gauge length. Shearing strain is the change in angle (expressed in radians) between two reference lines originally at right angles. When the term is used alone, it usually refers to linear strain in the direction of the applied stress.

strain aging—Aging induced by cold work. *See* AGING.

strain-hardening—An increase in hardness and strength caused by plastic deformation at temperatures below the recrystallization range.

stress—Force per unit area. True stress denotes stress determined by measuring force and area at the same time. Conventional stress, as applied to tension and compression tests, is force divided by original area. *Nominal* stress is stress computed by simple elasticity formulae.

stress-corrosion cracking—Failure by cracking under the combined action of corrosion and stress, either external (applied) or internal (residual). Cracking may be either intergranular or transgranular, depending on the metal and the corrosive medium.

stress relieving—Heating to a suitable temperature, holding long enough to reduce residual stresses and then cooling slowly enough to minimize the development of new residual stresses.

stretcher strains—Elongated markings that appear on the surfaces of some materials when they are deformed just past the yield point. These markings lie approximately parallel to the direction of maximum shear stress and are the result of localized yielding. *See also* LÜDERS LINES.

strip—A sheet of metal whose length is many times its width.

sub-boundary structure (subgrain structure)—A network of low-angle boundaries (usually with misorientations of less than one degree) within the main grains of a microstructure.

subcritical annealing—An annealing treatment in which a steel is heated to a temperature below the A_1 temperature and then cooled slowly to room temperature.

subgrain—A portion of a crystal or grain slightly different in orientation from neighboring portions of the same crystal. Generally, neighboring subgrains are separated by low-angle boundaries.

substitutional solid solution—A solid solution in which the solvent and solute atoms are located randomly at the atom sites in the crystal structure of the solution.

substrate—The layer of metal underlying a coating, regardless of whether the layer is base metal.

sulfide spheroidization—A stage of overheating in which sulfide inclusions are partly or completely spheroidized.

sulfur print—A macrographic method of examining distribution of sulfide inclusions.

supercooling—Cooling to a temperature below that of an equilibrium phase transformation without the transformation taking place.

superheating—(1) Heating a phase to a temperature above that of a phase transformation without the transformation taking place. (2) Heating molten metal to a temperature above the normal casting temperature to obtain more complete refining or greater fluidity.

surface-hardening—A generic term covering several processes applicable to a suitable ferrous alloy that produce, by quench-hardening only, a surface layer that is harder or more wear-resistant than the core. There is no significant alteration of the chemical composition of the surface layer.

The processes commonly used are *induction-hardening*, *flame-hardening* and *shell-hardening*. Use of the applicable specific process name is preferred.

taper section—A section made at an acute angle to a surface of interest, thereby achieving a geometrical magnification of depth. A sectioning angle of 5° 43′ achieves a depth magnification of 10:1.

teeming—Pouring molten metal from a ladle into ingot molds. The term applies particularly to the specific operation of pouring either iron or steel into ingot molds.

temper brittleness—A reversible increase in the ductile–brittle transition temperature in steels heated in, or slowly cooled through, the temperature range from about 375 to about 575°C.

temper rolling—Light cold rolling of sheet steel. The operation is performed to improve flatness, to minimize the formation of stretcher strains, and to obtain a specified hardness or temper.

tempering—In heat treatment, reheating hardened steel to some temperature below the A_1 temperature for the purpose of decreasing hardness and/or increasing toughness. The process also is sometimes applied to normalized steel.

tensile strength—In tensile testing, the ratio of the maximum force sustained to the original cross-sectional area.

texture—In a polycrystalline aggregate, the state of distribution of crystal orientations. In the usual sense, it is synonymous with preferred orientation, in which the distribution is not random.

thermal analysis—A method of studying transformations in metal by measuring the temperatures at which thermal arrests occur.

time–temperature–transformation (TTT) diagram. *See* ISOTHERMAL TRANSFORMATION (IT) DIAGRAM.

toughness—Capacity of a metal to absorb energy and deform plastically before fracturing.

transcrystalline—Same as INTRACRYSTALLINE.

transformation ranges (transformation temperature ranges)—Those ranges of temperature within which austenite forms during heating and transforms during cooling. The two ranges are distinct, sometimes overlapping but never coinciding. The limiting temperatures of these ranges depend on the composition of the alloy and on the rate of change of temperature, particularly during cooling. *See* TRANSFORMATION TEMPERATURE.

transformation temperature—The temperature at which a change in phase occurs. The term is sometimes used to denote the limiting temperature of a transformation range. The following symbols are used:

A_1—The temperature of the eutectoid transformation.

A_3—The temperature at which proeutectoid ferrite begins to separate from austenite under conditions of slow cooling.

A_{cm}—The temperature at which proeutectoid cementite begins to separate from austenite under conditions of slow cooling.

M_f—The temperature at which transformation of austenite to martensite finishes during cooling.

M_s—The temperature at which transformation of austenite to martensite starts during cooling.

transgranular—Same as INTRACRYSTALLINE.

transition temperature (ductile–brittle transition temperature)—An arbitrarily defined temperature that lies within the temperature range in which metal fracture characteristics (as usually determined by tests of notched specimens) change rapidly, such as from primarily fibrous (shear) to primarily cleavage.

triple point—The intersection of the boundaries of three adjoining grains, as observed in a section.

twin—Two portions of a crystal having a definite orientation relationship; one may be regarded as the parent, the other as the twin. The orientation of the twin is either a mirror image of the orientation of the parent across a "twinning plane" or an orientation that can be derived by rotating the twin portion about a "twinning axis."

twin, annealing—A twin produced as the result of heat treatment.

twin, crystal—A portion of a crystal in which the lattice is a mirror image of the lattice of the remainder of the crystal.

twin, deformation—A twinned region produced by a shearlike distortion of the parent crystal structure during deformation. In ferrite, deformation twins form on {211} planes.

upper yield stress—*See* YIELD POINT.

upset—(1) The localized increase in cross-sectional area resulting from the application of pressure during mechanical fabrication or welding. (2) That portion of a welding cycle during which the cross-sectional area is increased by the application of pressure.

vacancy—A type of structural imperfection in which an individual atom site is temporarily unoccupied.

veining—A type of sub-boundary structure that can be delineated because of the presence of a greater-than-average concentration of precipitate or solute atoms.

Walloon process—An early two-hearth process for making wrought iron by refining cast iron. The conversion proper was carried out in a hearth furnace known as a *finery*; reheating for forging was carried out in a second hearth furnace known as a *chafery*.

weld—A union made by welding.

weld bead—A deposit of filler metal from a single welding pass.

weldability—Suitability of a metal for welding under specific conditions.

welding—Joining two or more pieces of material by applying heat or pressure, or both, with or without filler metal, to produce a localized union through fusion or recrystallization across the interface.

wetting agent—A surface-active agent that produces wetting by decreasing the cohesion within the liquid.

Widmanstätten structure—A structure characterized by a geometric pattern resulting from the formation of a new phase on certain crystallographic planes in the parent phase. The orientation of the lattice in the new phase is related crystallographically to the orientation of the lattice in the parent phase.

wootz—A carbon steel containing 1 to 1.6% C produced by melting a bloomery iron or an inhomogeneous steel with charcoal in a crucible. The process originated in India as early as the 3rd century A.D.

work-hardening—Same as STRAIN-HARDENING.

wrought iron—An iron produced by direct reduction or ore or by refining molten cast iron under conditions where a pasty mass of solid iron with included slag is produced. The iron has a low carbon content.

wüstite—The oxide of iron of lowest valence which exists over a wide range of compositions that do not quite include the stoichiometric composition FeO.

yield point—The first stress in a material less than the maximum obtainable stress at which an increase in strain occurs without an increase in stress. Also known as UPPER YIELD STRESS.

yield strength—The stress at which a material exhibits a specified limiting deviation from the proportionality of stress to strain. The deviation is expressed in terms of strain. Also known as PROOF STRESS.

INDEX